To my Father and
the memory of my Mother

Jeremy Gray

Linear Differential Equations and Group Theory from Riemann to Poincaré

Birkhäuser
Boston · Basel · Stuttgart

Library of Congress Cataloging in Publication Data

Gray, Jeremy, 1947–
   Linear differential equations and group theory from Riemann to Poincaré.

   Bibliography: p.
   Includes index.
      1. Differential equations, Linear—History.
   2. Groups, Theory of—History. I. Title.
   QA372.G68   1985   515.3'54'09   85-22807
   ISBN 0-8176-3318-9

CIP-Kurztitelaufnahme der Deutschen Bibliothek

Gray, Jeremy:
Linear differential equations and group theory from Riemann to Poincaré /
Jeremy Gray. – Boston ; Basel ; Stuttgart : Birkhäuser, 1986.
      ISBN 3-7643-3318-9 (Basel...)
      ISBN 0-8176-3318-9 (Boston)

© 1986 Birkhäuser Boston, Inc.
Cover design: Justin Messmer, Basel
Printed in Germany
ISBN 0-8176-3318-9
ISBN 3-7643-3318-9

Introduction

Mathematicians often speak of the unity of their subject, whether to praise it or lament its passing. This book traces the emergence of such a unity from its nineteenth century origins in the history of linear ordinary differential equations, especially those closely connected with elliptic and modular functions. In its later chapters the book is concerned with the impact of group theoretical and geometrical ideas upon the problem of understanding the nature of the solutions to a differential equation. So far as is possible the mathematics is developed from scratch, and no more than an undergraduate knowledge of the subject is presumed. It is my hope that this historical treatment will re-acquaint many mathematicians with a rich and important area of mathematics perhaps known only to specialists today.

The story begins with the hypergeometric equation

$$x(1 - x) \frac{d^2y}{dx^2} + (c - (a + b + 1)x) \frac{dy}{dx} - aby = 0$$

studied by Gauss in 1812. This equation is important in its own right as a linear ordinary differential equation for which explicit power-series solutions can be given and, more importantly, their inter-relations examined. Gauss's work in this direction was extended by Riemann in a paper of 1857, in which the crucial idea of analytically continuing the solutions around their singularities in the complex domain was first truly understood. This approach, which may be termed the monodromy approach, gave a thorough global understanding of the solutions of the hypergeometric equation. It was extended in 1865 by Fuchs to those n^{th} order linear ordinary differential equations none of whose solutions have essential singularities, a class he was able to characterize. Fuchs's work revealed that, for technical reasons, equations other than the hypergeometric would be less easy to understand globally. He suggested, however, that a sub-class could be isolated, consisting of differential equations all of whose solutions were algebraic functions, and this problem was tackled in the 1870's by many mathematicians: Schwarz (for the hypergeometric equation), Fuchs, Gordan, and Klein (for the 2nd order equation), Jordan (for the nth order). The methods may be described as geometric (Schwarz and Klein), invariant-theoretic (Fuchs and Gordan), and group-theoretic (Jordan), and of these the group-theoretic was the most strikingly successful. It arose from the monodromy approach by taking all the monodromy transformations, i.e. all analytic transformations of a basis of solutions under analytic continuations around all paths in the domain, and considering this set in totality as a group. In the case at hand this group, which is evidently composed of linear transformations, is finite if and only if the solutions are all algebraic.

The hypergeometric equation is also important because it contains many interesting equations as special cases, notably Legendre's equation

$$(1 - k^2) \frac{d^2y}{dk^2} + \frac{1 - 3k^2}{k} \frac{dy}{dk} - y = 0$$

which is satisfied by the periods

$$K = \int_0^1 \frac{dx}{\sqrt{[(1 - x^2)(1 - k^2x^2)]}} \quad \text{and} \quad K' = \int_1^{1/k} \frac{dx}{\sqrt{[(1 - x^2)(1 - k^2x^2)]}}$$

of elliptic integrals considered as functions of the modulus k^2.

This equation was studied by Legendre, Gauss, Abel, Jacobi, and Kummer. It is useful in the study of the transformation problem of elliptic integrals: given a prime p and a modulus k^2, find a second modulus λ^2 for which the associated periods L and L' satisfy $\frac{L}{L'} = p\frac{K}{K'}$. The transformation problem also yields a polynomial relation between k^2 and λ^2 of degree p + 1, but Jacobi showed that, when p = 5, 7, and 11, the degree can be reduced to p. This mysterious observation was discussed by Galois in 1832 in terms of what would now be called the Galois group PSL(2; $\mathbb{Z}/p\mathbb{Z}$) and shown to depend on the existence of sub-groups of small index (in fact, p) in these groups. Galois's work was taken up by Betti, Hermite, Kronecker, and Jordan in the 1850's and 1860's and a connection was discovered between the general quintic equation and the transformation problem at the prime p = 5, which enabled Hermite to solve the quintic equation by modular functions. Analogous questions at higher primes remained unsolved until, in 1879, Klein explained the special significance of the Galois group PSL(2; $\mathbb{Z}/7\mathbb{Z}$).

Klein's work grew out of an attempt to reformulate the theory of modular functions and modular transformations without using the theory of elliptic functions. This had been almost completely achieved by Dedekind in 1878, when, inspired by some notes of Riemann's and Fuchs's considerations of the monodromy transformations of Legendre's equation, he constructed a theory of modular transformations based on the idea of the lattice of periods of an elliptic function. Klein enriched Dedekind's approach with an explicit use of group-theoretic ideas and an essentially Galois-theoretic approach to the fields of rational and modular functions. Both men relied on the theory of the hypergeometric equation at technical points in their arguments.

Klein's theory divided into the case when the genus of a certain Riemann surface was zero, when it connected with his earlier work on differential equations and with Hermite's theory of the quintic equation, and the cases when the genus was greater than zero.

The latter cases related to the study of higher plane curves, and in particular the case of the group PSL(2; $\mathbb{Z}/7\mathbb{Z}$) (when the genus is 3) related to plane quartics and their 28 bitangents. These curves had been studied projectively by Plücker and Hesse, and function-theoretically by Riemann, Roch, Clebsch, and Weber. Klein was able to give an account of their work in the spirit of his new theoretical formulations, and began to develop a systematic theory of modular functions

Meanwhile, and at first quite independently, Poincaré began in 1880 to develop a more general theory of Riemann surfaces, discontinuous groups (such as PSL(2; $\mathbb{Z}$)), and differential equations. At first he seems not to have known of the work of Schwarz or Klein, but to have picked up the initial idea only from a paper of Fuchs. Soon he learned, in correspondence with Klein, of what had already been done, but there is nonetheless a marked contrast between the novelty of Poincaré's work and the impeccably educated approach of Klein. The group-theoretic and Riemann-surface theoretic aspects of the theory were developed jointly, but the connection with differential equations remained Poincaré's concern.

This presentation ends in 1882 with the publication of Poincaré's and Klein's papers on the Fuschsian theory (a choice of name that Klein hotly denounced) and with Klein's collapse from nervous exhaustion. The less complete Kleinian theory of 1883-4 is scarcely discussed.

It can be argued that the history of mathematical ideas is a history of problems, methods, and results. These terms are, of course, not precise - a problem may be to find the method leading to an already known or intuited result, or to explain why a method works. But still one may sit gingerly on the edge of this Procrustean bed and argue that results in mathematics are, more or less, truths. Such-and-such numbers are prime, such-and-such geometries are possible, such-and-such functions exist. Problems are a more varied class of objects. They may be existence questions, or they may be more theoretical or methodological. They may arise outside mathematics, or so deep inside it that only specialists can raise them. Finally the methods are the means employed to formulate the theory. These methods are often quite personal, and may be subjected to two kinds of test. The first is their critical scrutiny by other mathematicians as to their mathematical validity, itself a matter having an historical dimension. The second is a matter of taste or style, leading mathematicians to adopt or reject methods for their own use.

If this trichotomy is applied to the subject at hand it suggests that the hypergeometric question raised the problem: understand the global relationships of the solutions given locally by power-series. Power-series and their convergence were dealt with carefully by Gauss and did not pose a problem, but the global question did. Riemann gave results in terms of monodromy, and in terms of the associated Riemann surfaces defined by the solutions. The monodromy method was not itself problematic, and could be used to formulate problems about, for example, Legendre's equation. This led to the successful elaboration of new results about modular functions. But the Riemann surface approach constructed functions transcendentally and by means of a doubtful use of Dirichlet's principle, so it raised methodological problems. These problems were avoided by Dedekind, Klein, and Poincaré, but tackled directly by Schwarz and C. A. Neumann. The transcendental approach was also rejected for a while in the study of algebraic curves, and Clebsch, Brill, and M. Noether developed a more strictly algebraic approach to the theorems of Abel, Riemann, and Roch.

The study of the transformation problem was explicitly raised as a methodological problem, that of emancipating the theory of modular equations from the theory of elliptic functions. As such it was solved by Dedekind, and that solution re-formulated by Klein. By now the formulation was dependent on the transcendental theory of functions. Hurwitz in 1881 freed the theory of modular functions from that embrace, but Poincaré and Klein chose to base their theories of automorphic functions firmly on Riemannian ideas.

This brief sketch is intended to illuminate the importance, for the conduct of the history of mathematics, of the state of the subject as a theoretically organized body of knowledge. Mathematics is not just a body of results, each one attached to a technical argument, but an intricate system of theories and a historical process. Histories which proceed from problems to results and leave out the methods by which those results were achieved, omit a crucial aspect of the process of this theoretical development. The gain in space, essential if a broad period is to be described, is made only by risking making the reasons for the development unintelligible. The vogue for histories of modern mathematics has, with notable exceptions, been too willing to leave out the details of how it was all done, and thus leave unexplored the question of why it was done. Similarly, all but the best histories of foundational topics have

concentrated naively on the foundations without appreciating the status
of foundational enquiry within the broader picture of mathematical
discovery. It is probably not true that mathematics can be built from
the bottom up, but is it certainly false that its history can be told
in that way.

For reasons of space and ignorance I have made no serious attempt
to ground this account in any questions of a social-historical kind,
although I believe such enquiries are most important. However, the
trichotomy of problems, methods, and results would accommodate itself
to such an approach. Problems may evidently be socially determined,
either from outside the subject altogether or within the development
of competing mathematical schools. Methods are naturally historically
and socially specific in large part, but results are more objective.
The formulation here adopted at least suggests what aspects of the
history of modular and automorphic functions may be treated more
sociologically while still securing an objectivity for the mathematical
results.

I do not claim that this study establishes any conclusions that
can be stated at an abstract level. Rather, it represents an attempt
to try out a methodology for exploring certain past events.

One striking observation can, however, be made at once. Only
Gauss was concerned to find any scientific applications for his results
in the theory of differential equations. Riemann, who was deeply
interested in physics, sought no role there for his P-functions.
Klein, who made great claims for the importance of mathematics in
physics, and Poincaré, who later did work of the greatest importance
in astronomy and electro-magnetism, likewise confined their researches

to the domain of pure mathematics. The Berlin school, led by
Weierstrass and Fuchs, showed less interest in physics anyway, but during
the period 1850-1880 it is clear that none of the works here discussed
were inspired by scientific concerns. It may be a fastidious comtemporary
preference for not claiming that large purposes stand behind narrow
papers, and it may be that a real concern for physics is shown by some
of these mathematicians elsewhere in their work - although their published
work is largely free of such claims. It is more likely that we are
confronted with an emerging speciality - pure mathematics - perhaps
being practised, in Germany at least, by a professeriat disdainful of
applications. Several of Klein's remarks about physics suggest such a
divorce was taking place, and an examination of physicist's work in
the period might suggest a comparable lack of interest in mathematics.
On the other hand, this period marks the introduction of the tools of
elliptic function theory into that curious hybrid applied mathematics
(notably into the study of heat diffusion and Lamé's equation, and the
motion of the top). It would be quite important to examine the
literature with an eye to the distinction between pure and applied
science, say between dynamics and electromagnetic theory, and to see
what the role of mathematics was in each of them.

I might add that the unity of mathematics, which is often invoked
in a rather imprecise way, was brought home to me very vividly as I
worked on the various aspects of this story. It is not just the
striking harmony of group theory and geometry in geometric function
theory which is here on display, but the many parallels between the
theory of the differential equations and the polynomial equations
derived from problems in elliptic functions. Nor is this a static
unity imposed, as it were, after the discoveries have been made.

In particular, the different ways in which transformations may be performed, or groups act, is a guiding thread for many mathematicians. Klein justifiably took pleasure in seeing how the regular solids could be found to lie so often at the heart of different problems, as indeed they still do, and I hope some of that pleasure has been conveyed here. But perhaps the reader will find the continuation of Klein's work into the world of non-Euclidean geometry an even greater delight.

Unity would be a terrible thing if it did not respect individual detail, and many specific items have been presented here because, finally, I could not keep them out. This is particularly true of the algebraic curves in Chapter V . Sometimes details have been included so that I may try to explain the mathematics. Perhaps, historically, this was unwise, and the jaded reader should skip such passages, I have in mind much of Chapter I, II.2, the end of II.3 and III.3. On the other hand, mathematics should not be made unduly mysterious. I have written numerous Exercises to help elucidate the mathematics. Two books may serve as excellent companions: on linear differential equations Poole's book of that name [1960]; and on modular functions the last Chapter of Serre's A Course of Arithmetic [1978]. Two good books on automorphic functions and Fuchsian groups are Lehner's Short Course... [1966] and Ford's classic Automorphic Functions [1929] which is certainly still worth reading. The reader seeking more information on elliptic functions can still scarcely do better than to consult the famous book by Whittaker and Watson [1973].

Observations on the text

I conclude with some observations.

References have generally been given in the following forms:
Shakespeare [1603] refers to the entry in the bibliography under
Shakespeare for that date, which is usually the date of first
publication. The symbols a, b, ... distinguish between works published
in the same year. Well known works are referred to by name, so the
above would be given as Hamlet.

Notation is largely as it appears in the original works under
discussion, so x and y are generally complex variables, but some
simplifications have been introduced. Mathematical comments of an
anachronistic kind which I have felt it necessary to make on occasion
are usually set off in square brackets [   ].

References to classic works, such as the German Encyklopädie
der Mathematischen Wissenschaften are relatively few. This is
because the classics are Berichte (reports) not Geschichte (histories)
a useful distinction carefully observed by their authors, but which
has reasonably restricted them to brief historical remarks. I have
tended to use these extensive reference works as guides, but to report
directly on what they had led me to find.

Conjectures and opinions of my own are usually stated as such;
the personal pronoun always indicates that a personal view is being
expressed.

I have followed recent German practice in using the symbol :=, as
in X:=Y or Y=:X, to mean 'X is defined to be Y'.

Acknowledgements

I would like to thank my supervisors David Fowler and Ian Stewart for all their help and encouragement during the years. An earlier incarnation of this work formed my doctoral thesis at the University of Warwick, and I would also like to thank my examiners, Ivor Grattan-Guinness and Rolf Schwarzenberger, for their helpful comments. I thank the Niedersächsiche Staatsbibliothek in Göttingen for permission to quote from the unpublished letter from Jordan to Klein, and the Académie des Sciences for permission to quote from the unpublished papers of Poincaré in their possession.

Last but certainly not least, my thanks to the Open University typists for their excellent job of typing this: Christine, Frances, Kim, Michelle, Norma, Shirley and Sue. My apologies to them, and the reader, for the mistakes my poor proof reading have allowed to slip through.

Jeremy Gray
The Open University
Milton Keynes
MK7 6AA
England.

Chapter One. Hypergeometric equations and modular equations

1.1 Euler and Gauss

1.2 Jacobi and Kummer

Chapter Two. Lazarus Fuchs

Fuchs a student of Kummer and Weierstrass.

2.1 Fuchs's early papers

His approach to differential equations a blend of
Riemannian techniques and those of the Berlin
school.

He characterized those equations whose solutions are
everywhere locally of the form $(x - x_0)^\alpha \sum_{n=-k}^{\infty} a_n (x - x_0)^n$,
as being of the form

$$\frac{d^n y}{dx^n} + \frac{F_{\rho-1}(x)}{\psi(x)} \frac{d^{n-1} y}{dx^{n-1}} + \frac{F_{2(\rho-1)}(x)}{\psi(x)^2} + \ldots + \frac{F_{n(\rho-1)}(x)}{\psi(x)^n} y = 0$$

where ρ is the number of finite singular points
$a_1, a_2, \ldots, a_\rho$, $\psi(x) := (x - a_1)(x - a_2) \ldots (x - a_\rho)$,
and $F_s(x)$ is a polynomial in x of degree at most s.

Such equations are 'of the Fuchsian class'. Fuchs
gave rigorous methods for solving linear differential
equations, and a careful analysis of their singular
points in terms of the monodromy matrices and their
eigenvalues; the case of repeated eigenvalues is

introduced. The 'indicial' equation for the

exponents. *

Amongst the equations of the Fuchsian class are those
all of whose solutions are algebraic. Fuchs asked how
these might be characterised.

Chapter Three. Algebraic Solutions to a Differential Equation

3.1 Schwarz

3.2 Generalizations

3.3 Klein's solution

3.4 The solutions of Gordan and Fuchs

3.5 Jordan's solution

Four passages which contain technical details which may

not be to everyone's taste have been marked with a

star*.

Some principal figures, showing their age in 1870.

Niels Hendrik Abel, 1802 - 1829, (d.).

Enrico Betti, 1823 - 1892, (47).

Jean-Claude Bouquet, 1819 - 1885, (51).

Francesco Brioschi, 1824 - 1897, (46).

Charles-Auguste Briot, 1817-1882, (53).

Augustin Louis Cauchy, 1789 - 1857, (d.).

Arthur Cayley 1821 - 1895, (49).

Rudolf Friedrich Alfred Clebsch, 1833 - 1872, (37).

Richard Dedekind, 1831 - 1916, (39).

Peter Gustav Lejeune Dirichlet, 1805 - 1859, (d.).

Ferdinand Gotthold Max Eisenstein, 1823 - 1852, (d.).

Leonhard Euler, 1707 - 1783, (d.).

Ferdinand Georg Frobenius, 1849 - 1917, (21).

Immanuel Lazarus Fuchs, 1833 - 1902, (37).

Evariste Galois, 1811 - 1832, (d.).

Carl Friedrich Gauss, 1777 - 1855, (d.).

Paul Gordan, 1837 - 1912, (33).

Georges Henri Halphen 1844 - 1889, (26).

Charles Hermite, 1822 - 1901, (48).

Ludwig Otto Hesse, 1811 - 1874, (59).

Adolph Hurwitz, 1859 - 1919, (11).

Carl Custav Jacob Jacobi, 1804 - 1851, (d.).

Camille Jordan, 1838 - 1922, (32).

Christian Felix Klein, 1849 - 1925, (21).

Leopold Kronecker, 1823 - 1891, (47).

Ernst Eduard Kummer, 1810 - 1893, (60).

Adrien Marie Legendre, 1752 - 1833, (d.).

Carl Neumann, 1832 - 1925, (38).

Charles Emile Picard, 1856 - 1941, (14).

Julius Plücker, 1801 - 1868, (d.).

Henri Poincaré, 1854 - 1912, (16).

Victor Alexandre Puiseux, 1829 - 1883, (50).

Georg Friedrich Bernhard Riemann, 1826 - 1866, (d.).

Gustav Roch, 1836 - 1866, (d.).

Ludwig Schläfli, 1814 - 1895, (56.).

Hermann Amandus Schwarz, 1843 - 1921, (27).

Jacob Steiner, 1796 - 1863, (d.).

Heinrich Weber, 1843 - 1913, (27).

Karl Theoder Wilhelm Weierstrass, 1815 - 1897, (55).

CHAPTER I   HYPERGEOMETRIC EQUATIONS AND MODULAR TRANSFORMATIONS

This chapter does three things. It gives a short account of the work of Euler, Gauss, Kummer, and Riemann on the hypergeometric equation, with some indication of its immediate antecedents and consequences. It therefore looks very briefly at some of the work of Gauss, Legendre, Abel and Jacobi on elliptic functions, in particular at their work on modular functions and modular trans-formations. It concludes with a description of the general theory of linear differential equations supplied by Cauchy and Weierstrass. There are many omissions, some of which are rectified elsewhere in the literature[1]. The sole aim of this chapter is to provide a setting for the work of Fuchs on linear ordinary differential equations, to be discussed in Chapter II, and for later work on modular functions, discussed in Chapter IV and V.

1.1 *Euler and Gauss*

Euler gave two accounts of the differential equation

$$x(1-x) \frac{d^2y}{dx^2} + [\gamma - (\alpha + \beta + 1)x] \frac{dy}{dx} - \alpha\beta y = 0 \qquad (1.1.1)$$

and the power series which represents one solution of it:

$$y = 1 + \frac{\alpha\beta}{1 \cdot \gamma} x + \frac{\alpha(\alpha+1)\beta(\beta+1)}{1.2.\gamma.(\gamma+1)} x^2 + \dots , \qquad (1.1.2)$$

now known as the hypergeometric equation and the hypergeometric series   respectively. Here y is a real valued function of a real variable x and α, β , and γ are real constants. The earlier account occupies four chapters of the Institutiones Calculi Integralis [1769, Vol. II, Part I, Chs. 8-11], the later one is a paper [1794] presented to the St. Petersburg Academy of Science in 1778 and published in 1794.

In the paper Euler demonstrated that the power series satisfies
the differential equation, and conversely, that the method of
undetermined coefficients yields the power series as a solution to
the differential equation. In the Institutiones he also considered
the slightly more general equation

$$x^2(a + bx^n) \frac{d^2y}{dx^2} + x(c + ex^n) \frac{dy}{dx} + (f + gx^n)y = 0 \qquad (1.1.3)$$

which has two solutions of the form

$$y = Ax^\lambda + Bx^{\lambda+n} + Cx^{\lambda+2n} + \ldots$$

where λ satisfies $\lambda(\lambda + 1)a + \lambda c + f = 0$. When the two values of
λ obtained from this equation differ by an integer Euler derived a
second solution containing a logarithmic term[2]. The substitution
$x^n = u$ reduces Euler's equation to

$$n^2u^2(a + bu) \frac{d^2y}{du^2} + [(a + bu)n(n - 1)u + nu(c + eu)] \frac{dy}{du} + (f + gu)y = 0$$

$$(1.1.4)$$

which would be of the hypergeometric type if f were zero and one
count divide throughout by u.

Euler, following Wallis [1655], used the term 'hypergeometric'
to refer to a power series in which the nth term is a(a + b)...
(a + (n - 1)b). The modern use of the term derives from Johann
Friedrich Pfaff, who devoted his Disquisitiones Analyticae (vol. I,
1797) to functions expressible by means of the series. Pfaff's
purpose was to solve the hypergeometric equation and various transforms of it in closed form, and he gave many examples of how this
could be done. His view of power series expansions of a function
seems to have been the typical view of his day; namely that they were
a means to the end of describing the function in terms of other,
better understood, functions,

and the view that a large class of functions can be understood
using power series without a reduction to closed form being possible
is one of the characteristic advances made by Pfaff's student Gauss.

Gauss

Much has been made of Gauss's legendary ability to calculate
with large numbers, and mention of this will be made below, but Gauss
was also a prodigious manipulator with series of all kinds. He
himself said many of his best discoveries were made at the end of
lengthy calculations, and much of his work on elliptic functions moves
in a sea of formulae with an uncanny sense of direction. Of course,
Gauss's skills as a reckoner with numbers are unusual, even amongst
mathematicians, whereas the ability to think in formulae is much more
common; one is struck as often by the technical power of great
mathematicians as by their profundity. Nonetheless, some mathematicians
have the ability more than others. It is present in a high degree
in Euler, Gauss, and Kummer, but much less in Klein or Poincaré.
Certainly it enabled Gauss to leave the circumscribed eighteenth
century domain of functions and move with ease into the large class
of functions known only indirectly. This move confronts all who take
it with the question: when is a function 'known'?, to which there
were broadly speaking two answers. One was to develop a theory of
functions in terms of some characteristic traits which can be used to
mark certain functions out as having particular properties. Thus one
might seek integral representations for the functions, and be able
to characterize those functions for which such-and-such a kind of
representation is possible. The second answer is to side-step the
question and to regard the inter-relation of functions given in

power series as itself the answer. Of course, most mathematicians adopted a mixture of the two approaches depending on their own success with a given problem. We shall see that Gauss was happy to publish a work of the second kind on the hypergeometric series, and that the work of the Berlin school led by Weierstrass regarded the study of series as a corner-stone of their theory of functions, as Lagrange had earlier. On the other hand, Riemann and later workers, chiefly Klein and Poincaré, sought more geometric answers. Fuchs, although at Berlin, adopted an interestingly ambiguous approach.

It is almost impossible to describe Gauss. Gifted beyond all his contemporaries he was doubly isolated: by the startling novelty of his vision, and by a quirk of history which produced no immediate successors to the French and German (rather, Swiss) mathematicians who had dominated the eighteenth century. In 1800 Lagrange was 64, Laplace 51, Legendre 48, Monge 54. Contact with them would have been difficult for Gauss because of the Napoleonic war, and perhaps dis- tasteful, given his conservation disposition. His teachers, Pfaff (then 35) and Kaestner (81) were not of the first rank, nor, unsurprisingly, were his contemporaries Bartels and W. Bolyai. By the time there were young mathematicians around with whom he could have conversed (Jacobi, Abel, or the generation of Cauchy and Fourier) he had become confirmed in a life-long avoidance of mathe- maticians. His contact was with astronomers, notably Bessel, in whose subject he worked increasingly. Only the tragic figure of Eisenstein caught Gauss's imagination towards the end of his life[3].

Gauss, it is said, wrote much but published little. However, May [1972, 300] found that Gauss published 323 works in his lifetime, many on astronomy - *matura sed non pauca?* Nonetheless, it is true that he published almost nothing of his work on elliptic functions, the vast store of discoveries left in the <u>Nachlass</u> account for most of the disparity between what he knew and what he saw fit to print. There is no space here for an adequate account of Gauss's approach to elliptic functions, and only one aspect can be discussed, which bears most closely on the themes of this chapter[4].

Gauss's elliptic functions

Gauss discovered for himself the arithmetico-geometric mean (agm) when he was 15. It is defined as follows for positive numbers a_0 and b_0:

$a_1 = \frac{1}{2}(a_0 + b_0)$,   their arithmetic mean, and

$b_1 = \sqrt{(a_0 b_0)}$,   their geometric mean. The iteration

of this process, defining

$a_{n+1} = \frac{1}{2}(a_n + b_n)$,

$b_{n+1} = \sqrt{(a_n b_n)}$, for $n \geq 1$,

produces two sequences $\{a_n\}$ and $\{b_n\}$ which in fact converge to the same limit, α, called the agm of a_0 and b_0. Convergence follows from the inequality $a_{n+1} - b_{n+1} < \frac{1}{2}(a_n - b_n)$. Gauss wrote $M(a, b)$ for the agm of a and b. Plainly $M(\lambda a, \lambda b) = \lambda M(a, b)$, and Gauss considered various functions of the form $M(1, x)$. For example $M(1, 1 + x) = M(1 + \frac{x}{2}, \sqrt{(1 + x)})$, so setting $x = 2t + t^2$ he obtained power series expansions with undetermined coefficients for M in terms of x and then in terms of t, from which the coefficients could be calculated. They display no particular pattern, but various manipulations led Gauss to the dramatic series for the reciprocal of $M(1 + x, 1 - x)$:

$$y: = M(1 + x, 1 - x)^{-1} = 1 + \frac{1}{4} x^2 + \frac{9}{64} x^4 + \frac{25}{256} x^6 + \ldots$$

$$= 1 + (\frac{1}{2})^2 x^2 + (\frac{1.3}{2.4})^2 x^4 + (\frac{1.3.5}{2.4.6})^2 x^6 + \ldots$$

$$(1.1.5)$$

As a function of x, y satisfies the differential equation

$(x^3 - x) \frac{d^2 y}{dx^2} + (3x^2 - 1) \frac{dy}{dx} + xy = 0$, and Gauss found another

linearly independent solution $M(1, x)^{-1}$.

The connection with elliptic integrals was discovered by Gauss

on 30 May 1799, as he tells us in his diary [1917, entry 98]. He

considered the lemniscatic integral

$$\int \frac{dx}{(1 - x^4)^{\frac{1}{2}}} \text{, for which } \int_0^1 \frac{dx}{(1 - x^4)^{\frac{1}{2}}} =: \frac{\omega}{2},$$

and showed that $M(1, \sqrt{2}) = \frac{\pi}{\omega}$ "to eleven places"[5]. This alerted him

to the possibility of making a much more general discovery applicable

to any complete elliptic integral, and he was able to claim such a

result almost exactly a year later [diary entries 105, 106 May 1800].

He took

$$\int_0^\pi \frac{d\phi}{(1 - k^2 \cos^2 \phi)^{1/2}}$$

expanded the denominator as a power series in $\cos^2 \phi$, used the known

integrals

$$\int_0^\pi \cos^{2n} \phi \, d\phi = \frac{1.3 \ldots (2n - 1)\pi}{2.4 \ldots 2n} \text{, and found}$$

$$\int_0^\pi \frac{d\phi}{(1 - k^2 \cos^2 \phi)^{1/2}} = \frac{\pi}{M(1 + k, 1 - k)} = \frac{\pi}{M(1,(1 - k^2)^{1/2})} \quad (1.1.6)$$

A hypergeometric equation is readily obtained from the equation for

$M(1 + x, 1 - x)^{-1}$ by the substitution $x^2 = z$, when it becomes

$$z(1 - z) \frac{d^2y}{dz^2} + (1 - 2z) \frac{dy}{dz} - \frac{1}{4} y = 0. \qquad (1.1.7)$$

This is a form of Legendre's equation and is a special case of the general hypergeometric equation, as will be seen, in which $\alpha = \beta = \frac{1}{2}$, $\gamma = 1$. Gauss has obtained an expression for a period of the elliptic integral

$$\int_0^\pi \frac{d\phi}{(1 - k^2\cos^2\phi)^{1/2}}$$

as a function of the modulus k. The term 'period' derives from the analogy with the integral

$$\int_0^1 \frac{dx}{(1 - x^2)^{1/2}}$$

which is discussed further below.

In 1827, Gauss made a study of a function more or less inverse to M, expressing the modulus as a function of the quotient of the periods. These later discoveries will be described in due course, for they remained unpublished until 1866 [Gauss Werke III, 470–480], by which time others had made them independently.

The hypergeometric equation

Gauss only published the first part of his study of the hyper-geometric equation, [1812a] in 1812. The second part, [1812b], found amongst the extensive Nachlass, follows on from the first, in numbered paragraphs (§§38–57).

Gauss's published paper is not remarkable by Gauss's standards, but even so it has several claims to fame: it considers x as a complex variable; and it contains the earliest rigorous argument for the convergence of a power series and a study of the behaviour of the function at a point on the boundary of the circle of convergence, as well as a thorough examinaton of continued fraction expansions for certain quotients of hypergeometric functions. Part two is given over to finding several solutions of the hypergeometric equation and the relationships between them, and will be of more concern to us in the sequel.

In part I Gauss observed that the series

$$1 + \frac{\alpha\beta}{1.\gamma} x + \frac{\alpha(\alpha + 1)\beta(\beta + 1)}{1.2\ \gamma(\gamma + 1)} x^2 + \ldots$$

is a polynomial if either $\alpha - 1$ or $\beta - 1$ is a negative integer, and is not defined at all if γ is a negative integer or zero (this case he excluded). In all other cases the series is convergent for $x = a + bi$, by the ratio test, provided that $a^2 + b^2 < 1$.

He gave, following Pfaff [1797], a list of functions which can be represented by means of hypergeometric functions and then introduced the idea of contiguous functions[6] (Section 2, §7): $F(\alpha, \beta, \gamma, x)$ is contiguous to any of the six functions $F(\alpha \pm 1, \beta \pm 1, \gamma \pm 1, x)$ obtained from it by increasing or decreasing one coefficient by 1. He obtained (§14) 15 equations connecting $F(\alpha, \beta, \gamma, x)$ with each of the fifteen pairs of its different contiguous functions by systematically permuting the α's, β's, γ's, $(\gamma - 1)$'s etc. and comparing coefficients. As an example, the fifteenth equation is

$$0 = \gamma(\gamma - 1 - (2\gamma - \alpha - \beta - 1)x)F(\alpha, \beta, \gamma; x)$$
$$+ (\gamma - \alpha)(\gamma - \beta)xF(\alpha, \beta, \gamma + 1; x)$$
$$- \gamma(\gamma - 1)(1 - x)F(\alpha, \beta, \gamma - 1; x).$$

As Klein remarked [1894, 16] these establish that any three contiguous functions satisfy a linear relationship with rational

functions for coefficients. They can be worked up to give linear
relationships over the rational functions between any three functions
of the form $F(\alpha \pm m, \beta \pm n, \gamma \pm p, x)$, where m, n, and p are
integers. Gauss's purpose in introducing contiguous functions was
to obtain partial fractions for quotients of hypergeometric
functions, e.g.

$$\frac{F(\alpha, \beta + 1, \gamma + 1; x)}{F(\alpha, \beta, \gamma; x)} ,$$

and hence for several familiar elementary functions. Observe that

$$F = F(\alpha, \beta, \gamma; x), \frac{dF}{dx} = \frac{\alpha\beta}{\gamma} F(\alpha + 1, \beta + 1, \gamma + 1; x)$$

and

$$\frac{d^2F}{dx^2} = \frac{\alpha(\alpha + 1)\beta(\beta + 1)}{\gamma(\gamma + 1)} F(\alpha + 2, \beta + 2, \gamma + 2; x)$$

are contiguous in the obvious generalized sense, the relationship
between them being, essentially, the differential equation itself.
Gauss made this observation at the start of the second, unpublished
paper.

In the third and final section of the published paper Gauss
considered the question of the value of $F(\alpha, \beta, \gamma, 1)$, i.e.
of $\lim\limits_{x \to 1} F(\alpha, \beta, \gamma, x)$, at least for real a, b, and c. He introduced
the gamma function in a somewhat modified form which is, perhaps,
more intuitively acceptable, by defining
$\Pi(k, z) := \frac{1.2...k.k^z}{(z + 1)(z + 2)...(z + k)}$, where k is a positive integer,
and $\Pi(z) := \lim\limits_{k \to \infty} \Pi(k, z)$. The limit certainly exists for $Re(z) \geq 0$,
and Π satisfies the functional equation $\Pi(z + 1) = (z + 1)\Pi(z)$,
with $\Pi(0) = 1$, from which it follows that $\Pi(n) = n!$ for positive
integral n; Π may be called (Gauss's) factorial function, it is

infinite at all negative integers. (In the usual notation, due to
Legendre [1814], $\Pi(z) = \Gamma(z + 1)$.) The factorial function enabled
Gauss to write

$$F(\alpha, \beta, \gamma, 1) = \frac{\Pi(k, \gamma - 1)\Pi(k, \gamma - \alpha - \beta - 1)}{\Pi(k, \gamma - \alpha - 1)\Pi(k, \gamma - \beta - 1)} F(\alpha, \beta, \gamma + k, 1),$$

and since $\lim_{k \to \infty} F(\alpha, \beta, \gamma + k, 1) = 1$, he obtained (§23)

$$F(\alpha, \beta, \gamma, 1) = \frac{\Pi(\gamma - 1)\Pi(\gamma - \alpha - \beta - 1)}{\Pi(\gamma - \alpha - 1)\Pi(\gamma - \beta - 1)}.$$

This expression is meaningful provided $\alpha + \beta - \gamma < 0$.

The gamma function also enabled him to attend to certain
integrals, for example Euler integrals of the first kind, in Legendre's
terminology:

$$\int_0^x z^{\lambda - 1}(1 - z^\mu)^\nu \, dz,$$

which vanishes at $z = 0$. This Gauss expressed as
$\frac{x^\lambda}{\lambda} F(-\nu, \frac{\lambda}{\nu}, \frac{\lambda}{\mu} + 1, x^\mu)$. When $x = 1$ the definite integral

$\int_0^1 z^{\lambda - 1}(1 - z)\, dz$ is equal to $\dfrac{\Pi(\frac{\lambda}{\mu})\Pi(\nu)}{\lambda\Pi(\frac{\lambda}{\mu} + \nu)}$. Gauss commented (§27)

"Whence many relations, which the illustrious Euler could only get
with difficulty, fall out at once". As examples of spontaneous
results Gauss considered the lemniscatic integrals:[7]

$$A = \int_0^1 \frac{dx}{(1 - x^4)^{1/2}}, \quad B = \int \frac{x^2\, dx}{(1 - x^4)^{1/2}} \quad \text{and showed}$$

$$A = \frac{\Pi(\frac{1}{4})\Pi(-\frac{1}{2})}{\Pi(-\frac{1}{4})}, \quad B = \frac{\Pi(\frac{3}{4})\Pi(-\frac{1}{2})}{3\Pi(\frac{1}{4})} = \frac{\Pi(-\frac{1}{4})\Pi(-\frac{1}{2})}{4\Pi(\frac{1}{4})}, \quad \text{so } AB = \frac{\pi}{4}.$$

The published paper virtually concludes with a study of a function ψ.
Gauss attributed to Euler the equation (now usually called Stirling's serie

$$\log \Pi(z) = (z + \frac{1}{2})\log z - z + \frac{1}{2}\log 2\pi + \frac{B_1}{1.2.z} - \frac{B_2}{3.4.z^3} + \frac{B_3}{5.6.z^5} + \dots \, ,$$

where B_1, B_2, B_3, etc. are the Bernoulli numbers, which are

defined by the expansion $\dfrac{x}{e^x - 1} = 1 - \dfrac{x}{2} + \sum \dfrac{(-1)^{k+1} B_k x^{2k}}{(2k)!}$

(so $B_1 = \frac{1}{6}$, $B_2 = \frac{1}{30}$, $B_3 = \frac{1}{42}$). He introduced

$\Psi(z): = \log z + \dfrac{1}{2z} - \dfrac{B_1}{2z^2} + \dfrac{B_2}{4z^4} - \dfrac{B_3}{6z^6} + \dots$ and showed $\dfrac{d}{dz}\Pi(z) = \Pi(z).\Psi(z)$,

so Ψ is the logarithmic derivative of Π. The last paragraph is a 2

page tabulation of z, $\log \Pi(z)$, and $\Psi(z)$ where $0 \leq z \leq 1$, z increases

in steps of 0.01, and the tabulated values are given to 18 decimal

places, thus surpassing Legendre's 12-place tables [1814] of $\log$

$\Gamma(z)$ over the same range. The paper frequently carries calculations

of specific values of functions to over 20 decimal places; such

calculations were both easy and congenial to Gauss.

Gauss began the second and unpublished part of the paper,

Determinatio series nostrae per Aequationem Differentialem Secundi

Ordinis, by observing that P: $= F(\alpha, \beta, \gamma; x)$ is a solution of the

differential equation

$$(x - x^2)\frac{d^2P}{dx^2} + (\gamma - (\alpha + \beta + 1)x)\frac{dP}{dx} - \alpha\beta P = 0. \tag{1.1.8}$$

To find a second linearly independent solution he set $1 - y = x$, when

the equation becomes

$$(y - y^2)\frac{d^2P}{dy^2} + (\alpha + \beta + 1 - \gamma - (\alpha + \beta + 1)y)\frac{dP}{dy} - \alpha\beta P = 0$$

which is the first equation with γ replaced by $\alpha + \beta + 1 - \gamma$. It,

therefore, has a solution $F(\alpha, \beta, \alpha + \beta + 1 - \gamma, 1 - x)$, and the

differential equation in general has solutions of the form

$$MF(\alpha, \beta, \gamma, x) + NF(\alpha, \beta, \alpha + \beta + 1 - \gamma, 1 - x), \qquad (\S39)$$

where M and N are constants.

Other solutions may arise which do not at first appear to be of this type, but, he remarked, any three solutions must satisfy a linear relationship with constant coefficients. This fact was of most use to him when transforming the differential equation by means of a change of variable. For instance, the substitution $P = x^{1-\gamma}P'$ transforms the differential equation into

$$(x - x^2) \frac{d^2P'}{dx^2} + (2 - \gamma - (\alpha + \beta + 3 - 2\gamma)x) \frac{dP'}{dx} - (\alpha + 1 - \gamma)(\beta + 1 - \gamma)P = 0$$

which has the general solution:

$$P' = MF(\alpha+1-\gamma, \quad \beta+1-\gamma, \quad 2-\gamma; x) + NF(\alpha+1-\gamma, \quad \beta+1-\gamma, \quad \alpha+\beta+1-\gamma; 1-x).$$

But Gauss was able to show

$$F(\alpha, \beta, \alpha + \beta + 1 - \gamma; 1 - x) = \frac{\Pi(\alpha + \beta - \gamma)\Pi(-\gamma)}{\Pi(\alpha - \gamma)\Pi(\beta - \gamma)} F(\alpha, \beta, \gamma; x)$$

$$+ \frac{\Pi(\alpha + \beta - \gamma)\Pi(\gamma - 2)}{\Pi(\alpha - 1)\Pi(\beta - 1)} x^{1-\gamma}(1 - x)^{\gamma-\alpha-\beta}F(1 - \alpha, 1 - \beta, 2 - \gamma; x)$$

He considered various substitutions to obtain a variety of equations relating the hypergeometric functions. For instance ($\S47$), setting $x = \frac{y}{y - 1}$ gives a new equation for P and y, whence, setting $P = (1 - y)^{\mu}P'$, an equation for P' and y. From this equation he deduced, when $\mu = \alpha$, $F(\alpha, \beta, \gamma, x) = (1-y)^{\alpha}F(\alpha, \gamma-\beta, \gamma, y)$

$$= (1-x)^{-\alpha}F(\alpha,\gamma-\beta,\gamma,\tfrac{-x}{1-x})$$

and when $\mu = \beta$, $\quad F(\alpha, \beta, \gamma, x) = (1-x)^{-\beta}F(\beta,\gamma-\alpha,\gamma,\tfrac{-x}{1-x})$.

The substitutions he considered are of two types: the following transformations of x: $x = 1 - y$, $x = \frac{1}{y}$, $x = \frac{y}{y-1}$, $x = \frac{y-1}{y}$, and these transformations of P: $P = x^{\mu}P'$, $P = (1 - x)^{\mu}P'$ for particular values of μ. These gave him several solutions to the original equation in terms of functions like $F(-, -, -, x)$ and $F(-, -, -, 1-x)$

etc., possibly multiplied by powers of x and $1 - x$, and also some
linear identities between these solutions. He also gave an
impressive calculation to illustrate the linear dependence of the
three solutions

$$P = F(\alpha, \beta, \gamma; x), \quad Q = x^{1-\gamma}F(\alpha + 1 - \gamma, \beta + 1 - \gamma, 2 - \gamma; x) \text{ and}$$

$$R = F(\alpha, \beta, \alpha + \beta + 1 - \gamma, 1 - x):$$

$$R = F(\alpha, \beta, \gamma)P + F(\alpha + 1 - \gamma, \beta + 1 - \gamma, 2 - \gamma)Q, \text{ where}$$

$$F(\alpha, \beta, \gamma) = \frac{\Pi(\alpha + \beta - \gamma)\Pi(-\gamma)}{\Pi(\alpha - \gamma)\Pi(\beta - \gamma)}.$$

The paper concluded with a discussion of certain special cases that
can arise when α, β, and γ are not independent, for example when
$\beta = \alpha + 1 - \gamma$, and the quadratic change of variable $x = 4y - 4y^2$
can be made.

Gauss made a very interesting observation at this point. The
equation has as one solution in this case $F(\alpha, \beta, \alpha + \beta + \frac{1}{2},$
$4y - 4y^2) = F(2\alpha, 2\beta, \alpha + \beta + \frac{1}{2}, y)$. If, he said, y is replaced by
$1 - y$, this produces $F(\alpha, \beta, \alpha + \beta + \frac{1}{2}, 4y - 4y^2)$
$= F(2\alpha, 2\beta, \alpha + \beta + \frac{1}{2}, 1 - y)$, and one is led to the seeming paradox
$F(2\alpha, 2\beta, \alpha + \beta + \frac{1}{2}, y) = F(2\alpha, 2\beta, \alpha + \beta + \frac{1}{2}, 1 - y)$ which equation
is certainly false (§55). To resolve the paradox he distinguished
between F as a function, which satisfies the hypergeometric equation,
and F as the sum of an infinite series. The latter is only defined
within its circle of convergence, but the former is to be understood
for all continuous changes in its fourth term, whether real or
imaginary, provided the values 0 and 1 are avoided. This being so,
he argued one would no more be misled than one would infer from arc $\sin \frac{1}{2}$
$= 30^{\circ}$ and that also $\sin 150^{\circ} = \frac{1}{2}$ that $30^{\circ} = 150^{\circ}$, for a (many-valued) function
may have different values even though its variable has taken the same
value, whereas a series may not.

Gauss here confronted the question of analytically continuing a function outside its circle of convergence. It was his view that the solutions of the differential equation exist everywhere but at 0, 1, (and ∞, although he avoided the expression). However, their representation in power series is a local question, and the same function may be represented in different ways. In particular, the series expression may not be recaptured if the variable is taken continuously along some path and restored to its original value.

Because he here talked of continuous change in the variable in the complex number plane, one may thus infer that Gauss here is truly discussing analytic continuation, and not merely the plurality of series solutions at a given point. In later terminology, used by Cauchy and Riemann in the 1850's, a function is <u>monodromic</u> if its analytic continuations always yield a unique value for the function at each point, and such considerations are called 'monodromy questions'. Gauss is therefore the first to raise the monodromy problem in the question of differential equations, albeit in the unpublished part of his paper. One may reasonably speculate that it was connected in his mind with the linear relations that exist between any three solutions at a point, but he does not say so explicitly[8].

In these papers, Gauss introduced a large class of functions of a complex variable which were defined by the hypergeometric equation and were capable of various expressions in series. The main direction of his research was in studying relationships between the series, which in turn provided information about the nature of the functions under consideration.

1.2 *Jacobi and Kummer*

Gauss published only a small part of his work on elliptic functions, and made no mention of it in his paper on the hypergeometric series. Kummer, the next author to discuss the series significantly, did so with a view to using them to explore the new functions, which by then had been announced publicly. So before discussing his work it will be necessary to look briefly at the theory of the elliptic functions as began by Euler and Legendre and developed by Abel and, in particular, by Jacobi.[9]

Elliptic functions

The equation $k(1 - k^2) \frac{d^2y}{dk^2} + (1 - k^2) \frac{dy}{dk} + ky = 0$, now called Legendre's equation for elliptic integrals of the second kind, was studied by Euler in [1733], because of its connection with the rectification of the ellipse, and again in [1750], when he found the solution for which $y = 1$ when $k = 0$ to be

$$y = 1 + \alpha k^2 + \beta k^4 \ldots + \log k(\gamma k^2 + \delta k^4 + \ldots)$$

(for suitable α, β, γ, δ, $\ldots$). This is interesting, for it shows that logarithmic terms may be expected near a singular point of the equation, a matter discussed more fully in Chapter II.

The equation

$$\frac{\partial}{\partial \mu} (1 - \mu^2) \frac{\partial U}{\partial \mu} + \frac{1}{(1 - \mu^2)} \frac{\partial^2 U}{\partial \phi^2} + n(n + 1)U = 0$$

was obtained by Laplace [1782] in a study of potential theory . When ϕ is absent the equation becomes Legendre's equation for elliptic integrals of the first kind, (1.2.1) below. Legendre used this equation (with ϕ absent) in the course of his own work on potential theory, [1793], and introduced the differential equations for elliptic

integrals of the first and second kinds in [1785], and, more

influentially, in his Exercises de Calcul intégral, [1814] and Traité des

Fonctions Elliptiques et des Intégrales Eulériennes [1825, vol 1,

Ch. 13] where he derived the equation

$$k(1 - k^2) \frac{d^2Q}{dk^2} + (1 - 3k^2) \frac{dQ}{dk} - kQ = 0 \qquad (1.2.1)$$

for the periods 4K and 2iK' as functions of the modulus k, by

differentiating the complete integral $\int_0^{\pi/2} \frac{d\phi}{\sqrt{(1 - k^2 \sin^2 \phi)}}$

with respect to k. Here

$$K: = \int_0^1 \frac{dx}{\sqrt{[(1 - x^2)(1 - k^2x^2)]}} \quad \text{and} \quad K': = \int_0^1 \frac{dx}{\sqrt{[(1 - x^2)(1 - k'^2x^2)]}}$$

and $k'^2: = 1 - k^2$ is the so-called complementary modulus.

Legendre's elliptic integrals are real, and the modulus c = sin θ

lies between 0 and 1, [Traité, Ch. 5]. So they define single-valued

functions of their upper end-points which have single-valued inverse

functions. Legendre's tables and his acoompanying comments make it clear

he regarded the problem of inversion as solved (in this case). In

the Exercises de calcul intégral ... [1811] he remarked (p. 380):

"The same formulae serve to solve the inverse problem, that is to say,

to determine the amplitude φ when one knows the function F(c, φ)."

In the Traité (p. 383) he said; "Thus being given an arbitrary value

of the angle ψ, one can find the corresponding value of the time t and

reciprocally." (Quoted in Krazer [1909, 55n].)

Once the upper end-point is allowed to vary in the complex plane,

the presence of the periods means that the integral

$$u = \int_0^\phi \frac{d\phi}{\sqrt{(1 - k^2 \sin^2 \phi)}} = \int_0^x \frac{dx}{\sqrt{[(1 - x^2)(1 - k^2 x)]}} \; ,$$

where $x = \sin \phi$, only defines an infinitely many-valued 'function'
of the upper end-point, for the path of integration may loop several
times around ± 1 and $\pm 1/k$. Jacobi (and Abel independently) had the idea
of studying instead the inverse functions $\phi = am(u)$, $x = x(u) = \sin am \; u$.
Jacobi was from the first particularly interested in the transformation
problem for this function, the change of modulus from k^2 to λ^2:

$$\frac{dy}{\sqrt{[(1 - y^2)(1 - \lambda^2 y^2)]}} = \frac{dx}{M\sqrt{[(1 - x^2)(1 - k^2 x^2)]}} \; . \qquad (1.2.2)$$

This is connected to the problem of relating the inverse functions
$\sin am \; u$ and $\sin am \; nu$ for odd integers n, which was suggested to him
by the evident analogy with $u = \int_0^{\sin u} \frac{dx}{\sqrt{(1 - x^2)}}$. But whereas the
equation connecting $\sin n \; \theta$ has n roots, that between $\sin am \; nu$ and $\sin$
$am \; u$ has n^2 when n is odd (the case n even is a little more
complicated). Jacobi explained this in terms of the double periodicity of $\sin am$, which, he showed, satisfied
$\sin am(u + 4K) = \sin am(u + 2iK') = \sin am \; u$, whence $4K$ and $2iK'$ are
the periods.

In his _Fundamenta Nova_ [1829] Jacobi, inspired by a paper of Abel's,
gave explicit rational transformations of the kind he sought, $y = \frac{U(x)}{V(x)}$,
and when U was a polynomial of order p said the transformation was of
order p. To take one of his examples, corresponding to the transformation of order 3, if the periods of the integral taken with
modulus λ are Λ and Λ', then one asks for $\frac{\Lambda'}{\Lambda} = 3 \frac{K'}{K}$. Jacobi found
that the substitution

$$y = \frac{x(a + a'x^2)}{1 + b'x^2} = \frac{U(x)}{V(x)}$$

where a = 1 + 2α, a' = α^2, and b' = α(2 + α) produces a complicated

expression for $\dfrac{dy}{\sqrt{[(1 - y^2)(1 - \lambda^2 y^2)]}}$ of the form $\dfrac{P(x)dx}{\sqrt{Q(x)}}$, where P(x)

is of degree 4 and Q(x) of degree 6. But, for suitable λ, P(x) occurs

squared as a factor of Q(x) and the expression in y reduces to

$\dfrac{dx}{M\sqrt{[(1 - x^2)(1 - k^2 x^2)]}}$ for some constant M. In this case a $= \sqrt{(\dfrac{k}{\lambda})} \cdot \dfrac{b'}{k}$

a' = $\sqrt{(\dfrac{k^3}{\lambda})}$, so, setting $4\sqrt{k}$ = u and $4\sqrt{\lambda}$ = v, he obtained this equation

connecting the moduli k and λ:

$$u^4 - v^4 + 2uv(1 - u^2 v^2) = 0. \quad [1829 \ \S13 = 1969, \ I, \ 74]$$

This equation, and the others like it for transformations of

higher order, will be discussed in Chapter IV when their significance

as polynomial equations will be discussed in the context of the

emerging Galois theory. Jacobi also derived differential equations

connecting k and λ, and since it was these equations which

interested Kummer, they will be presented here. Jacobi argued [Fund.

Nova §32-34=1969, I, 129-138] that if Q = aK + bK' and

Q' = a'K + b'K' are two solutions of Legendre's equations then their

quotient satisfies

$$d\left(\frac{Q'}{Q}\right) = -\frac{\pi}{2}\left(\frac{ab' - a'b}{k(1 - k^2)}\right)\frac{dk}{Q^2}.$$

The same equations hold for the periods Λ and Λ' taken with respect

to a different modulus λ

$$(\lambda - \lambda^3)\frac{d^2 L}{d\lambda^2} + (1 - 3\lambda^2)\frac{dL}{d\lambda} - \lambda L = 0.$$

So, if λ is obtained from k by a transformation of order n, i.e.

$\dfrac{\Lambda'}{\Lambda} = \dfrac{nK'}{K}$, then $\Lambda = \dfrac{K}{M}$, and the corollary of Legendre's equation implies

that

$$\frac{ndk}{k(1 - k^2)K^2} = \frac{d\lambda}{\lambda(1 - \lambda^2)\Lambda^2},$$

from which it follows that

$$M^2 = \frac{1}{n} \frac{\lambda(1 - \lambda^2)}{k(1 - k^2)} \frac{dk}{d\lambda}$$

If n is fixed and M is regarded as a function of k then Legendre's equations can be used to eliminate M from this equation, and the result is [10]

$$3\left(\frac{d^2\lambda}{dk^2}\right)^2 - 2 \frac{d\lambda}{dk} \cdot \frac{d^3\lambda}{dk^3} + \left(\frac{d\lambda}{dk}\right)^2 \cdot \left\{ \left[\frac{1 + k^2}{k - k^3}\right]^2 - \left[\frac{1 + \lambda^2}{\lambda - \lambda^3}\right]^2 \left(\frac{d\lambda}{dk}\right)^2 \right\} = 0.$$

(Jacobi's equation 12)   (1.2.3)

This equation, which may be called Jacobi's differential equation for the moduli, was one of the targets of Kummer's work. It has amongst its particular integrals the quotients of solutions of Legendre's equations:

$$\frac{a'K + b'K'}{aK + bK'} = \frac{\alpha'\Lambda + \beta'\Lambda'}{\alpha\Lambda + \beta\Lambda'}.$$

The second part of Gauss's paper on the hypergeometric series raises two main types of question. First, it would be useful to have a systematic account of the solutions obtained by the various substitutions, and of the nature of the substitutions themselves. Second, it would be instructive to connect the hypergeometric functions with the newer functions in analysis, especially in complex analysis, such as the elliptic functions. It is striking that Kummer's 1836 paper [Kummer 1836 = Coll Papers, II] sets itself both these tasks and resolves them while, moreover, observing Gauss's restrictions where the work would otherwise be too difficult (for example, by considering only real coefficients).

Kummer

Ernst Eduard Kummer (1810-1893) had studied Mathematics at Halle,
after first intending to study Protestant theology - a common enough
false start - and in 1836 was a lecturer at the Liegnitz Gymnasium.
His earliest work was in function theory, but from the mid-1840's
onwards he concerned himself with algebraic number theory, which he
came to dominate, and he is also remembered for his quartic surface
with 16 nodal points. Leopold Kronecker was one of his students at
Liegnitz, and with Kronecker and Weierstrass Kummer dominated the Berlin
school of mathematics from 1856 until his retirement in 1883. He was
a gifted teacher and organizer of seminars; he concerned himself
greatly with the fortunes of his many students, and his students were
correspondingly devoted to him. He was also a man of great charm, and
he had a great appetite for administration, being dean of the University
of Berlin twice, rector once, and perpetual secretary of the physics-
mathematics section of the Berlin Academie from 1863 to 1878. Although
he never attended a lecture by Dirichlet, he considered him to have been
his real teacher, and this is perhaps reflected in the topics which he
came to study most closely[11].

Kummer alluded briefly to other discussions of the hypergeometric
equation at the start of his long paper [1836]. Of Gauss's paper he
remarked: "But this work is only the first part of a greater work as
yet unpublished, and wants comparison of hypergeometric series in which
the last element x is different. This will therefore be the principal
purpose of the present work; the numerical application of the discovered
formulae will preferably be made to elliptic transcendents, to which
in great part the general series corresponds". To elucidate the first
problem, Kummer sought the most general transformation there could be
between two hypergeometric equations.

He considered two hypergeometric equations:

$$\frac{d^2y}{dx^2} + p\,\frac{dy}{dx} + qy = 0, \tag{1.2.4}$$

and

$$\frac{d^2v}{dz^2} + P\,\frac{dv}{dz} + Qv = 0, \tag{1.2.5}$$

where x and z are real, $z = z(x)$ is a function of x, and $y = w.v$ where $w = w(x)$ is a function to be determined. Eliminating y and v he showed

$$w^2 = c.(e^{\int Pdz - \int pdx})\,\frac{dx}{dz}. \tag{1.2.6}$$

He now remarked that if z was known as a function of x then w would also be known as a function of x, so he eliminated w and found[12]

$$2\,\frac{d^3z}{dx^3}\left(\frac{dz}{dx}\right)^{-1} - 3\left(\frac{d^2z}{dx^2}\right)^2\left(\frac{dz}{dx}\right)^{-2} - \left(2\,\frac{dP}{dz} + P - 4Q\right)\frac{dz}{dx} - \left(2\,\frac{dp}{dx} + p - 4q\right) = 0, \tag{1.2.7}$$

and so, he said, the only difficulty was to solve this equation and determine z as a function of x. Since Kummer stipulated that the transformations between the transcendental functions should be algebraic, he required algebraic solutions of (1.2.7.). The general problem of finding all algebraic solutions to a differential equation appeared to him to be impossible, but in this case he had shown in an earlier paper [1834] that the general solution of (1.2.7) was of the form

$$A\phi(x)\,\psi(z) + B\phi(x)\,\psi_1(z) + C\phi_1(x)\,\psi(z) + D\phi_1(x)\,\psi_1(z) = 0,$$

where $\phi(x)$ and $\phi_1(x)$ are independent solutions of (1.2.4) and $\psi(z)$ and $\psi_1(z)$ are independent solutions of (1.2.5), so

$$A\phi(x) + B\phi_1(x) = A'w\psi(z) + B'w\psi_1(z). \tag{1.2.8}$$

Indeed, equation (1.2.6)

implies $\dfrac{ce^{-\int pdx}\,dx}{(A\phi(x) + B\phi_1(x))^2} = \dfrac{e^{-\int Pdz}\,dz}{(A'\psi(z) + B'\psi_1(z))^2}$,

whence $\dfrac{A\phi(x) + B\phi_1(x)}{C\phi(x) + D\phi_1(x)} = \dfrac{A'\psi(z) + B'\psi_1(z)}{C'\psi(z) + D'\psi_1(z)}$ (1.2.9)

is the general solution of (1.2.7).

Kummer's analysis of the cases when certain relationships exist between α, β, and γ, and accordingly quadratic changes of variable are possible, must be looked at only briefly. It led him in fact to the same paradox as Gauss, a seemingly impossible equation between

 $F(-, -, -, x)$ and a sum of two $F(-, -, -, \frac{1}{x})$

for certain α's, β's and γ's. In this case either one side converges or the other but not both. Kummer took more or less Gauss's (un-published) view that the equality meant the same function was being represented in two ways, one valid when x < 1, the other when x > 1. But Kummer's discussion was confined to real x, and so lacks the concept of continuous change in x connecting the two branches. For Kummer x = 1 is a genuine barrier; the series fail to converge and are, so to speak, kept apart. When in Chapter VII he allowed x to become complex, he did not return to this problem, so one cannot infer that he was aware of monodromy considerations. This claim is made for him by Klein [1967, 267] and Biermann, [1973, 523] and several workers risk implying it when they connect Kummer's 24 solutions with the question of monodromy. While they are making an entirely permissible interpretation of Kummer's results, it is not one made by Kummer himself. The honour of discovery must go to Gauss, and the first to grasp the significance of the idea was Riemann.

Kummer's 24 solutions

In section II of his paper Kummer produced a set of 24 solutions to the hypergeometric equation which, in some sense, are the complete solution to the equation. To be precise, if the variable is allowed to be complex, they provide not only sets of bases for the solutions everywhere, but a description of their analytic continuation on the complex sphere. For this reason they are central to the insights of Riemann and Schwarz, as we shall see. However, in Kummer's work the variable is only real, and he regarded them as the best way to obtain solutions valid near the singular points in a variety of convenient forms. Let the coefficients in (1.2.4) and (1.2.5) be α, β, γ and α', β', γ'.

Kummer imposed the following simplifying requirements: z is to be a function of x alone, there are to be no relationships between α, β, and γ or α', β', and γ', but α', β', and γ' are to be linear combinations of α, β, and γ with constant coefficients[13]. Under these restrictions

Kummer found $z = \dfrac{ax + b}{cx + d}$ (48 = <u>Coll.Papers</u> II 84)

In §6 Kummer noted that for such a function z of x
$$2\left(\frac{d^3 z}{dx^3}\right)\left(\frac{dx}{dz}\right) - 3\left(\frac{d^2 z}{dx^2}\right)^2 \left(\frac{dx}{dz}\right)^2 = 0,$$ simplifying (1.2.7) considerably. It becomes

$$\frac{Ax^2 + Bx + C}{x^2(1-x)^2} = \frac{(ad-bc)^2(A'(ax+b)^2+B'(ax+b)(cx+d)+C'(cx+d)^2)}{(ax+b)^2(cx+d)^2((c-a)x+d-b)^2}$$
$$(1.2.10)$$

There can be no common factors either side since α, β, γ are arbitrary so the two numerators can only differ by a constant factor m (say). There are precisely six solutions to the two equations that arise for m by equating the numerators and denominators separately:

TABLE 2

21.
$$\begin{cases} F = Ax^{1-\gamma}F(\alpha-\gamma+1, \beta-\gamma+1, 2-\gamma, x) + BF(\alpha, \beta, \alpha+\beta-\gamma+1, 1-x), \\[4pt] A = \dfrac{\Pi(\gamma-1)\Pi(\alpha-\gamma)\Pi(\beta-\gamma)}{\Pi(1-\gamma)\Pi(\alpha-1)\Pi(\beta-1)}, \qquad B = \dfrac{\Pi(\alpha-\gamma)\Pi(\beta-\gamma)}{\Pi(\alpha+\beta-\gamma)\Pi(-\gamma)}, \end{cases}$$

22.
$$\begin{cases} F = A_1 x^{1-\gamma}F(\alpha-\gamma+1, \beta-\gamma+1, 2-\gamma, x) \\[2pt] \qquad + B_1(1-x)^{\gamma-\alpha-\beta}F(\gamma-\alpha, \gamma-\beta, \gamma-\alpha-\beta+1, 1-x), \\[4pt] A_1 = \dfrac{\Pi(1-\gamma)\Pi(-\alpha)\Pi(-\beta)}{\Pi(1-\gamma)\Pi(\gamma-\alpha-1)\Pi(\gamma-\beta-1)}, \qquad B_1 = \dfrac{\Pi(-\alpha)\Pi(-\beta)}{\Pi(\gamma-\alpha-\beta)\Pi(-\gamma)}, \end{cases}$$

23.
$$\begin{cases} F = A_2 F(\alpha, \beta, \alpha+\beta-\gamma+1, 1-x) \\[2pt] \qquad + B_2(1-x)^{\gamma-\alpha-\beta}F(\gamma-\alpha, \gamma-\beta, \gamma-\alpha-\beta+1, 1-x), \\[4pt] A_2 = \dfrac{\Pi(\gamma-1)\Pi(\gamma-\alpha-\beta-1)}{\Pi(\gamma-\alpha-1)\Pi(\gamma-\beta-1)}, \qquad B_2 = \dfrac{\Pi(\gamma-1)\Pi(\alpha+\beta-\gamma-1)}{\Pi(\alpha-1)\Pi(\beta-1)}, \end{cases}$$

24.
$$\begin{cases} F = A_3(1-x)^{-\alpha}F\!\left(\alpha, \gamma-\beta, \alpha-\beta+1, \dfrac{1}{1-x}\right) \\[2pt] \qquad + B_3(1-x)^{-\beta}F\!\left(\beta, \gamma-\alpha, \beta-\alpha+1, \dfrac{1}{1-x}\right), \\[4pt] A_3 = \dfrac{\Pi(\gamma-1)\Pi(\alpha-\gamma)\Pi(-\beta)}{\Pi(1-\gamma)\Pi(\alpha-1)\Pi(\gamma-\beta-1)}, \qquad B_3 = \dfrac{\Pi(-\beta)\Pi(\alpha-\gamma)}{\Pi(\alpha-\beta)\Pi(-\gamma)}, \end{cases}$$

25.
$$\begin{cases} F = A_4(1-x)^{-\alpha}F\!\left(\alpha, \gamma-\beta, \alpha-\beta+1, \dfrac{1}{1-x}\right) \\[2pt] \qquad + B_4(1-x)^{-\beta}F\!\left(\beta, \gamma-\alpha, \beta-\alpha+1, \dfrac{1}{1-x}\right), \\[4pt] A_4 = \dfrac{\Pi(\gamma-1)\Pi(\beta-\gamma)\Pi(-\alpha)}{\Pi(1-\gamma)\Pi(\beta-1)\Pi(\gamma-\alpha-1)}, \qquad B_4 = \dfrac{\Pi(-\alpha)\Pi(\beta-\gamma)}{\Pi(\beta-\alpha)\Pi(-\gamma)}, \end{cases}$$

26.
$$\begin{cases} F = A_5(1-x)^{-\alpha}F\!\left(\alpha, \gamma-\beta, \alpha-\beta+1, \dfrac{1}{1-x}\right) \\[2pt] \qquad + B_5(1-x)^{-\beta}F\!\left(\beta, \gamma-\alpha, \beta-\alpha+1, \dfrac{1}{1-x}\right), \\[4pt] A_5 = \dfrac{\Pi(\gamma-1)\Pi(\beta-\alpha-1)}{\Pi(\beta-1)\Pi(\gamma-\alpha-1)}, \qquad B_5 = \dfrac{\Pi(\gamma-1)\Pi(\alpha-\beta-1)}{\Pi(\alpha-1)\Pi(\gamma-\beta-1)}. \end{cases}$$

TABLE 1

1) $F(\alpha, \beta, \gamma, x)$,

2) $(1-x)^{\gamma-\alpha-\beta}F(\gamma-\alpha, \gamma-\beta, \gamma, x)$,

3) $x^{1-\gamma}F(\alpha-\gamma+1, \beta-\gamma+1, 2-\gamma, x)$,

4) $x^{1-\gamma}(1-x)^{\gamma-\alpha-\beta}F(1-\alpha, 1-\beta, 2-\gamma, x)$,

5) $F(\alpha, \beta, \alpha+\beta-\gamma+1, 1-x)$,

6) $x^{1-\gamma}F(\alpha-\gamma+1, \beta-\gamma+1, \alpha+\beta-\gamma+1, 1-x)$,

7) $(1-x)^{\gamma-\alpha-\beta}F(\gamma-\alpha, \gamma-\beta, \gamma-\alpha-\beta+1, 1-x)$,

8) $x^{1-\gamma}(1-x)^{\gamma-\alpha-\beta}F(1-\alpha, 1-\beta, \gamma-\alpha-\beta+1, 1-x)$,

9) $x^{-\alpha}F\left(\alpha, \alpha-\gamma+1, \alpha-\beta+1, \dfrac{1}{x}\right)$,

10) $x^{-\beta}F\left(\beta, \beta-\gamma+1, \beta-\alpha+1, \dfrac{1}{x}\right)$,

11) $x^{\alpha-\gamma}(1-x)^{\gamma-\alpha-\beta}F\left(1-\alpha, \gamma-\alpha, \beta-\alpha+1, \dfrac{1}{x}\right)$,

12) $x^{\beta-\gamma}(1-x)^{\gamma-\alpha-\beta}F\left(1-\beta, \gamma-\beta, \alpha-\beta+1, \dfrac{1}{x}\right)$,

13) $(1-x)^{-\alpha}F\left(\alpha, \gamma-\beta, \alpha-\beta+1, \dfrac{1}{1-x}\right)$,

14) $(1-x)^{-\beta}F\left(\beta, \gamma-\alpha, \beta-\alpha+1, \dfrac{1}{1-x}\right)$,

15) $x^{1-\gamma}(1-x)^{\gamma-\alpha-1}F\left(\alpha-\gamma+1, 1-\beta, \alpha-\beta+1, \dfrac{1}{1-x}\right)$,

16) $x^{1-\gamma}(1-x)^{\gamma-\beta-1}F\left(\beta-\gamma+1, 1-\alpha, \beta-\alpha+1, \dfrac{1}{1-x}\right)$,

17) $(1-x)^{-\alpha}F\left(\alpha, \gamma-\beta, \gamma, \dfrac{x}{x-1}\right)$,

18) $(1-x)^{-\beta}F\left(\beta, \gamma-\alpha, \gamma, \dfrac{x}{x-1}\right)$,

19) $x^{1-\gamma}(1-x)^{\gamma-\alpha-1}F\left(\alpha-\gamma+1, 1-\beta, 2-\gamma, \dfrac{x}{x-1}\right)$,

20) $x^{1-\gamma}(1-x)^{\gamma-\beta-1}F\left(\beta-\gamma+1, 1-\alpha, 2-\gamma, \dfrac{x}{x-1}\right)$,

21) $x^{-\alpha}F\left(\alpha, \alpha-\gamma+1, \alpha+\beta-\gamma+1, \dfrac{x-1}{x}\right)$,

22) $x^{-\beta}F\left(\beta, \beta-\gamma+1, \alpha+\beta-\gamma+1, \dfrac{x-1}{x}\right)$,

23) $x^{\alpha-\gamma}(1-x)^{\gamma-\alpha-\beta}F\left(1-\alpha, \gamma-\alpha, \gamma-\alpha-\beta+1, \dfrac{x-1}{x}\right)$,

24) $x^{\beta-\gamma}(1-x)^{\gamma-\alpha-\beta}F\left(1-\beta, \gamma-\beta, \gamma-\alpha-\beta+1, \dfrac{x-1}{x}\right)$.

$$c = 0 \quad = b = a - d, \qquad m = a^6$$

$$c = 0 \quad = d - b = a + b, \; m = a^6$$

$$a = 0 \quad = d = c - b, \qquad m = b^6 \tag{1.2.11}$$

$$a = 0 \quad = d - b = c + d, \; m = b^6$$

$$c-a = 0 = b = c + d, \qquad m = a^6$$

$$c-a = 0 = d = a + b, \qquad m = b^6$$

for which the substitutions are precisely $z = x$, $z = 1 - x$, $z = \dfrac{1}{x}$,

$z = \dfrac{1}{1 - x}$, $z = \dfrac{x}{x - 1}$, $z = \dfrac{x - 1}{x}$ respectively.

In the first case, there are four possible substitutions for
α', β', γ', in place of α, β, γ

$$(\alpha', \; \beta', \; \gamma') = (\alpha, \; \beta, \; \gamma)$$

$$= (\gamma - \alpha, \; \gamma - \beta, \; \gamma)$$

$$= (\alpha - \gamma + 1, \; \beta - \gamma + 1, \; 2 - \gamma)$$

$$= (1 - \alpha, \; 1 - \beta, \; 2 - \gamma) \qquad (50 = \underline{\text{Coll.Papers}} \text{ II, 86})$$

as can be seen from the equation $Ax^2 + Bx + C = A'x^2 + B'x + C'$ upon

replacing A by $(\alpha - \beta)^2$, A' by $(\alpha' - \beta')^2$, and so on.

Furthermore, there are 4 substitutions in α, β, γ for each of the
other substitutions of z for x giving rise to 24 solutions to the
differential equations, which Kummer listed in §8. They have become
known as Kummer's 24 solutions to the hypergeometric equation, and
they are displayed in Table 1 [52, 53 = $\underline{\text{Coll.Papers}}$, II, 88, 89].

Independently of Gauss, Kummer pointed out that several of these
substitutions could be spotted without going through the argument
above. For example (1.2.7) is unaltered by the substitution of
$\gamma' - \alpha'$ for α', $\gamma' - \beta'$ for β', so from the solution $F(\alpha', \beta', \gamma', z)$
the solution $(1 - z)^{\gamma'-\alpha'-\beta'} F(\gamma' - \alpha', \gamma' - \beta', \gamma', z)$ is obtained.

Kummer's analysis presents a complete answer to the first problem: what are the allowable changes of variable for the general hypergeometric equation? It immediately raises the second question: what are the solutions themselves? As Kummer pointed out (§9), there are many relationships between the solutions. Some, in any case, are equal to others, for example

$$F(\alpha, \beta, \gamma, x) = (1 - x)^{\gamma-\alpha-\beta}F(\gamma - \alpha, \gamma - \beta, \gamma, x)$$

$$= (1 - x)^{-\alpha}F(\alpha, \gamma - \beta, \gamma, \frac{x}{x-1}).$$

In fact, the six families of four solutions rearrange themselves into six different families of four equal solutions thus: 1,2,17, and 18; 3,4,19 and 20; 5,6,21,22; 7,8,23,24; 9,12,13,15; and 10,11,14,16. So to find all the linear relations between the twenty-four solutions it is enough to consider the six different ones 1,3,5,7,13,14. Of these, 5 and 7 converge or diverge exactly when 13 and 14 diverge or converge respectively. Kummer here restricted x to be real, but the observations is valid for complex x. The problem is thus reduced to finding the relations between the following triples: 1,3,5; 1,3,7; 1,3,13; 1,3,14. As an example of Kummer's results, the relationship between 1,3, and 5 is:

$$F(\alpha, \beta, \gamma, x) = \frac{\Pi(\gamma-1)\Pi(\alpha-\gamma)\Pi(\beta-\gamma)}{\Pi(1-\gamma)\Pi(\alpha-1)\Pi(\beta-1)} F(\alpha - \gamma + 1, \beta - \gamma + 1, 2 - \gamma, x)x^{1-}$$

$$+ \frac{\Pi(\alpha-\gamma)\Pi(\beta-\gamma)}{\Pi(\alpha+\beta-\gamma)\Pi(-\gamma)} F(\alpha, \beta, \alpha + \beta - \gamma + 1, 1 - x),$$

where Π stands for Gauss's factorial function. He listed the relationship that arise in §11 (see Table 2). They all arise from evaluating $F(-, -, -, x)$ at 0 or 1, and hence obtaining equations for the A's and B's.

Chapter 6 of Kummer's paper is his analysis of the transcendental

functions which can be represented by the hypergeometric functions. He

found that $F(\frac{1}{2}, \frac{1}{2}, 1, c^2)$ was already well known in analysis, for

$$F(\tfrac{1}{2}, \tfrac{1}{2}, 1, c^2) = \frac{1}{\pi} \int_0^1 u^{-\frac{1}{2}} (1-u)^{-\frac{1}{2}} (1 - c^2 u)^{-\frac{1}{2}} du$$

$$= \frac{2}{\pi} \int_0^{\pi/2} \frac{d\phi}{(1 - c^2 \sin \phi^2)^{\frac{1}{2}}}$$

$$= F^1(c),$$

an elliptic integral of the first kind, in Legendre's terminology

[1825, 11], and $F(-\frac{1}{2}, \frac{1}{2}, 1, c^2) = \frac{2}{\pi} \int_0^{\pi/2} (1 - c^2 \sin \phi^2)^{1/2} d\phi$

$$= E^1(c),$$

an elliptic integral of the second kind. Legendre had shown that

$F^1(c)$ satisfies $(1 - c^2) \dfrac{d^2 y}{dc^2} + \dfrac{1 - 3c^2}{c} \dfrac{dy}{dc} \cdot y = 0$, and that $E^1(c)$

satisfies $(1 - c^2) \dfrac{d^2 y}{dc^2} + \dfrac{1 - c^2}{c} \dfrac{dy}{dc} + y = 0$. These equations were now

identified by Kummer as special cases of the hypergeometric equation

(§29). In the course of deriving them Legendre had deduced his famous

relationship between the periods of elliptic integrals:

$$F^1(c) \, E^1(b) + F^1(b) \, E^1(c) - F^1(b) \, F^1(c) = \frac{\pi}{2} \text{ (where } b^2 = 1 - c^2\text{)}.$$

Kummer derived this result as a consequence of his theory (§30). As

has been remarked, his interest in such matters had been awakened by

the related equations for the transformations of elliptic functions

found by Jacobi, which had been the subject of his [1834].

Kummer concluded the paper with an unremarkable study of what

happens when x is allowed to be complex but α, β, and γ stay real,

largely devoted to the study of special functions. The first significant

advance on Kummer's work was to be made by Riemann, who went at once
to a general discussion of the complex case. This will be discussed
next, after a brief sketch of its context for since Riemann's ideas
about complex function theory form a coherent whole and some of his
ideas about Riemann surfaces will be encountered again later on, a
brief synopsis of them has been provided which may serve as a context
for his theory of functions defined by differential equations. The
contemporary theory of differential equations themselves is described
at the end of the chapter. Material which Riemann expounded in lectures,
and which seems only to have reached a significant audience well after
his death, is discussed in Appendix 2.

1.3 *Riemann's approach to complex analysis.*

The central theme of both Riemann's mathematics and his physics is
that of the complex function, understood geometrically. Riemann sought
to prise the independent variable of a complex function off the complex
plane and free it to roam over a more general surface. This removed
an unnecessary constraint upon mathematicians' attitudes to such functions,
and opened the way to a topological study of their properties. To
accomplish this Riemann gave a purely local definition of a complex
function, so that it may be equally well regarded as defined on a patch
of surface or on a part of the plane. He then sought global restrictions
determining the nature of function under certain given conditions. The
same dialectic between local and global properties can be found in
other of his works not particularly concerned with complex variables,
for example in his work on the foundations of geometry. It differs
considerably from the emphasis on convergent power series and the theory
of analytic continuation[14] of his influential contemporary Karl
Weierstrass. Where Weierstrass worked outwards from a function defined

by an infinite series on a disc towards the complete function, and used analytic tools, Riemann sought to anchor his global ideas in the specifics of a given problem. There is an interesting contrast, within the study of differential equations, between Riemann and Lazarus Fuchs, an exponent of the Berlin school whose work is considered in the next chapter. First, however, Riemann's general view of the theory of functions of a complex variable will be considered.

The papers [1851, 1857c] have become famous for several reasons. They introduced what are now called Riemann surfaces, in the form of domains spread out over the complex plane. They presented enough tools to classify all compact orientable surfaces, and so gave a great impetus to topology[15]. They provided a topological meaning for an otherwise unexplained constant which entered into Abel's work on Abelian integrals (see Chapter V), and more generally gave a geometric framework for all of complex analysis. They are thus the first mature, though obscure, papers in the study of topology of manifolds and are equally decisive for the development of algebraic geometry and the geometric treatment of complex analysis. Riemann's own use of these ideas in his study of Abelian functions and integrals on algebraic curves is perhaps the greatest indication of their profundity, and will be described below (Chapter V). This chapter concentrates on the implications of his idea of a complex function for the theory of differential equations, which Riemann presented in an earlier paper of 1857, the "Beiträge zur Theorie der durch Gauss'sche Reihe $F(\alpha, \beta, \gamma, x)$ darstellbaren Functionen" [1857a]. His theory of functions will be expounded in this section, and its use in [1857a] in section 1.4.

In his inaugural dissertation, [1851], Riemann defined a function of
a   complex variable in this way (§1): "A complex variable w is called a
function of another complex variable z if it varies with the other in
such a way that the value of the derivative $\frac{dw}{dz}$ is independent of the
value of the differential dz." This is equivalent to the modern
definition of an analytic function w as being differentiable at z_0
(as a function of z) in such a way that $\frac{dw}{dz}$ at z_0 is independent of
the path of z as it tends to z_0. The function defines a surface in
$\mathfrak{C}^2$, which Riemann said may be considered as spread out over the
complex z-plane. It follows from the definition that infinitesimal
neighbourhoods of z and w(z) are conformally equivalent, as Riemann
showed in §3, unless the corresponding variations in z and w cease to
have a finite ratio to one another which he said it was tacitly assumed
they did not. Riemann observed in a footnote that his matter had been
thoroughly discussed by Gauss in [Gauss 1822]. Furthermore, if w = u + vi
the functions u and v satisfy the equations $\frac{\partial u}{\partial x} = \frac{\partial v}{\partial y}$, $\frac{\partial v}{\partial x} = -\frac{\partial u}{\partial y}$ (the
Cauchy-Riemann equations) and consequently separately satisfy

$$\frac{\partial^2 u}{\partial x^2} + \frac{\partial^2 u}{\partial y^2} = 0, \quad \frac{\partial^2 v}{\partial x^2} + \frac{\partial^2 v}{\partial y^2} = 0.$$ Riemann remarked (§§2-4) that these
equations can be used to study the individual properties of u and v
and thus the complex function w = u + vi.

However, he said, it is not necessary to assume z lies in $\mathfrak{C}$. It
may lie in some finite domain T, having a boundary, spread out over
$\mathfrak{C}$, and covering the plane several times. The different parts of surface
covering each region of the plane are joined together at points, but
not along lines. Under these conditions the number of times a point
is covered is completely determined with the boundary and interior are
specified, but the form of the covering may be different. More

precisely, the surface winds around various branch points at which
the various 'leaves' (Flächentheile, literally 'pieces of surface')
are interchanged in cycles. These may be considered as copies of parts
of the plane (cut by a line emanating from the branch point) and joined
up according to a certain rule.

A point at which m leaves are interchanged was said by Riemann to
have order m - 1. Once these branch points are determined, so is the
surface (up to a finite number of different shapes deriving from the
arbitrariness in the choice of original leaf in each cycle). Functions
can be defined on T provided there is not a line of exceptional points
(in which case the function would not be differentiable). In a footnote
Riemann observed that this restriction on the set of singular points
did not derive from the idea of a function, but from the conditions under
which the integral calculus can be applied. He gave as an example of
a function discontinuous everywhere in the (x, y) plane the function
which takes the value 1 when x and y are commensurable and the value 2
otherwise. Dirichlet [1829] had given the example of a function which
takes the value c on the rationals and d ≠ c on the irrationals.

The connectivity of T is important and Riemann proceeded as
follows. Two parts of a surface were said to be connected if any point
of one can be joined to any point of the other by a curve lying entirely
in the surface. Strictly, this defines the modern concept of path
connected, but the notions of connectedness and path connectedness
agree here, since Riemann's surfaces are manifolds. A boundary cut
[Querschnitt] is a curve which joins two boundary points without cutting
itself. A (bounded) connected surfaces was said by Riemann to be simply
connected if any boundary cut makes it disconnected; it then falls into
two simply connected pieces. It was said to be n-fold connected if it

can be made simply connected by n − 1 suitably chosen boundary cuts;
the number n is well-defined and independent of the choice of cuts [§6].
The purpose of introducing these concepts was to generalize Cauchy's
theorem on contour integration to T. In general the integral of a
complex function taken around a closed curve in T which contains no
poles of the function does not vanish, and Cauchy's theorem is true only
when T is simply connected[16]. From this observation the whole of the
elementary theory of analytic functions can be generalized to functions
satisfying the Cauchy-Riemann equations locally on T. For instance,
such functions, w , are infinitely differentiable, and locally one-to-one
except near branch point. Near a branch point z' of order (n − 1) the
function becomes one-to-one if $(z - z')^{1/n}$ is taken as the new variable.
The image of T is again a surface, S, and the inverse of w is an analytic
function z = z(w) [§15].

Riemann was prepared to use power series or Fourier series methods
to express a function locally. He described such methods in his [1857c]
as standard techniques. Where he differed from Weierstrass was in his
emphasis on geometrical reasoning, which was avoided in Berlin. In
particular Riemann was prepared to use Dirichlet's principle to
guarantee the existence of functions without seeking them necessarily
in any other form. For him a function was known once a certain topological property was known about it (the connectivity of the surface
it defined) and once certain discrete facts were known (the nature
of its singular points). The careful separation of these two kinds
of data is characteristic of Riemann, and even more so is his brilliant
yoking together of the two. It occurs again, for example, in his
study of the zeta function $\zeta(s) = \Sigma n^{-s}$ (Re(s) > 2) [1859]. In a more
diffuse form the polarity of continuous and discrete haunts nearly

all his work. It also manifests itself as an interplay between global
and local properties, and it is in this guise that it appears in his
work on differential equations, where it will be seen that he
immediately looked for the number of leaves and for the branch points
of the solution functions.

1.4 Riemann's P-functions.

The paper discussed in this section presents Riemann's analysis,
[1857a], of the hypergeometric functions as P-function (defined
below). Riemann's remarkable extension of the theory of the hypergeometric equation as given in his lecture notes of 1858/9 will be
discussed in Appendix 2.

It will be recalled that Kummer had found 24 solutions to the
hypergeometric equation in the form of a hypergeometric series in
x, $\frac{1}{x}$, $1 - x$ etc, possibly multiplied by some powers of x or $(1 - x)$.
Each solution was therefore presented in a form which restricted
it to a certain domain and the relationship between overlapping
solutions was given. In [1857a] Riemann observed that this method
of passing from the series to the function it represents depends on the
differential equation itself. It would be possible, he said, to study
the solutions expressed as definite integrals, although the theory was
not yet sufficiently developed (a task soon to be taken up by Schläfli
[1870] and by Riemann himself in lectures [1858/59, Nachträge 69-94]).
However, he proposed to study the hypergeometric equation according to
his new, geometric methods, which were essentially applicable to all
linear differential equations with algebraic coefficients - a rather
dramatic claim.

Riemann began by specifying geometrically the functions he intended to study. Any such function P is to satisfy three properties:

1. It has three distinct branch points at a, b, and c, but each branch is finite at all other points:

2. A linear relation with constant coefficients exists between any three branches P', P", P''' of the function: $c'P' + \bar{c}"P" + c'''P''' = 0$;

3. There are constants α and α', called the exponents, associated with the branch point a, such that P can be written as a linear combination of two branches $P^{(\alpha)}$ and $P^{(\alpha')}$ near a, $(z-a)^{-\alpha} P^{(\alpha)}$ and $(z-a)^{-\alpha'} P^{(\alpha')}$ are single valued, and neither zero nor infinite at a. Similar conditions hold at b and c with constants β, β' and γ, γ' respectively.

To eliminate troublesome special cases Riemann futher assumed that none of $\alpha - \alpha'$, $\beta - \beta'$, $\gamma - \gamma'$ are integers, and that, furthermore, the sum $\alpha + \alpha' + \beta + \beta' + \gamma + \gamma' = 1$.

He denoted such a function of z

$$P \left\{ \begin{matrix} a & b & c \\ \alpha & \beta & \gamma \\ \alpha' & \beta' & \gamma' \end{matrix} \quad z \right\} \quad \text{or} \quad P \left(\begin{matrix} \alpha & \beta & \gamma \\ \alpha' & \beta' & \gamma' \end{matrix} \right) \text{when } (a, b, c) = (0, \infty, 1).$$

The first and third conditions express the nature of a P-function, as Riemann called them, in terms of the singularities at a, b, and c: for example, $P^{(\alpha)}$ is branched like $(x-a)^{\alpha}$. The second condition expresses the global relationship between the leaves and says that there are at most two linearly independent determinations of the function under analytic continuation of the various separate branches. It is not immediately clear that this information specifies a function exactly; in fact it turns out that it defines P up to a constant multiple.

When a, b, and c take the value 0, ∞, and 1 respectively, as they may be assumed to do without loss of generality, the analogy between P-functions and hypergeometric functions becomes clear. There are two linearly independent solutions of the hypergeometric equation at each singular point, they are branched according to certain expressions in α, β, and γ, and any three solutions are linearly dependent. Riemann showed that information of this kind about the solutions determines the differential equation completely. This goes some way to explain the great significance of the equation, and to illuminate the difficulties we shall see others were to experience in generalizing from it to other equations.

To investigate the global behaviour of P functions under analytic continuation Riemann argued that it is enough to specify their behaviour under circuits of the branch points. When two linearly independent branches P' and P'', say, are continued analytically in a loop around the branch point a in the positive (anti-clockwise) direction they return as two other branches, $\widetilde{P}'$ and $\widetilde{P}''$, say. But then

$$\widetilde{P}' = a_1 P' + a_2 P''$$

$$\widetilde{P}'' = a_3 P' + a_4 P''$$

for some contants a_1, a_2, a_3, a_4, so in some sense the matrix

$A = \begin{pmatrix} a_1 & a_2 \\ a_3 & a_4 \end{pmatrix}$ describes what happens at a. Let B and C be the matrices which describe the behaviour of P', P'' under analytic continuation around b and c respectively[17]. A circuit of a and b can be regarded as a circuit of c in the opposite direction, so

$$CBA = \begin{pmatrix} 1 & 0 \\ 0 & 1 \end{pmatrix}.$$

Any closed path can be written as a product of loops around a, b, or c in the same order, or, as Riemann remarked; "the coefficients of A, B, and C completely determine the periodicity of the function." The choice of the word 'periodicity' ('Periodicität') is interesting, suggesting a connection with doubly-periodic functions and the moduli('Periodicitäts-moduln') of Abelian functions via the use of closed loops.

The idea of using such a matrix to describe how an algebraic function is branched had been introduced by Hermite [1851], but it seems that Riemann is the first to have considered products of such matrices. We shall see that he effectively determined the monodromy group of the hypergeometric equation, i.e., the group generated by the matrices A, B, C. The term 'monodromy group' was first used by Jordan [Traité, 278] and its subsequent popularity derives from its successful use by Jordan and Klein, to be discussed in Chapter III.

For definiteness Riemann supposed a = 0, b = ∞, c = 1, and chose branches P^{α}, $P^{\alpha'}$, P^{β}, $P^{\beta'}$, P^{γ}, $P^{\gamma'}$ as in (3) above. A circuit around a in the postive direction returns P^{α} as $e^{2\pi i\alpha}P^{\alpha}$ and $P^{\alpha'}$ as $e^{2\pi i\alpha'}P^{\alpha'}$ so

$$A = \begin{pmatrix} e^{2\pi i\alpha} & 0 \\ 0 & e^{2\pi i\alpha'} \end{pmatrix}$$

To express the effect on P^{α} and $P^{\alpha'}$ of a circuit around b = ∞ he replaced them by their expressions in terms of P^{β} and $P^{\beta'}$, conducted the new expressions around ∞, and then changed them back into P^{α} and $P^{\alpha'}$, by writing

$$P^{\alpha} = \alpha_{\beta}P^{\beta} + \alpha_{\beta'}P^{\beta'}$$

$$P^{\alpha'} = \alpha'_{\beta} P^{\beta} + \alpha'_{\beta'} P^{\beta'}$$

or more briefly[18]

$$\begin{pmatrix} P^{\alpha} \\ P^{\alpha'} \end{pmatrix} = B' \begin{pmatrix} P^{\beta} \\ P^{\beta'} \end{pmatrix}$$

Then $B = B' \begin{pmatrix} e^{2\pi i \beta} & 0 \\ 0 & e^{2\pi i \beta'} \end{pmatrix} B'^{-1}$

By similarly conducting P^{α} and $P^{\alpha'}$ around $c = 1$, Riemann found

$C = C' \begin{pmatrix} e^{2\pi i \gamma} & 0 \\ 0 & e^{2\pi i \gamma'} \end{pmatrix} C'^{-1}$, where C' is the matrix relating P^{α}

and $P^{\alpha'}$ to P^{γ} and $P^{\gamma'}$:

$$\begin{pmatrix} P^{\alpha} \\ P^{\alpha'} \end{pmatrix} = C' \begin{pmatrix} P^{\gamma} \\ P^{\gamma'} \end{pmatrix}$$

Since CBA = I, it follows on taking determinants that $\det(C) \det(B) \det(A) = 1 = e^{2\pi i (\alpha + \alpha' + \beta + \beta' + \gamma + \gamma')}$, so $\alpha + \alpha' + \beta + \beta' + \gamma + \gamma'$ must be an integer. Riemann assumed it was in fact 1.

Riemann showed that the entries in B' and C' could be expressed in terms of the six coefficients $\alpha, \ldots, \gamma'$. In the published paper he contented himself with expressions relating the various ratios $\frac{\alpha_{\gamma}}{\alpha'_{\gamma}}$, $\frac{\alpha_{\beta}}{\alpha'_{\beta}}$, etc. but the precise values of $\alpha_{\beta}, \ldots, \alpha'_{\gamma'}$ can be also be determined; Riemann had obtained them himself in July of the previous year. It is most convenient to do so by obtaining the differential equation which the P-function satisfies and thence comparing its various branches with various branches of the 24 solutions obtained by Kummer. To do this, one must first see that the nine quantities a, b, c, $\alpha, \ldots, \gamma'$ define P up to a constant multiple.

Riemann did this in §4 of his paper. The method involves showing from the definition of a P-function that any two with the same a, b, c, α, ...,γ' have a constant quotient.

The same argument also shows that given two branches P' and P" of a P-function, whose quotient is not a constant, any other P-function having the same branch points and the same exponents can be expressed linearly as a sum C'P' + C"P".

Riemann went on to study P-functions whose exponents α, ..., γ' and $\tilde{\alpha}$, ..., $\tilde{\gamma}$' say, differ, by integers. These are Gauss's contiguous functions transformed to Riemann's setting.

In this way he could determine the differential equation the P-function satisfies. For, P = y, $P_1 = \frac{dy}{dx}$, and $P_2 = \frac{d^2y}{dx^2}$ are three such P-functions and, Riemann showed, they satisfy a linear relationship with coefficients certain rational functions in x. When γ = 0 he found explicitly that P satisfies the hypergeometric equation in this form:

$$(1 - z) \frac{d^2y}{dlogz^2} - (A + Bz) \frac{dy}{dlogz} + (A' - B'z)y = 0.$$

Riemann could therefore connect his P-functions with the functions F(α, β, γ, x) of Gauss:

$$F(a, b, c, z) = const. \ P^{\alpha} \begin{pmatrix} 0 & a & 0 \\ 1-c & b & c-a-b \end{pmatrix} z \end{pmatrix}.$$

So Riemann had shown that the branching data of the P-function determined its monodromy relations (later generations would say determine the monodromy group of the equation explicitly) and moreover he had established that the hypergeometric equation is the only second order linear equation whose solutions satisfy the geometric conditions of his three postulates.

Riemann gave neither the general form of the differential equation

satisfied by $P \left\{ \begin{matrix} a & b & c \\ \alpha & \beta & \gamma & z \\ \alpha' & \beta' & \gamma' \end{matrix} \right\}$ - a task later accomplished by

Papperitz [1889] - nor a full discussion of the P-function as an Euler

integral with suitable paths for integration - matters taken up by

Pochhammer [1889], Jordan [1915, 251], and Schläfli [1870].

Riemann concluded by illuminating the relationship between P and

F, its hypergeometric series representation. Since α and α' may be

interchanged,

$$P \left(\begin{matrix} \alpha & \beta & \gamma \\ & & & z \\ \alpha' & \beta' & \gamma' \end{matrix} \right) = P \left(\begin{matrix} \alpha' & \beta & \gamma \\ & & & z \\ \alpha & \beta' & \gamma' \end{matrix} \right) \text{, there are}$$

8 P functions for each hypergeometric series in z, say

$$P \left(\begin{matrix} \alpha & \beta & \gamma \\ & & & z \\ \alpha' & \beta' & \gamma' \end{matrix} \right) = z^{\alpha}(1 - z)^{\gamma} F(\beta + \alpha + \gamma, \beta' + \alpha + \gamma, \alpha - \alpha' + 1, z).$$

There are six choices of variable, so 48 representations of a function

as a P function.

Concluding comments

The most immediate difference between Riemann and his predecessors

is the relative lack of computation. Rather than starting from a hyper-

geometric series he began with a P-function having ∞^2 branches and 3

branch-points. To be sure, any two linearly independent branches have

expansions as hypergeometric series, and any three branches are linearly

dependent, but the argument employed by Riemann inverted that of Gauss

and Kummer. His starting point was the set of solutions, functions which

are shown to satisfy a certain type of equation. Their starting point

was the equation from which a range of solutions are derived. Riemann

showed that a very small amount of information, the six exponents at the

branch points, entirely characterizes the equation and defines the
behaviour of the solutions. The hypergeometric equation is special
in this respect, as Fuchs [1865] was able to explain, and consequently
the task of generalizing the theory to cope with other differential
equations was to be quite difficult.

The crucial step in Riemann's argument is his use of matrices to
capture the behaviour of the solutions in the neighbourhood of a branch
point. These monodromy matrices enable one to study the global nature
of the solutions on analytic continuation around arbitrary paths, and
were introduced by Hermite in response to Puiseux's [1850, 1851].
Puiseux had considered the effect on a branch of an algebraic function
of analytically continuing it around one of its branch-points, and
also the effect of integration an algebraic function over a closed
path containing a branch point. Cauchy also reported on this work
[Cauchy, 1851]. Naturally, the question of influence arises. Riemann
was sparing with references especially to contemporaries. Quick to
mention Gauss, especially in works Gauss was to examine, and always willing to acknowledge his mentor Dirichlet, he otherwise usually contented
himself with general remarks of a historical kind which might set the
scene mathematically. Yet, as Brill and Noether said [1892-93, 283]:
Riemann"... everywhere betrayed an exact knowledge of the literature..."
and, they continued, (p. 286) "... evidently the fundamental researches
of Abel, Cauchy, Jacobi, Puiseux provide the starting point" for his
research on algebraic curves. We must suppose Riemann was one of those
mathematicians who absorbed the work of others and then re-derived it
in his own way. He did in fact mention Cauchy fleetingly when discussing
the development of a function in power series, but never Puiseux by

name[19]. However, he tells us that his first work in Abelian functions was carried out in 1851-52 [1857c, 102], and that would fit well with a contemporary study of related French work.

The connection between Riemann's study of the hypergeometric equation itself and those of Gauss and Kummer was, fortunately, a matter on which he chose to be explicit. In the Personal Report, [1857b], on his [1857a], which he submitted to the Göttinger Nachrichten, he wrote. "The unpublished part of the Gauss's study on this series, which has been found in his Nachlass, was already supplemented in 1835 by the work of Kummer contained in the 15th volume of Crelle's Journal". This also makes clear that Riemann had spent some of 1856 looking at the Gauss treasure trove, while working on his own ideas. However, we know from his lectures, see for example [Werke 379-390], that Riemann considered the hypergeometric equation from the standpoint of the linear differential equations with algebraic coefficients, which gave the level of generality he most wanted to attain (see Appendix 2).

1.5 Cauchy's theory of the existence of solutions of a differential equation.

Cauchy's study of differential equations has been considered frequently by historians of mathematics, notably Freudenthal [1971], Kline [1972] and Dieudonné [1978]. I follow here the recent and thorough discussion of C. Gilain [1977].

Cauchy was critical of his contemporaries' use of power series methods to solve differential equations, on the grounds that the series so obtained may not necessarily converge, nor, if it does converge, need it represent the solution function. He offered two methods for solving such equations rigorously. The first, given in lectures at the Ecole Polytechnique in 1823 and published in 1841, [Cauchy 1841] is a method of approximation by difference equations. It is nowadays known as the Cauchy-Lipschitz method, and it applies to equations

$$\frac{dy}{dx} = f(x, y)$$

where f and $\frac{\partial f}{\partial y}$ are bounded, continuous functions on some rectangle. He suggested that this theory might be extended to systems of such equations.

The second method [1835] treats a system of first order equations by passing to a certain partial differential equation. Candidates for the solution arise as power series, and Cauchy established their convergence by his "calcul des limites" or, in modern terminology, the method of majorants. In this method a series is shown to be convergent if its nth term is less in modulus than the nth term of a series known to be convergent; Cauchy usually took a geometric series for the majorizing series.

Gilain [1977, 14] notes that the second method supplanted the first in the minds of nineteenth-century mathematicians, and that both were taken to establish the same theorem, although in fact they apply to different types of equation (only the second is analytic). This is in keeping with a tendency throughout the later nineteenth century to regard variables as complex rather than real. Cauchy's preference in this matter is well known, indeed he is the principal founder of the theory of functions of a complex variable, but it was often, if tacitly, understood that function theory really meant complex function theory.

Since the eighteenth century had regarded functions as (piece-wise)
analytic, the critical spirit of the nineteenth century at first
provided a rigorous theory of analytic functions. Cauchy's own example
of e^{-1/t^2}, for which the Maclaurin series at $t = 0$ is identically
zero, showed that such a theory would necessarily be complex
analytic (solving his second objection to the older solution methods
for differential equation referred to above).

Cauchy's approach was extended by Briot and Bouquet [1856b] to
consider the singularities of the solution functions. These can arise
when the coefficients of the equation are singular, but may also occur
elsewhere if the equation is non-linear. French mathematicians of the
1850's were much concerned with the nature of an analytic function near
its singular points; Puiseux and Cauchy had already studied the singularities of algebraic functions in 1850 and 1851. Briot and Bouquet began
by simplifying Cauchy's existence proof using the method of majorants,
and then considered cases where it breaks down because $f(y, z)$ becomes
undetermined in the equation $\frac{dy}{dz} = f(y, z)$. At such points the solution
function may have a branch point and also a pole, even an essential
singularity. If it has a branch point it is not single-valued, or
"monodrome" in their terminology. But unless it has an essential
singularity at the point in question (which may be taken to be $z = 0$)
it still has a power series expansion of the form

$$z^\alpha \sum_{-k}^{\infty} a_n z^n, \quad a^{-k} \neq 0.$$

Briot and Bouquet made a particular study [1856c] of equations of
the form $F(u, \frac{du}{dz}) = 0$, where F is a polynomial, which have elliptic
functions as their solutions. For such equations the solutions have
finite poles but no branch points. The nature of the singular points

of an analytic function was to remain obscure for a long time.
Neuenschwander [1978a, 143] has established that the point $z = \infty$ was
treated ambiguously by Briot and Bouquet, and the nature of an essential
singularity remained obscure until the Casorati-Weierstrass theorem
was published. Near an essential singularity the function has a
series expansion

$$\sum_{n=-\infty}^{\infty} a_n z^n$$

but neither f nor 1/f can be defined at the singular point; for example,
e^z near $z = \infty$, can be regarded as $e^{1/t} = \sum_{n=-\infty}^{0} \frac{t^n}{(-n)!}$ near $t = 0$ and has
an essential singularity at $t = 0$. In France, Laurent [1843] had drawn
attention to these points, however they had by then received a much more
thorough treatment in Germany at the hands of Karl Weierstrass.

Weierstrass developed his theory independently of both Laurent and
Cauchy. In his [1842] he studied the system of differential equations

$$\frac{dx_i}{dt} = G_i(x_1,\ldots,x_n) \qquad 1 \leq i \leq n,$$

where the G_i are rational functions and obtained solutions in the form
of power series convergent on a certain domain. To show convergence he
took power series expansions of functions $\tilde{G}_i$, which are obtained from the
G_i by replacing the coefficients a_{ij} of G_i by positive numbers
$\alpha_{ij} \geq |a_{ij}|$, and dominated the $\tilde{G}_i$ by suitably chosen geometric progressions
Similar methods are still used today. This paper of Weierstrass was
left unpublished until 1894, although he lectured on related topics involvi
his Abelian functions at Berlin in 1863. In these lectures Weierstrass
connected the singular points with the domain of validity of a power
series expansion, and introduced the idea of the analytic continuation
of a function outside its circle of convergence. The matter is
discussed further below, in Chapters II and V .

Exercises on Chapter I

The theory of elliptic functions is a vast topic which provides an origin for many of the topics discussed in this book. However, it is not necessary to know anything about that theory to appreciate the story, nor is it difficult to find it out. These exercises cover most of the salient features of the elementary theory of elliptic integrals and functions, and outline the accompanying historical developments, which are discussed in more detail in Houzel [1978].

1. Show that the formula for the element of arc of the ellipse $\frac{x^2}{a^2} + \frac{y^2}{b^2} = 1$ as a function of x, obtained by using the formula $ds^2 = dx^2 + dy^2$, is $ds^2 = \left(\frac{a^2 - e^2 x^2}{a^2 - x^2}\right) a^2 dx^2$, $e^2 = \frac{a^2 - b^2}{a^2}$.

This integrand cannot in fact be integrated in terms of elementary functions. Failure to do so drove mathematicians to try various other solutions. Newton (1669) gave a power series expansion for the arc-length in terms of x, Euler (1733) and MacLaurin (1742) expansions in terms of e^2.

2. Suppose $x = a \sin \phi$, $y = b \cos \phi$. Show that
$$ds^2 = (a^2 - e^2 \sin^2 \phi) d\phi^2.$$

The curious term 'lemniscatic' derives from the study of a curve, introduced by Jakob Bernoulli in 1694, which can be written in polar coordinates as $r^2 = a^2 \cos 2\phi$.

3. Show that the Cartesian equation of the lemniscate is $(x^2 + y^2)^2 = a^2(x^2 - y^2)$, and that its element of arc in terms of r, using the formula $ds^2 = dr^2 + r^2 d\phi^2$, is $ds^2 = \left(\frac{a^4}{a^4 - r^4}\right) dr^2$.

4. The expression for the arc-length of the lemniscate recalls that
for the circle, $ds^2 = \dfrac{a^2}{(a^2 - r^2)}\, dr^2$. In 1752, Euler considered
the evident integration of the differential equation

$$\frac{dx}{(1 - x^2)^{\frac{1}{2}}} = \frac{dy}{(1 - y^2)^{\frac{1}{2}}}, \text{ which yields}$$

$$x^2 + y^2 = c^2 + 2xy(1 - c^2)^{\frac{1}{2}}$$

where c is an arbitrary constant. Prove that this is a solution
of the differential equation and show that when it is interpreted
in terms of the integrals $\displaystyle\int_0^{\sin\phi} \frac{dx}{(1 - x^2)^{\frac{1}{2}}} = \phi$, etc, for which

$$\int_0^y \frac{dy}{(1 - y^2)^{\frac{1}{2}}} = \int_0^x \frac{dx}{(1 - x^2)^{\frac{1}{2}}} + \int_0^c \frac{dc}{(1 - c^2)^{\frac{1}{2}}},$$

one obtains

$$\sin(\phi + \theta) = \sin\phi \cos\theta + \cos\phi \sin\theta.$$

5. The formulae in (4) are an <u>algebraic addition theorem</u> for the
integrals, for c is presented as an algebraic function of x and y,
and $\sin(\theta + \phi)$ as an algebraic function of $\sin\theta$ and $\sin\phi$. Euler
went on to consider the lemniscatic integrals, for which a
duplication formula had been found earlier by Fagnano. Fagnano
considered the substitutions

$$r^2 = \frac{2u^2}{1 + u^4} \text{ and } u^2 = \frac{2v^2}{1 - v^4} - \text{ show that they lead to}$$

$$\frac{dr^2}{1 - r^4} = \frac{2du^2}{1 + u^4} \text{ and } \frac{du^2}{1 + u^4} = \frac{2dv^2}{1 - v^4}.$$

So $\dfrac{dr}{(1 - r^4)^{\frac{1}{2}}} = \dfrac{2dv}{(1 - v^4)^{\frac{1}{2}}}$, from which it follow that if

$$\int_0^R \frac{dr}{(1 - r^4)^{\frac{1}{2}}} = S \text{ then}$$

$$\int_0^V \frac{dv}{(1 - v^4)^{\frac{1}{2}}} = S/2, \text{ so V is half-way along the arc from 0}$$

to R, where V is a known (algebraic) function of R.

6. Euler found that the general solution of the differential equation

$$\frac{dx}{(1 - x^4)^{\frac{1}{2}}} = \frac{dy}{(1 - y^4)^{\frac{1}{2}}}$$

was $x^2 + y^2 = c^2 + 2xy(1 - c^4) - c^2x^2y^2$ for an arbitrary constant c.

Verify this. Euler was not satisfied with this rather artifical method of solving the equation, and was much impressed by the young Lagrange's direct method, which I omit.

7. Deduce from (4) and (5) that sin is periodic, with period

$$2\pi = 4 \int_0^1 \frac{dt}{(1 - t^2)^{\frac{1}{2}}} .$$

8. The conclusions to be drawn from the above exercises are that

$$u = \int_0^x \frac{dt}{(1 - t^2)^{\frac{1}{2}}}$$

does not define a single valued function of x, for u(x) is only well-defined up to a multiple of 2π, but that $x = x(u)$ is a well-defined function (the sine function), which moreover is periodic; $\sin(u) = \sin(u + 2\pi n)$, and satisfies an algebraic addition theorem.

What about the similar integral

$$v = \int_0^x \frac{dt}{(1 - t^4)^{\frac{1}{2}}} \; ?$$ Deduce from

Euler's algebraic addition theorem that it is also periodic, with period $2w = 4 \int_0^1 \frac{dt}{(1 - t^4)^{\frac{1}{2}}}$. Whenever x is real and the path of integration is real, it defines a single-valued function of v, as Legendre was clearly aware. The decisive step is to let x become complex.

Write down the equation for sin 2u as a function of sin u, and for sin nu, say when n is odd.

9. Deduce from (5) or (6) that there are 4 values of x(u) defined by

$$u = \int_0^x \frac{dt}{(1 - t^4)^{\frac{1}{2}}}$$ for a given value of x(2u).

It may be shown that when n is odd there are n^2 solutions, and so the problem arises: explain the "extra" solutions. Gauss seems to have seen at once that the answer was to admit complex periods.

10. Set t = it in $v = \int_0^x \frac{dt}{(1 - t^4)^{\frac{1}{2}}}$,

and deduce a connection between x(v) and x(iv): x(iv) = ix(v).

Gauss called the function x(v) the sinus lemniscaticus, sl(v). Since he had shown sl(u) = sl(u + m2w), m $\in \mathbb{Z}$, and sl(iv) = isl(v), he could, by the addition theorem, define sl(u + iv), and deduce that it was doubly periodic, indeed sl(z + 2(m + ni)w) = sl(z) for any complex z and any 'Gaussian integer' m + ni. The periods are 2w and 2iw. The n values of z such that sin nx has a given value lie along the interval [0, 2π]. The n^2 values of z such that sl(z) has a given value are distributed regularly in the square [0, 2w] × [0, 2w].

11. Show, by differentiating under the integral with respect to (k^2),

that $K = \int_0^{\pi/2} (1 - k^2 \sin^2\theta)^{-\frac{1}{2}} d\theta = \frac{\pi}{2} F(\frac{1}{2}, \frac{1}{2}, 1, k^2)$

and $K' = \int_0^{\pi/2} (1 - k'^2 \sin^2\theta)^{-\frac{1}{2}} d\theta = \frac{\pi}{2} F(\frac{1}{2}, \frac{1}{2}, 1, k'^2)$,

where $1 = k^2 + k'^2$, both satisfy the same hypergeometric equation.

12. Show that the substitution $t^2 = \sin^2\theta + k^2 \cos^2\theta$ establishes

$$K' = \int_1^{1/k} [(t^2 - 1)(1 - k^2 t^2)]^{-\frac{1}{2}} dt.$$

13. Show that

$$E: = \int_0^{\pi/2} (1 - k^2 \sin^2\theta)^{\frac{1}{2}} d\theta = \frac{\pi}{2} F(\frac{1}{2}, -\frac{1}{2}, 1, k^2),$$

and that they satisfy

$$EK' + E'K - KK' = (K'\frac{dK}{dk^2} - K\frac{dK'}{dk^2})$$

Show that $\lim_{x \to 1} 2x(1 - x)(K'\frac{dK}{dk^2} - K\frac{dK'}{dk^2}) = $

$\pi \lim_{x \to 1} x(1 - x)\frac{dK}{dk} = \frac{\pi}{2}$, and deduce Legendre's relation: $EK' + E'K - KK' = \frac{\pi}{2}$.

14. Deduce some of Gauss's realizations of the hypergeometric series,

e.g.:

(i) $(t + u)^n = t^n F(-n, \beta, \beta, \frac{-u}{t})$

(ii) $\log(1 + t) = t F(1, 1, 2, -t)$

(iii) $e^t = \lim_{k \to \infty} F(1, k, 1, \frac{t}{k}) = \lim_{k \to \infty} \{1 + t F(1, k, 2, \frac{t}{k})\}$

$= \lim_{k \to \infty} \{1 + t + \frac{1}{2} t^2 F(1, k, 3, \frac{t}{k})\}$ etc.

(iv) $\quad e^t + e^{-t} = \lim_{k,k' \to \infty} 2F(k, k', \frac{1}{2}, \frac{t^2}{4kk'})$

(v) $\quad \sin t = \lim_{k,k,' \to \infty} t\, F(k, k', \frac{3}{2}, -\frac{t^2}{4kk'})$

(vi) $\quad t = \sin t\; F(\frac{1}{2}, \frac{1}{2}, \frac{3}{2}, \sin^2 t)$

[Gauss wrote: k, k' 'numeros infinite magnos', infinitely great numbers.]

15. Show that Gauss's function M(1, x) is a continuous and indeed a differentiable function of x.

16. The Bernoulli numbers were defined on p.11 by an expansion like

$$\frac{x}{e^x - 1} = \sum \frac{b_n}{n!} x^n.$$

Show that $b_0 = 1$, $b_1 = -\frac{1}{2}$, and other $b_n = 0$ if n is odd and greater than 1, by showing

$$\frac{x}{e^x - 1} = \frac{-x}{2} + \frac{x}{2} \cdot \frac{e^x + 1}{e^x - 1}$$

$$= \frac{-x}{2} - \frac{x}{2} \cdot \left(\frac{e^{-x} + 1}{e^{-x} - 1}\right).$$

Use the expansion

$$x = \left(\sum \frac{b_n}{n!} x^n\right)\left(x + \frac{x^2}{2!} + \frac{x^3}{3!} + \dots\right)$$

to express b_n in terms of $b_0, \dots, b_{n-1}$. Deduce that b_n is rational.

17. There is an infinite product expansion for sine, due to Euler:

$$\sin \pi z = \pi z\, \Pi(1 - \frac{z^2}{n^2})$$

Find its logarithmic derivative, which is an infinite series for $\pi z \cot \pi z$, and show from the definition of cot in terms of exponentials, that $\pi z \cot \pi z = \sum_{0}^{\infty} \frac{(2\pi)}{(2m)!} B_{2m} z^{2m}$. Deduce that

$\zeta(2k): = \sum_{n} \frac{1}{n^{2k}} = -\frac{(2\pi)^{2k}}{2(2k)!} B_{2k}$, which gives the exciting result

that $\frac{1}{(2\pi)^{2k}} \zeta(2k)$ is rational.

18. The only meremorphic maps of the Riemann sphere to itself are of

the form

$$z \to \frac{az + b}{cz + d}$$

These maps are called Möbius, or fractional linear, transformations.

Show that they form a group.

19. Show that Möbius transformations preserve the cross-ratio of four

points; i.e. if $z_i \to z_i'$, $i = 1, 2, 3, 4$, then

$$(z_1, z_2, z_3, z_4) = (z_1', z_2', z_3', z_4'), \text{ where}$$

$$(z_1, z_2, z_3, z_4) = \frac{(z_1 - z_2)(z_3 - z_4)}{(z_1 - z_4)(z_3 - z_2)}.$$

20. Deduce from Question 19 that Möbius transformations send circles

to circles.

21. Show that, if $(z_1, z_2, z_3, z_4) = \lambda$, then the possible values of

any permutations of z_1, z_2, z_3, z_4 are $\lambda, \frac{1}{\lambda}, 1 - \lambda, \frac{1}{1 - \lambda}, 1 - \frac{1}{\lambda}$,

$\frac{\lambda}{\lambda - 1}$.

22. Show that there is a subgroup, K, of S_4, the group of all

permutations of the symbols (z_1, z_2, z_3, z_4), which preserves the

cross-ratio (z_1, z_2, z_3, z_4), is of order 4, and is normal in S_4.

23. Summarize your findings in Q's 21 and 22 by showing that there is a

homomorphism from S_4 onto S_3 with kernel K.

24. Obtain Papperitz's result [1889]: the differential equation satisfied by

$$P \left\{ \begin{array}{ccc} a & b & c \\ \alpha & \beta & \gamma \\ \alpha' & \beta' & \gamma' \end{array} \; z \right\} \text{ is}$$

$$\frac{d^2w}{dz^2} + \sum \frac{1 - \alpha - \alpha'}{z - a} \frac{dw}{dz} + \sum \frac{\alpha\alpha'(a - b)(a - c)}{(z - a)} \cdot \frac{w}{(z - a)(z - b)(z - c)}$$

$$= 0$$

by showing firstly that the equation is satisfied by two linearly independent determinations of P, say P_1 and P_2, so it is of the form

$$\frac{d^2w}{dz^2} + p \frac{dw}{dz} + qw = 0$$

where $p = -\dfrac{P_2 P_1'' - P_1 P_2''}{P_2 P_1' - P_1 P_2'}$

and $q = \dfrac{P_2' P_1'' - P_1' P_2''}{P_2 P_1' - P_1 P_2'}$

(Hint: write down a suitable determinant).

Now deduce that $p = \dfrac{1 - \alpha - \alpha'}{z - a} + u_1(z)$ near $z = a$ from the monodromy relation of $z = a$, and thence that

$$p = \frac{1 - \alpha - \alpha'}{z - a} + \frac{1 - \beta - \beta'}{z - b} + \frac{1 - \gamma - \gamma'}{z - c} + u_2(z)$$

where $u_2(z)$ is analytic at infinity. Since p is analytic at infinity, $p = \dfrac{2}{z} + 0(z^{-2})$; deduce that $u_2(z) = 0$, and $\alpha + \alpha' + \beta + \beta' + \gamma + \gamma' = 1$. Conduct a similar argument to find q; $q(z) = 0(z^{-4})$.

25. Which of the following differential equations are hypergeometric?

(i)   Legendre's equation: $(1 - x^2) \dfrac{d^2y}{dx^2} - 2x \dfrac{dy}{dx} + n(n + 1)y = 0$;

(ii)   Bessel's equation: $x^2 \dfrac{d^2y}{dx^2} + x \dfrac{dy}{dx} + (x^2 - n^2)y = 0$;

(iii) A second order linear differential equation with constant

coefficients.

26. (i)   Show that, when n is not an integer, two independent

solutions of Legendre's equation are

$$y_1 = P_n(x) = F(n + 1, -n, 1, \frac{1 - x}{2})$$

and $y_2 = P_n(-x)$.

(ii)   Find the monodromy group of the equation by finding the

monodromy relations at $x = +1$ and $x = -1$.

27. Show that the monodromy matrices of a differential equation of the

form $y'' + py = 0$ all have determinant 1, by considering

$$D: = \begin{pmatrix} y_1' & y_1 \\ y_2' & y_2 \end{pmatrix},$$
where y_1 and y_2 are a basis of solutions. This

means that the monodromy group of a second order linear equation

can always be made to lie in $PSL(2; ¢)$. If the equation has order

n the same method shows that the group may be made to lie in

$PSL(n; ¢)$.

28. (Riemann's calculation of the ratios $\alpha_\beta / \alpha'_\beta$, etc.)

(i)   Show $\alpha_\gamma e^{2\pi i \gamma} P^\gamma + \alpha_{\gamma'} e^{2\pi i \gamma'} P^{\gamma'} = (\alpha_\beta e^{-2\pi i \beta} P^\beta + \alpha_{\beta'} e^{-2\pi i \beta'} P^{\beta'}) e^{-2\pi i \alpha}$

by considering the LHS as the result of analytically continuing

$P^\alpha = \alpha_\gamma P^\gamma + \alpha_{\gamma'} P^{\gamma'}$ on a positive circuit around $c = 1$ and the

RHS as the result of analytically continuing $P^\alpha = \alpha_\beta P^\beta + \alpha_{\beta'} P^{\beta'}$

on a negative circuit around $a = 0$ and $b = \infty$.

(ii) Use $\alpha_\beta P^\beta + \alpha_{\beta'} P^{\beta'} = \alpha_\gamma P^\gamma + \alpha_{\gamma'} P^{\gamma'}$ to deduce

$\alpha_\gamma \sin(\sigma - \gamma) \pi e^{\pi i \gamma} P^\gamma + \alpha_{\gamma'} \sin(\sigma - \gamma') \pi e^{\pi i \gamma'} P^{\gamma'} = \alpha_\beta \sin(\sigma + \alpha + \beta) \pi e^{-\pi i (\alpha + \beta)} P^\beta$

$+ \alpha_{\beta'} \sin(\sigma + \alpha + \beta') \pi e^{-\pi i (\alpha + \beta')} P^{\beta'}$.

(iii)   Eliminate $P^{\gamma'}$ from the equation in (ii) and its counter-
part in which α has been replaced by α', to obtain two
equations of the form

$$C'P^{\gamma} = C''P^{\beta} + C'''P^{\beta'}$$

and deduce

$$\frac{\alpha_{\gamma}}{\alpha'_{\gamma}} = \frac{\alpha_{\beta}\sin(\alpha+\beta+\gamma')\pi e^{-\alpha\pi i}}{\alpha'_{\beta}\sin(\alpha'+\beta+\gamma')\pi e^{-\alpha'\pi i}} = \frac{\alpha_{\beta'}\sin(\alpha+\beta'+\gamma')\pi e^{-\alpha\pi i}}{\alpha'_{\beta'}\sin(\alpha'+\beta'+\gamma')\pi e^{-\alpha'\pi i}}$$

(iv)   Set $\sigma = \gamma$ and deduce

$$\frac{\alpha_{\gamma'}}{\alpha'_{\gamma'}} = \frac{\alpha_{\beta}\sin(\alpha+\beta+\gamma)\pi e^{-\alpha\pi i}}{\alpha'_{\beta}\sin(\alpha'+\beta+\gamma)\pi e^{-\alpha'\pi i}} = \frac{\alpha_{\beta'}\sin(\alpha+\beta'+\gamma')\pi e^{-\alpha\pi i}}{\alpha'_{\beta'}\sin(\alpha'+\beta'+\gamma)\pi e^{-\alpha'\pi i}}$$

These equations are consistent because $\alpha + \alpha' + \ldots + \gamma' = 1$,
and $\sin s\pi = \sin(1 - s)\pi$, as Riemann remarked. Riemann
also showed that

$$\alpha_{\beta} = \frac{\sin(\alpha+\beta'+\gamma')\pi}{\sin(\beta'-\beta)\pi} \qquad\qquad \alpha_{\beta'} = -\frac{\sin(\alpha+\beta+\gamma)\pi}{\sin(\beta'-\beta)\pi}$$

$$\alpha'_{\beta} = \frac{\sin(\alpha'+\beta'+\gamma)\pi}{\sin(\beta'-\beta)\pi} \qquad\qquad \alpha'_{\beta'} = -\frac{\sin(\alpha'+\beta+\gamma')\pi}{\sin(\beta'-\beta)\pi}$$

$$\alpha_{\gamma} = \frac{\sin(\alpha+\beta'+\gamma')\pi e^{(\alpha'+\gamma)\pi i}}{\sin(\gamma'-\gamma)\pi} \qquad\qquad \alpha_{\gamma'} = \frac{-\sin(\alpha+\beta+\gamma)\pi e^{(\alpha'+\gamma')\pi i}}{\sin(\gamma'-\gamma)\pi}$$

$$\alpha'_{\gamma} = \frac{\sin(\alpha'+\beta+\gamma')\pi e^{(\alpha+\gamma)\pi i}}{\sin(\gamma'-\gamma)\pi} \qquad\qquad \alpha'_{\gamma'} = -\frac{\sin(\alpha'+\beta'+\gamma)\pi e^{(\alpha+\gamma')\pi i}}{\sin(\gamma'-\gamma)\pi}$$

29. The uniqueness of a P-function with given data.

$$P = P\begin{pmatrix} a & b & c \\ \alpha & \beta & \gamma & z \\ \alpha' & \beta' & \gamma' \end{pmatrix}, \text{ and let } P_1 \text{ be another P}$$

function with the same data. To show $\dfrac{P_1}{P}$ is constant, show

(i) $P^{\alpha}P^{\alpha'}_1 - P^{\alpha'}P^{\alpha}_1 = (\det B)(P^{\beta}P^{\beta'}_1 - P^{\beta'}P^{\beta}_1) = (\det C')(P^{\gamma}P^{\gamma'}_1 - P^{\gamma'}P^{\gamma}_1).$

(ii) $(P^{\alpha}P^{\alpha'}_1 - P^{\alpha'}P^{\alpha}_1)z^{-\alpha-\alpha'}$ is single valued and finite at $z = 0$,

as is $(P^{\beta}P^{\beta'}_1 - P^{\beta'}P^{\beta}_1)z^{\beta+\beta'}$ at $z = \infty$,

and $(P^{\gamma}P^{\gamma'}_1 - P^{\gamma'}P^{\gamma}_1)(1-z)^{-\gamma-\gamma'}$ at $z = 1$.

Deduce $(P^\alpha P_1^{\alpha'} - P^{\alpha'} P_1^\alpha) z^{-\alpha-\alpha'} (1-z)^{-\gamma-\gamma'}$ is continuous, single

valued, and behaves at infinity like

$$(P^\beta P_1^{\beta'} - P^{\beta'} P_1^\beta) z^{\alpha+\alpha'+\gamma+\gamma'} = (P^\beta P_1^{\beta'} - P^{\beta'} P_1^\beta) z^{\beta+\beta'-1} \text{ whence}$$

that $(P^\alpha P_1^{\alpha'} - P^{\alpha'} P_1^\alpha) z^{-\alpha-\alpha'} (1-z)^{-\gamma-\gamma'}$

is bounded everywhere and must be a constant; indeed must be

zero, its value at $z = \infty$. Deduce

$$\frac{P_1^{\alpha'}}{P^\alpha} = \frac{P_1^\alpha}{P^\alpha}$$

$$\frac{P_1^\beta}{P^\beta} = \frac{P_1^\beta}{P^{\beta'}} = \frac{\alpha_\beta P_1^\beta + \alpha_{\beta'} P_1^{\beta'}}{\alpha_\beta P^\beta + \alpha_{\beta'} P^{\beta'}} = \frac{P_1^\alpha}{P^\alpha}, \text{ and}$$

$$\frac{P_1^\gamma}{P^\gamma} = \frac{P_1^{\gamma'}}{P^{\gamma'}} = \frac{P_1^\alpha}{P^\alpha} .$$

It remains to ensure that P^α and $P^{\alpha'}$ do not simultaneously

vanish for some $z \neq 0, 1, \infty$. This follows from an analogous

consideration of $(P^\alpha \frac{dP^{\alpha'}}{dz} - P^{\alpha'} \frac{dP^\alpha}{dz}) z^{-\alpha-\alpha'+1} (1-z)^{-\gamma-\gamma'+1}$, which

is a non-zero constant (provided $\alpha \neq \alpha'$, as has been assumed).

CHAPTER II   LAZARUS FUCHS

Introduction

In the years 1865, 1866, and 1868 Lazarus Fuchs published three
papers, each entitled "Zur Theorie der Linearen Differentialgleichungen
mit veränderlichen Coefficienten"("On the theory of linear differential
equations with variable coefficients"). These will be surveyed in this
chapter. In them he characterized the class of linear differential
equations in a complex variable x all of whose solutions have only
finite poles and possibly logarithmic branch points. So, near any
point x_0 in the domain of the coefficients, the solutions become
finite and singled-valued upon multiplication by a suitable power
of $(x-x_0)$ unless it involves a logarithmic term. This class came to
be called the Fuchsian class, and equations in it equations of the
Fuchsian type. As will be seen, it contains many interesting equations,
including the hypergeometric. In the course of this work Fuchs
created much of the elementary theory of linear differential equations
in the complex domain: the analysis of singular points; the nature of
a basis of n linearly independent solutions to an equation of degree n
when there are repeated roots of the indicial equation; explicit forms
for the solution according to the method of undetermined coefficients.
He investigated the behaviour of the solutions in the neighbourhood of
a singular point, much as Riemann had done, by considering their
monodromy relations - the effect of analytically continuing the
solutions around the point - and, like Riemann, he did not explicitly
regard the transformations so obtained as forming a group. One problem
which he raised but did not solve was that of characterizing those
differential equations all of whose solutions are algebraic. It became
very important, and is discussed in the next chapter.

Fuchs's career may be said to have begun with these papers. Born in Moschin near Posen in 1833, he went to Berlin University in 1854 and started his graduate studies under Kummer, writing his dissertation on the lines of curvature of a surface (1858), and then papers on complex roots of unity in algebraic number theory[1]. Just as Fuchs was the first to present his thesis to the new Berlin University under Kummer's direction, so his friend Koenigsberger was the first to present one under Weierstrass's. But where Kummer's lectures were always clear and attractive, Weierstrass at first was a disorganized lecturer, and Fuchs does not seem to have come under his influence until Weierstrass lectured on Abelian functions in 1863. Weierstrass's approach was based on his theory of differential equations, and that subject became Fuchs's principal field of interest, Fuchs in turn becoming its chief exponent at Berlin.

When Fuchs presented his Habilitationsschrift in 1865 Kummer was the principal referee and Weierstrass the second, but by 1870 Fuchs spoke to himself as "a pupil of Weierstrass" (letter to Casorati, quoted in Neuenschwander [1978b, p46]).

In 1866 Fuchs succeeded Arndt at Berlin, before going to Greifswald in 1868. He returned to Berlin in 1884 as Professor Ordinarius, occupying the chair vacated by Kummer. His friend Koenigsberger described him as irresolute and anxious, a temporizer and one easily persuaded by others, but humorous and unselfish. He played a weak role in the battle between Weierstrass and Kronecker that polarized Berlin in the 1880's, disappointing Weierstrass thereby, and was no match for Frobenius in the 1890's, but several of his students, notably Schlesinger, spoke warmly of him.

2.1 Fuchs's theory of linear differential equations

The study of complex functions and their singularities was a matter of active research at this time. The general method used in Berlin was the analytic continuation of power series expansions. If the function could be expanded as a series of the form $\sum_{n=0}^{\infty} a_n(x-a)^n$, convergent on some disc with centre $x = a$, then a was a non-singular point. If a term of the form $(x-a)^{\alpha}$ appeared in the expansion then a was a branch point, and it was a logarithmic branch point if $\log (x-a)$ appeared. No distinction was made between single-valued, many-valued and even infinitely many-valued 'functions'. All were referred to as functions and distinguished as single-valued or many-valued as appropriate (and that usage will be followed here). If a function was analytic in a neighbourhood of $x = a$ Fuchs and others called it finite and continuous; the term 'regular' introduced by Thomé (see below, p85) will sometimes be used here. The points at which a function became infinite were called its points of discontinuity, but the nature of these points was obscure. Several writers, for example Briot and Bouquet, equivocated over whether the function was defined at that point (and took the value ∞) or was not defined at all. However, Weierstrass had developed the theory of power series expansions[2] of the form $\sum_{n=k}^{\infty} a_n(x-a)^n$ in 1841, although he did not publish it until 1894, and on the basis of this theory he made a clear distinction between functions admitting a finite "Laurent" expansion, which are infinite at $x = a$, and those of the second kind, which are not defined at $x = a$. The behaviour of a function near such singular points depends dramatically on the kind of expansion they admit. In the former case its value tends so infinity, but in the second case the Casorati-Weierstrass theorem asserts that it comes arbitrarily close to any pre-assigned value. Neuenschwander [1978a, 162] has considered the history of this theorem and finds that

"Weierstrass .... had the Casorati-Weierstrass theorem already in the year 1863", the year he lectured on Abelian functions. So Fuchs picked the largest class of linear differential equations whose solutions can be defined at every point of the domain of the coefficients, including their singular points, logarithmic cases aside.

Of Fuchs's three papers considered here, the [1866] is the most important. Its first three sections are identical with those of the [1865] but thereafter it diverges to give a much more thorough analysis of the case where the monodromy matrices have repeated eigenvalues and logarithmic terms enter the solution. It was also published in a more accessible journal, Crelle rather than the Jahresbericht of the Berlin Gewerbeschule, and it was more widely read. [1868], regarded by Fuchs as the concluding paper in the series, goes into more detail in special cases, extends the method of inhomogeneous equations of the appropriate kind, and discusses accidental singular points - the unusual case when a singularity of the coefficients does not give rise to a singularity in the solution. Therefore the two earlier papers will be considered first.

When Fuchs published his first paper on the theory of linear differential equations in 1865, he quite correctly stated that the task facing mathematicians was not so much to solve a given differential equation by means of quadratures, but rather to develop solutions defined at all points of the plane. Since analysis, he said, teaches how to determine a function if its behaviour in a neighbourhood of its points of discontinuity and its multiple-valuedness can be ascertained, the essential task is to discover the position and nature of these points. This approach had been carried out, he observed, by Briot and Bouquet in their paper [1856b] on the solution of differential equations which are of the form $f(y, \frac{dy}{dx}) = 0$, where $f(y, \frac{dy}{dx})$ is a polynomial expression in y and $\frac{dy}{dx}$; and by Riemann in his 1857 lectures on Abelian functions. Fuchs proposed to investigate homogeneous linear ordinary differential equations. The conventions of his 1865 and 1866 papers will be followed fairly closely if such an equation is written as

$$\frac{d^n y}{dx^n} + P_1 \frac{d^{n-1}y}{dx^{n-1}} + \ldots + P_{n-1} \frac{dy}{dx} + P_n y = 0, \qquad (2.1.1)$$

where $P_1, \ldots, P_n$ are single-valued meromorphic functions of x in the entire complex x-plane, or on some simply-connected region $T \subset \mathbf{C}$. Fuchs called the points in T where one or more of the p_i were discontinuous <u>singular</u> points, a term he attributed to Weierstrass, and he assumed each p_i had only finitely many singular points[3].

Solutions near a non-singular point

Fuchs first considered a solution of (2.1.1) in a neighbourhood of a non-singular point x_0, and showed that it is a single-valued, finite, continuous function which is prescribed once values for y, $\frac{dy}{dx}$, $\ldots$, $\frac{d^{n-1}y}{dx^{n-1}}$ are given at $x = x_0$. His method was to replace (2.1.1) by a second differential equation known to have a solution in a neighbourhood of x_0 and so related to (2.1.1) that its solution guarantees the existence to a solution to (2.1.1). The question of convergence was thus handled by Cauchy's method of majorants, much as Briot and Bouquet had done.

The solutions obtained in this way to (2.1.1) are power series of the form $\sum a_r (x - x_0)^r$, so it follows at once that the singular points of the solution can only be located amongst[4] the singular points of the coefficients $p_1(x) \ldots, p_n(x)$. They are therefore fixed, and determined by the equation. This is in sharp contrast with the case of non-linear

differential equations, where the singular points may depend on
arbitrary constants involved in the solutions. (Fuchs later classified
those non-linear equations whose branch points are fixed, see his
[1884a] and also Ince [1926, Chs. 13,14] and Matsuda [1980, 11-18.])

Solutions near a singular point

To study the solutions in the neighbourhood of a singular point
Fuchs proceeded as Riemann had. With the singular points removed, T
becomes a multiply connected region, T', and if circles are drawn
enclosing each singular point and the circles joined by cuts to the
boundary of T a simply connected region T" is marked out. Inside T" a
single-valued, continuous, and finite solution y of (2.1.1) can be
found by the methods just described, which can be represented by an
infinitely many leaved surface spread over T', the leaves joining along
the cuts in a fashion to be determined by the method of monodromy
matrices.

The special case of the second-order equation

In order to make Fuchs's argument more immediately comprehensible
I shall specialize to the case of a second-order differential equation

$$\frac{d^2y}{dx^2} + p_1 \frac{dy}{dx} + p_2 y = 0. \tag{2.1.2}$$

Fuchs worked throughout with an equation of degree n, and I shall
indicate the more general results in Appendix 3.

Fuchs said that the two solutions y_1 and y_2 to (2.1.2) form a
fundamental system (i.e. a basis of solutions) if their Wronskian
$D := \begin{vmatrix} \frac{dy_1}{dx} & y_1 \\ \frac{dy_2}{dx} & y_2 \end{vmatrix}$ never vanishes in T', and he proved that the general

solution to (2.1.2) can always be written in the form $\eta = c_1 y_1 + c_2 y_2$ where c_1 and c_2 are constants. He also proved theorems about the freedom to choose the elements of a fundamental system, nowadays familiar from the theory of finite dimensional vector spaces but here developed ad hoc, as was often the case in the contemporary work of Weierstrass and Kronecker[5].

Fuchs next showed that the prescribed conditions on p_1, p_2 imply that at least one solution of the equation would be single-valued upon multiplication by a suitable power of $x - a_1$, where a_1 is a singular point. Let y_1 and y_2 be a fundamental system near a_1, and y_1 and y_2 the result of analytically continuing the solutions y_1 and y_2 once round a_1. There are then equations of the form

$$y_1 = \alpha_{11} y_1 + \alpha_{12} y_2$$

$$y_2 = \alpha_{21} y_1 + \alpha_{22} y_2$$

where the α's are constants, and $\alpha_{11}\alpha_{22} - \alpha_{12}\alpha_{21} \neq 0$.

Fuchs showed that the roots[6] of what he called the fundamental equation $\begin{vmatrix} \alpha_{11} - w & \alpha_{12} \\ \alpha_{21} & \alpha_{22} - w \end{vmatrix} = 0$ (2.1.3) are independent of the choice of fundamental system, and neither can be zero. Let w_1 be one of them, then y_1 can be chosen so that $y_1 = w_1 y_1$, and if $w_1 = e^{2\pi i r}1$, say, then $y_1(x - a_1)^{-r_1}$ has the required property of being single-valued near a_1. The case in which (2.1.3) has repeated roots was first discussed by Fuchs in his [1865] then more fully in the later papers, and will be described below (Appendix 3) when it will be seen that the general solution involves a logarithmic term. Indeed, if the transformations are

$$\tilde{y}_1 = w y_1$$

$$\tilde{y}_2 = y_1 + w y_2$$

(which is the general case if $w_1 = w_2 = w$) then the solutions must be of the form

$$y_1 = (x - a_1)^r \phi_1$$

$$y_2 = (x - a_1)^r \phi_2 + (x - a_1)^r \phi_1 \log(x - a_1),$$

where ϕ_1 and ϕ_2 are single-valued and holomorphic near a_1 and $w = e^{2\pi i r}$. So the general solutions contains $(x-a_1)^r \phi$, where ϕ is a polynomial in $\log(x - a_1)$ with holomorphic coefficients.

Equations of the 'Fuchsian' class

The natural question was then: what conditions must be placed on p_1 and p_2 so that all the solutions are regular, i.e. have only finite poles at the singular points a_1, ..., a_ρ, $a_{\rho+1} = \infty$, and are at worst logarithmic upon multiplication by a suitable power of $(x-a_i)$ or $\frac{1}{x}$? From now on Fuchs took ∞ to be a singular point, and T to be $\hat{\mathbb{C}}$: $= \mathbb{C} \cup \{\infty\}$, so the coefficient functions must be rational functions, i.e. quotients of polynomials in x. His solution to this question [1865, §4, 1866, §4] produced a class of differential equations readily characterized by the purely algebraic restrictions upon the coefficients p_i; this class has become known as the Fuchsian class, and such equations will henceforth be referred to as 'equations of the Fuchsian class'. Fuchs himself always preferred more modest circumlocutions. (The phrase "Fuchsian equation" refers to a much more general linear differential equation; see below, Chapter VI.)

To answer this question he took a fundamental system y^i_1, y^i_2 such that near a_i, $y^i_k = (x-a_i)^{r_{ik}} \phi^i_k$ ($k = 1, 2$) for some single-valued function ϕ^i_k. If I denote[7] the pair (y^i_1, y^i_2) by $\underline{y}^i$ then the $\underline{y}^i$ ($i > 1$) are related to $\underline{y}^1$ by equations $\underline{y}^1 = B_i \underline{y}^i$, B_i matrices of constants, and if the analytic continuation of $\underline{y}^1$ once around[8] a_1 produces $\underline{\tilde{y}}^1$ then $\underline{\tilde{y}}^1 = R_1 \underline{y}^1$. Likewise a circuit of a_i produces $\underline{\tilde{y}}^i = R_i \underline{y}^i$ and the analytic continuation of $\underline{y}^1$ around a_i produces $B_i R_i B_i^{-1} \underline{y}^1$. Since a circuit of the finite singular points $a_1, \ldots, a_\rho$ is simultaneously a circuit of $a_{\rho+1}$ in the opposite sense

$$R_1 (B_2 R_2 B_2^{-1}) \cdots (B_{\rho+1} R_{\rho+1} B_{\rho+1}^{-1}) = I, \text{ the identity matrix}$$

so $\det R_1 \cdots \det R_{\rho+1} = 1$.

But $\det R_i = w_{i1} w_{i2} = e^{2\pi i (r_{i1} + r_{i2})}$

so $\sum\limits_{i=1}^{\rho+1} (r_{i1} + r_{i2}) = k$ is an integer.

Fuchs's notation, and his use of the monodromy matrices, follows Riemann's here. He even wrote his matrices in brackets, $(R)_i$, (B), and so forth. It is likely that Kummer encouraged Fuchs to follow Riemann's ideas; Kummer had already commented most favourably on the thesis of Riemann's student Prym[9]. On the other hand his study of the eigenvalues of these matrices is in the thorough-going Berlin spirit.

The integer k was made precise by a technically complicated study of the analytic continuation of a fundamental system of solutions. The underlying idea is that one knows what happens near $x = \infty$ in two ways (from the monodromy and by transforming the equation) so one can obtain an equation for the sum $\sum r_{ij}$ and hence for k. The result is $k = \rho - 1$, and the same argument also showed that $\lim\limits_{x \to \infty} x p_1$ is finite. More work of this kind enabled Fuchs to find p_2 and therefore to show that the most general equation of the Fuchsian class (and the second order) has the form

$$\frac{d^2 y}{dx^2} + \frac{F_{\rho-1}(x)}{\psi} \frac{dy}{dx} + \frac{F_{2(\rho-1)}(x)}{\psi^2} y = 0, \tag{2.1.4}$$

where $\psi = (x-a_1) \ldots (x-a_p)$ and $F_s(x)$ is a polynomial in x of degree at most s. This is Fuchs's main Theorem, but as it applies to second-order equations.

Fuchs then looked again at solutions valid near the singular point $x = a_i$, which are necessarily of the form $u = (x-a)^{r_{ij}} \sum_{k=0}^{\infty} c_k (x-a_i)^k$, and showed that the r_{ij} satisfy $r(r-1) - r P_{i1}(a_i) - P_{i2}(a_i) = 0$, where $P_{im} = (x-a_i)^m F_m(\rho-1)\psi^{-m}$. This last equation was later called the determining fundamental equation (determinirende Fundamentalgleichung) by Fuchs [1868, 367 = 1904, 220] and subsequently the indicial equation by Cayley ([1883, 5]). The r's are related to the w's obtained earlier by the equations $w = e^{2\pi i r}$. Fuchs here showed how to obtain the solutions valid near a singular point a_i directly, without considering the monodromy relations, but he did not discuss the case when two of the r's differ by an integer, and a w is therefore repeated. He only discussed that situation in terms of w's and so of analytic continuation, never in terms of r's and the formal derivation of the solutions.

Having shown that any differential equation all of whose solutions are of the form required has necessarily the form of (2.1.4), Fuchs had next to show that, conversely, any equation of that form had solutions of that type. But this was a routine exercise in computing with power series and using the method of majorants to establish their convergence, so I omit the details.

The nth order equation

Since the description of the 2nd order equation was deliberately constructed to follow Fuchs's analysis of the nth order equation, it is possible simply to state his more general results. He found that if the fundamental equation for any singular point $\underline{a}$ had no repeated roots then the solutions near that point were all of the form $(x-a)^\alpha \phi(x-a)$, where ϕ is a holomorphic function, what he called a finite, single-valued, continuous function. But if the fundamental equation had a k-fold repeated root the general solution was of the form of a poly-nomial in $\log (x-a)$ of degree k-1 with holomorphic coefficients. The determining fundamental equation and the method of obtaining solutions was described quite generally. The general form of an nth order equation of the Fuchsian class was shown to be

$$\frac{d^n y}{dx^n} + \frac{F_{\rho-1}(x)}{\psi} \frac{d^{n-1} y}{dx^{n-1}} + \frac{F_{2(\rho-1)}(x)}{\psi^2} \frac{d^{n-2} y}{dx^{n-2}} + \ldots + \frac{F_{n(\rho-1)}(x)y}{\psi^n} = 0$$

$$(2.1.5)$$

where $\psi = (x-a_1) \ldots (x-a_\rho)$; the finite singular points are $a_1, a_2, \ldots, a_\rho$.

It is very interesting to note that exactly this equation, in the form it takes when ∞ is not a singular point, was given by Riemann in a lecture on 20 February 1857, when he stated that no solution of it becomes infinitely great to an infinitely great order (Werke XXI). It was not published, however, until 1876, a decade after Fuchs's independent re-discovery of it.

Inhomogeneous equations

Fuchs also showed [1868 §5] how his methods could deal with the inhomogeneous equation

$$P_0 \frac{d^n y}{dx^n} + P_1 \frac{d^{n-1} y}{dx^{n-1}} + \ldots + P_n y + q = 0 \qquad (2.1.6)$$

which he abbreviated to $Y + q = 0$.

He showed that every solution of $Y + q = 0$ and $Y = 0$ is a solution of the equation

$$q \frac{dY}{dx} - Y \frac{dq}{dx} = 0, \qquad\qquad (2.1.7)$$

and that conversely every solution of (2.1.7) is either a solution of $Y = 0$ or of the inhomogeneous equation $Y + q = 0$. Furthermore, if the equation $Y = 0$ is of the Fuchsian class then the equation $Y + q = 0$ will have all its solutions regular provided $\frac{d \log q}{dx}$ becomes finite and single-valued on being multiplied by a finite power of $(x-a)$, whenever a is a singular point of (2.1.6). The converse is also true, and the function q must be of the form $C(x-a_1)^{\mu_1} \ldots (x-a_\rho)^{\mu_\rho}$, where $a_1, \ldots, a_\rho$ are the finite singular points of the coefficient $p_1, \ldots, p_n$, and C, $\mu_1, \ldots, \mu_\rho$ are constants. The conditions on q are simply derived from insisting that (2.1.7) be an equation of the Fuchsian class.

Corollaries of Fuchs's work

Fuchs drew several conclusions from his work. One concerned with the problem of characterizing those differential equations whose solutions were algebraic and another the special role played by the hypergeometric equation.

In [1865, §7] and [1866, §6] he observed that the class of differential equations all of whose solutions are algebraic is contained within the Fuchsian class, for Puiseux had showed [1850, 1851] that in a neighbourhood of a singular point, a, an algebraic function admits an expansion of the form

$$\eta = \sum_0^{\mu-1} c_k (x-a)^{-r+k/\mu}$$

where μ and σ are integers. This implies that

$$\eta = (x-a)^{-r} . \phi + (x-a)^{-r+1/\mu} . \phi_1 + \ldots + (x-a)^{-r+\frac{\mu-1}{\mu}} \phi_{\mu-1} .$$

say, where the ϕ's are single-valued, continuous, and finite functions near a and each term $(x-a)^{-r+j/\mu}$ satisfies the conditions required to make η a solution to differential equation of the Fuchsian class. It would, Fuchs said, be an interesting problem to seek a precise characterization of the equations which have only algebraic solutions, but one which he was then unable to solve. He was able to solve the much simpler special case: when are the solutions always rational functions? The necessary and sufficient conditions are that the roots of the indicial equation are all real integers, for then the solutions are single valued in the entire complex plane. The question of when a differential equation has only algebraic solutions was taken up by many mathematicians in the 1870's and is the subject of the next chapter. It is a question with a long history. Liouville had made a thorough, if ultimately inconclusive, study of it in the 1830's and 1840's, which has recently been thoroughly analysed in Lützen's analysis of Liouville's early work [Lützen, 1984], where the even earlier work of Abel and later 19th century and some 20th century works are also discussed.

Fuchs also observed ([1865 §6, 1866 §6]) that the hypergeometric equation is of the Fuchsian class. It appears in the form of (2.1.4) as

$$\frac{d^2y}{dx^2} + \frac{f_0 + f_1 x}{(x-a_1)(x-a_2)} \frac{dy}{dx} + \frac{g_0 + g_1 x + g_2 x^2}{(x-a_1)^2 (x-a_2)^2} y = 0. \qquad (2.1.8)$$

If $a_1 = 0$, $a_2 = 1$ the indicial equations at 0, 1, and ∞ are

$$r(r-1) - f_0 r + g_0 = 0$$

$$r(r-1) + (f_0 + f_1)r + g_0 + g_1 + g_2 = 0 \text{ and}$$

$$r(r-1) + (2-f_1)r + g_2 = 0 \qquad \text{respectively.}$$

If the roots are denoted by α, α': 0, γ'; and β, β' respectively

where $\alpha+\alpha' + \beta+\beta' + \gamma' = 1$ in conformity with the determination of k in

this case, the differential equation takes Riemann's form $(1-x) \dfrac{d^2 y}{d \log x^2}$

$- [\alpha+\alpha' + (\beta+\beta')x] \dfrac{d \, y}{d \, \log \, x} + (\alpha\alpha' - \beta\beta')y = 0$, where it is assumed that

none of $\alpha - \alpha'$, $\beta - \beta'$, γ' are integers.

Fuchs was also able to give a significant characterization of the

hypergeometric equation. In the general differential equation of the

Fuchsian type, each function $F_{k(\rho-1)}$ contains $k\rho - k + 1$ constants, so

there are $\frac{1}{2}n(n+1)\rho - \frac{1}{2}n(n-1)$. There are n exponents at each of the

$\rho+1$ singular points (including ∞). So there are $n(\rho+1)$ in all, of

which only $n(\rho+1) - 1$ are arbitrary in virtue of Fuchs's equation

for k. Accordingly, if a set of given exponents is to determine the

equation, then

$$\frac{n(n+1)\rho}{2} - \frac{n(n-1)}{2} = n(\rho+1) - 1 \qquad (2.1.9)$$

so $\rho = 1 + \dfrac{2}{n}$ and either[10] $n = 1$, $\rho = 3$ or $n = 2$, $\rho = 2$. Accordingly,

if the first-order equations are excluded, the class of equations of

the Fuchsian type for which the exponents at the singular points deter-

mine the coefficients of the equation contains precisely the hyper-

geometric equation. This explains why Riemann's methods worked so well

for the Gaussian equation, for it is precisely in that case that his

initial data (the exponents) characterize not only the solution func-

tions but also the equation itself. Contrariwise, in all other cases,

the number of exponents is too few to characterize the equation, and

the excess numbers, called the auxiliary parameters, have since proved

quite intractable.

2.2 Generalizations of the hypergeometric equation.

After 1868, Fuchs turned his attention to finding applications and consequences of his theory, rather than to finding ways of extending it. Among the problems then arousing the greatest interest in mathematics was the investigation of hyper-elliptic functions. Weierstrass had published two important papers on Abelian functions and the inversion of hyper-elliptic integrals in Crelle's Journal für Mathematik of 1854 and 1856 (Weierstrass [1854, 1856]) which were instrumental in securing him an invitation to become associate professor at the University of Berlin. Liouville hailed the first papers as "one of those works that marks an epoch in science" (quoted in Bierman: [1976, 221]), and Weierstrass's influential lecture cycle also dealt largely with elliptic and Abelian functions[11]. Riemann's [1857c] had solved the inversion problem for an arbitrary integral of an algebraic function by means of the theory of θ-functions in several variables, and this work was also much discussed in Berlin (see Chapter V).

Fuchs first proposed, in his [1870a], to study the periods of hyper-elliptic integrals from the standpoint of his new theory, reserving the full treatment of Abelian integrals for a subsequent opportunity [1871 a, b]. These later works require a knowledge of the theory of θ-functions, and will not be discussed. It will be clearest to begin with the example that dominates the [1870] paper, the hyper-elliptic integral

$$\int \frac{dx}{[(x-k_1) \ldots (x-k_{n-1})(x-u)]^{\frac{1}{2}}} = f(z) \qquad (2.2.1)$$

Fuchs introduced this example in §7 to illustrate the more general integral $\int \frac{dx}{s}$, where $s^2 = \phi(x,u)$ and $\phi(x,u)$ is algebraic of degree n in x, "partly" as he said "to elucidate the preceeding and partly for later use". It certainly presents a simpler piece of analysis, and one not without applications. When n = 3 his equation for the periods has become famous as the Picard-Fuchs equation. Its significance in algebraic geometry is well described in Clemens [1980].

Fuchs's strategy for finding the periods of the integral (2.2.1) was straightforward. Suppose u is fixed for the moment, then the periods are found by cutting the x-plane along a curve, called the principal cut, joining u to k_1, k_1 to k_2, ..., k_{n-2} to k_{n-1}, and (if n is odd) k_{n-1} to ∞. If a, b are any two consecutive points of the sequence (k, k_1, ..., k_{n-1}, ∞) the period corresponding to the cut joining a to b is found by evaluating f(x) around a closed curve γ enclosing that part of the principal cut but no other points in {u, k_1, ..., k_{n-1}, ∞}. Precisely,

$$\eta = \frac{1}{2} \int_\gamma y \, dx, \quad y = [(x_1 - k_1) \ldots (x - k_{n-1})(x - u)]^{-\frac{1}{2}}$$

The first question is, how to arrange all this when u is allowed to vary? Fuchs showed that the periods were continuous, single-valued functions of u away from the branch points. Continuity is evident; to show they are single-valued Fuchs considered

$$s^2 = [(x - k_1) \ldots (x - k_n)(x - u)].$$

For each u, s is defined as a two-leaved Riemann surface T_u over the complex x-plane. The branch points are at u, and at k_1, k_2, ..., k_{n-1} and possibly ∞, all of which are independent of u.

Accordingly the principal cut is independent of u once it has reached k_1. It can furthermore be made to lie in the upper leaf of T_u for all u, for, as u makes a circuit σ around some k_λ the factor x - u of s becomes $e^{2\pi i}(x - u)$ if x is enclosed within σ while all the others remain unchanged, so s becomes −s. On the other hand, if x is not within σ, then x − u does not change and s remains unaltered. The circuit σ has transformed T_u into T'_u (the lower leaf) and outside σ upper leaves are joined to upper leaves because s did not change. But σ could be taken arbitrarily small, whence the desired result.

Of course, the periods are not single-valued functions of u, but Fuchs could now write down (§7) a sequence of equations connecting the periods before and after a circuit is made by u around any k_λ. This gave him the monodromy relations, at each singular point of the differential equation he sought, connecting the periods. To obtain the differential equation he observed that it must be of order n − 1 since there are n − 1 linearly independent periods. It therefore has the form

$$\beta_{n-1} \frac{d^{n-1}\eta}{dx^{n-1}} + \ldots \beta_0 \eta = 0, \tag{2.2.2}$$

whose solutions, $\eta_1 \ldots, \eta_{n-1}$, are the periods along the parts of the principal cut. After some calculation he was able to show that $\frac{\beta_{(n-1)-i}}{\beta_{n-1}}$ is necessarily of the form $\frac{F_{i(n-2)}(u)}{\psi^i(u)}$ where $F_{i(n-2)}$ is a polynomial of degree at most i(n−2) and $\psi(u) = (u - k_1) \ldots (u - k_{n-1})$. So the singular points are precisely $k_1, \ldots, k_{n-1}$ (and ∞ if and only if n is odd); none are accidental. The differential equation is of the Fuchsian type and all of its solutions are regular[12].

When $\phi(x,u) = A[(x - k_1) \ldots (x - k_{n-1})](x - u) = \psi(x)(x - u)$, the calculations to find $\beta_{n-1-i}/\beta_{n-1}$ are not too difficult. Fuchs did not work directly with the monodromy relations he had obtained in §7 but preferred the expressions involving partial derivatives of y with respect to u. The result is clear, even if the method chosen passed through some murky country; Fuchs found (§19, equation 4)

$\beta_{n-i} = \rho_0 \tau_{n-1} \psi^{(i-1)}(u)$, where ρ_0 is a constant,

$$\tau_{n-i} = \frac{1 - 2n + 2}{2\sigma_{n-i}(i-1)!},$$

and $\sigma_k = \dfrac{1.3 \ldots (2k-1)}{2^k}.$

The monodromy relations give directly (§13) that the roots of the indicial equation are rational numbers. This result remains true if the general integral

$$\int \frac{f(x)}{s}\, dx, \quad s^2 = \phi(x,u), \quad \phi \text{ of order } n \text{ in } x,$$

is considered, but the rest of the analysis is more complicated. Some of the periods may now depend linearly on others, so, if there are exactly p linearly independent ones the differential equation has order only p. It remains of the Fuchsian class, but the β's are much harder to find. There are n finite singular points (∞ is singular if and only if n is odd) which are actual, and there may also be accidental singularities, depending on $f(x)$. Fortunately the examples Fuchs gave are of the simpler type (2.2.1).

For the elliptic integral $n = 3$ and $\psi(x) = x(x-1)$,

$\beta_2 = -2u(u - 1)\rho_0$, $\beta_1 = -2(2u - 1)\rho_0$, $\beta_0 = -\frac{1}{2}\rho_0$, and the differential equation is

$$2u(u - 1)\frac{d^2\eta}{du^2} + 2(2u - 1)\frac{d\eta}{du} + \frac{\eta}{2} = 0 \qquad (2.2.3)$$

This takes Legendre's form on substituting $u = \dfrac{1}{k^2}$, $\eta = k\zeta$:

$$k(1 - k^2)\frac{d^2\zeta}{dk^2} + (1 - 3k^2)\frac{d\zeta}{dk} - k\zeta = 0.$$

For the hyper-elliptic integral of the first kind $n = 5$ and Fuchs obtained a differential equation equivalent, as he said, to the system of simultaneous differential equations deduced by Koenigsberger in the first volume of the Mathematische Annalen.

Fuchs concluded the paper with a detailed study (§21) of the case of the elliptic integral, which turned out to be important for his later work. He knew the singular points of (2.2.3) were at $u = 0, 1$, and ∞, and he knew the monodromy relations at each point; taking them in order he found:

at $u = 0$ the indicial equation is $r^2 = 0$, so one solution is

$$v_{01} = 1 + \sum_k \left(\frac{1.3\ldots2k-1}{2.4\ldots2k}\right)^2 u^k,$$

and substituting $\eta = v_{01}\int \zeta_0 du$ into (2.2.3) he found $\zeta_0 = -\frac{C}{u} + CG_0(u)$, where

$$G_0(u) = \frac{1 + v^2_{01}(u - 1)}{v^2_{01}u(u - 1)},$$

and C is an arbitrary constant.

Accordingly, a second solution of (2.2.3) valid near $u = 0$ is

$$v_{02} = H_0(u)v_{01} - v_{01}\log u, \text{ where}$$

$$H_0(u) = 4 \log 2 + \int_0^u G_0(u)du.$$

But the monodromy relations between η_1 (the period along $(u, 0)$) and η_2 (the period along $(0, 1)$) relate η_1 and η_2 to v_{01} and v_{02}. The monodromy relations for η_1 and η_2 are $\tilde{\eta}_1 = \eta_1$ and $\tilde{\eta}_2 = 2\eta_1 + \eta_2$. The relations for v_{01} and v_{02} are immediate from their explicit representations: $\tilde{v}_{01} = v_{01}$ and $\tilde{v}_{02} = -2\pi i v_{01} + v_{02}$. So the relationships $\eta_1 = c_{11}v_{01} + c_{12}v_{02}$, $\eta_2 = c_{21}v_{01} + c_{22}v_{02}$ simplify, because $c_{12} = 0$, $c_{11} = -\pi i c_{22}$, to become : $\eta_1 = c_{11}v_{01}$, $\eta_2 = c_{21}v_{01} - \frac{1}{\pi i}c_{11}v_{02}$. The constants c_{11}, c_{21} can be found from

$$\eta_1 = \int_u^0 \frac{dx}{[x(x-1)(x-u)]^{\frac{1}{2}}}, \quad \eta_2 = \int_0^1 \frac{dx}{[x(x-1)(x-u)]^{\frac{1}{2}}},$$

on letting $u \to 0$. It turns out that $\eta_1 = -\pi v_{01}$, $\eta_2 = \pi v_{01} - i v_{02}$. For $\eta_1(0) = -\pi = c_{11} v_{01}(0) = c_{11} \cdot 1$, and, as $u \to 0$

$$\eta_1(u) \simeq \int_0^1 \frac{1 - (1-x)^{\frac{1}{2}} dx}{[x(x-1)(x-u)]^{\frac{1}{2}}} + \int_0^1 \frac{dx}{[x(u-x)]^{\frac{1}{2}}}$$

$$= -2i\log 2 \qquad + -2i\log((u - 1)^{\frac{1}{2}} + i) + i\log u.$$

$H_0(0) = 4\log 2$, so $v_{02}(u) \simeq 4\log 2 - \log u$,

and $\eta_2(u) = c_{21} v_{01}(u) - i v_{02}(u)$ implies

$i\log u - 4i\log 2 + \pi = c_{21} - i(4\log 2 - i \log u)$, so $c_{21} = \pi$. An exactly similar calculation near $u = 1$ produces solutions v_{11}, v_{12} to (2.2.3) related to η_1 and η_2 by

$$\eta_1 = -\pi i \, v_{11} + v_{12}$$

$$\eta_2 = -2\pi i \, v_{11} - v_{12}$$

Finally near $u = \infty$, where the indicial equation is $(r - \frac{1}{2})^2 = 0$

$$v_{\infty 1} = (\frac{1}{u})^{\frac{1}{2}} [1 + \sum_{K=1}^{\infty} (\frac{1.3....(2K-1)}{2.4....2K})^2 (\frac{1}{u})^K]$$

and

$$v_{\infty 2} = v_{\infty 1} -4 \log 2 + \int_0^t \left(\frac{t + v_{\infty 1}^2 (t-1)}{v_{\infty 1}^2 (1-t)t} \right) dt - v_{\infty 1} \log u$$

$$\eta_1 = -\pi v_{\infty 1} - i v_{\infty 2}$$

$$\eta_2 = -\pi v_{\infty 1}$$

The substitution $u = \frac{1}{k^2}$, $x = z^2$ transforms the integral

$$\int \frac{dx}{\sqrt{[x(x-1)(x-u)]}} \quad \text{into} \quad 2k \int \frac{dx}{\sqrt{[(1-z^2)(1-k^2x^2)]}} .$$

So, if

$$K = \int_0^1 \frac{dz}{\sqrt{[(1-z^2)(1-k^2z^2)]}} \quad \text{and} \quad K'i = \int_1^{1/k} \frac{dz}{\sqrt{[(1-z^2)(1-k^2z^2)]}},$$

then $K = -\frac{1}{2k} n_2 = \frac{\pi}{2k} v_{\infty 1}$

and $K'i = \frac{1}{2k} (n_1 - n_2) = -\frac{i}{2k} v_{\infty 2}$,

which give the well-known power series for K and K'.

Fuchs's techniques were thus able to give him a new derivation of the solutions to Legendre's equation, and in principle they were capable of handling the analogues of Legendre's equation for the periods of hyper-elliptic integrals[13]. Legendre had also considered what he called elliptic integrals of the second kind, for which the periods are

$$J = \int_0^1 \frac{k^2 y^2 dy}{[(1-y^2)(1-k^2y^2)]^{\frac{1}{2}}}, \quad \text{and} \quad J'i = \int_1^{1/k} \frac{k^2 y^2 dy}{[(1-y^2)(1-k^2y^2)]^{\frac{1}{2}}}$$

J and J' satisfy the differential equation

$$(1-k^2)\frac{d^2 y}{dk^2} + \frac{1-k^2}{k} \frac{dy}{dx} + y = 0, \quad \text{Legendre [1828, I, 62]}.$$

They are connected to K and K' by Legendre's relation $KJ' - K'J = \frac{\pi}{2}$, Legendre [1825, I, 61].

As I shall discuss in Chapter IV, the J's behave in some ways surprisingly differently to the K's, and when Fuchs was led to investigate this at Hermite's request a decade later, the consequences for the theory of modular functions were to be significant.

Conclusion

In these papers Fuchs succeeded in characterizing those linear ordinary differential equations none of whose solutions have an essential singular point anywhere in the extended complex plane. In so doing he developed the theory of their singularities in terms of their monodromy matrices and their eigenvalues - the roots of the associated fundamental equation. He showed how these roots were connected to the exponents of the branch points, which themselves satisfy the indicial or determinantal fundamental equation. He gave a thorough analysis of the situation when one or more eigenvalues are repeated and logarithmic terms enter the solution. He indicated the special position occupied by the hypergeometric equation among equations of the Fuchsian class, and raised the problem of isolating those equations which have algebraic solutions. He followed Riemann in using monodromy matrices to study the analytic continuation of the solutions around the singular points, but did not rely on Dirichlet's principle to obtain the solutions globally. Rather he preferred to study the global forms using the analytic continuation of power series in the Weierstrassian style. His study of hyper-elliptic integrals is his most 'Riemannian' paper, later ones are couched more and more in the theory of power series. This shift of emphasis reflects a general tendency during the 1870's to avoid Dirichlet's principle, and to seek to obtain Riemann's results in other ways (see Chapter V p218 and Appendix 1). This tendency was most marked in Berlin.

This achievement raised three areas for future work. One, with which I shall be little concerned, is the problem of going outside the Fuchsian class to study equations whose solutions have essential singularities. This problem was taken up by Thomé in the 1870's, and later by Poincaré. Although power series solutions to such equations can be found formally, they frequently do not converge in the neighbourhood of singular points of the coefficients which are not of Fuchs's type. Instead they provide asymptotic solutions to the equation, as was first realized by Poincaré [1886]. There is a thorough discussion of this topic in Schlissel [1976/77].

The second area was the use of the methods of Riemann and Fuchs to explore equations of the Fuchsian class. This included using the methods to reformulate and add to the knowledge of familiar equations, such as Legendre's, and attempting to discuss new equations. By and large Fuchs took the former path, quite successfully as will be seen in Chapter IV, but attempts to delineate new equations with precision were less successful. The pioneer in this direction was Thomae[14]. The difficulty here is precisely Fuchs's discovery that the hypergeometric equation is the only equation of order greater than 1 for which the exponents at the branch points uniquely determine the coefficients.

The third area was the attempt to formulate a theory of differential equations whose coefficients were not rational functions but, say, elliptic functions. This class includes, for example, Lamé's equation, and was studied accordingly by several writers in the 1870's. It forms an intriguing stepping stone on the way to the general theory of differential equations with algebraic coefficients, and will be described in Chapter VI.

Fuchs turned his attention to non-linear equations and movable singularities in the 1880's and 1890's. For a brief account of that work and its consequences see Gray [1984a].

2.3 The new methods of Frobenius and others

In this section some developments are considered which lie to one side of the main story: Frobenius' simplification of Fuchs's work, his introduction of the idea of irreducibility of an equation, Thomé's study of equations with irregular singular points, and the introduction of Jordan's theory of canonical forms for matrices. These ideas will not be referred to again, and they are included purely for the sake of completeness.

Fuchs's arguments were, one might say, solidly contemporary adaptations of general methods for studying algebraic differential equations. Consequently they are long, and more difficult than they need be. The man who first proposed the simpler methods which have since become customary in treatments of linear ordinary differential equations was Georg Frobenius, in 1873. Frobenius, then 24, had not yet presented his <u>Habilitationsschrift</u> and even before he did so he was nominated by Weierstrass for a newly created Professorship Extraordinarius the next year. He only held that post for half a year before going to the Polytechnic in Zurich, one of a series of distinguished professors the Swiss recruited from Berlin; see K R Biermann [1973b, 96].

Frobenius' main aim in his paper [1873a] was, as he said, to find simpler methods for obtaining Fuchs's results. He found that this could be done by working directly with power-series. He considered the differential equation (where x and y are complex and $y^{(n)} = \dfrac{d^n y}{dx^n}$)

$$x^n p(x) \, y^{(n)} + \ldots + p_n(x)y = 0,$$

which he denoted $P(y) = 0$, in a neighbourhood of $x = 0$. The solutions were to be bounded near 0 when multiplied by a suitable power of x.

He assumed, as he may without loss of generality, that $p(x) = 1$ and the other p_i have power series expansions near $x = 0$.

For y, he substituted the power series $g(x, \rho) = \sum_{j=0}^{\infty} g_j x^{\rho+j}$, containing a parameter ρ, and observed that $P(x^\rho) = x^\rho f(x, \rho)$, where $f(x, \rho)$ is a polynomial in ρ:

$$f(x,\rho) = \rho(\rho-1)\ldots(\rho - n + 1)p(x) + \rho(\rho - 1)\ldots + (\rho - n + 2)p_1(x)$$

$$+ \ldots + p_n(x)$$

$$= \sum_j f_j(\rho)x^j$$

here each $f_j(\rho)$ is a polynomial function of ρ. Accordingly he obtained a recurrence relation for the g_j's:

$$g_0 f_0(\rho) = 0$$

$$g_1 f_0(\rho+1) + g_0 f_1(\rho) = 0,$$

$$\ldots ,$$

$$g_j f_0(\rho+j) + g_{j-1} f_1(\rho+j-1) + \ldots + g_0 f_j(\rho) = 0.$$

Since it is assumed that $g_0 \neq 0$ the equation $g_0(\rho) = 0$ determines ρ. Frobenius observed this, but preferred to let ρ remain a variable and determine the g_j as functions of ρ instead. Each g_j is a rational function,

$$g_j(\rho) = g_0(\rho)h_j(\rho)/f_0(\rho+1)\ldots f_0(\rho+j),$$

where $h_j(\rho)$ can be determined explicitly as a function of $f_j(\rho + k)$'s. If ρ is restricted to suitably small neighbourhoods of the roots of $f_0(\rho) = 0$ then the denominators of the g_j's vanish only at the roots of $f_0(\rho) = 0$, and $g_0(\rho)$ can be chosen so that the functions $g_j(\rho)$ are bounded. So, if the series $g(x, \rho) = \sum g_j x^{\rho+1}$ converges it represents a solution of the differential equation. Frobenius next established the necessary convergence, by means of the ratio test and a simple

majorizing argument, and showed explicitly that within that domain the

convergence is uniform. This established that the power series can be

differentiated term by term, and concluded the proof that the series

represents a solution of the differential equation.

In the next two sections of his paper Frobenius considered the

nature of the solutions which are obtained once ρ is a root of

$f_0(\rho) = 0$. He simplified Fuchs's treatment of repeated roots (when

polynomials in log x enter the solution) by exploiting the parameter

ρ. Since $P(g(x, \rho)) \equiv f_0(\rho)g(\rho)x^\rho$, if ρ_k is a k-fold root of

$f(\rho) = 0$, then differentiating both sides of this identity k times

with respect to ρ and setting $\rho = \rho_k$ yields $P(\dfrac{d^k}{d\rho^k} g(x, \rho_k)) = 0$. This

implies that $\dfrac{d^k}{d\rho^k} g(x, \rho_k) = g^{(k)}(x, \rho_k)$ is a solution of the differen-

tial equation $P(y) = 0$, and since $g(x, \rho) = x^\rho \sum g_j(\rho)x^j$, the solution

obtained is

$$g(k)(x,\rho_k) = x^{\rho_k} \sum (g_j^{(k)}(\rho_k) + kg_j^{(k-1)}(\rho_k) \log x + \frac{k(k-1)}{2!} g_j^{(k-2)}(\rho_k)(\log x)^2 +$$

$$\cdots + g_j(\rho_k)(\log x)^k)x^j$$

Consequently the solution to the differential equation has no logarith-

mic terms if the equation $f_0(\rho) = 0$ has no repeated roots. Frobenius

concluded the paper with other methods for obtaining the coefficients

$g_j(\rho)$.

Frobenius confined his first study of linear differential equations

to rederiving the results of Fuchs. Almost none of the results were

new: the indicial equation is to be found in Fuchs's [1865] and [1866],

as are the associated power series solutions. The concern about

uniform convergence is also present in the earlier work. What is new

is the method, involving the parameter ρ, for proving the convergence of the solutions obtained via the indicial equation. Frobenius was quite scrupulous in acknowledging Fuchs's work, but as his simpler methods drove out those of Fuchs, the comparison became blurred until he is sometimes remembered more for the results than the methods (see e.g Birkhoff [1973, 282], Hille [1976, 344], or Piaggio [1962, 109]).

Frobenius's new results were largely connected with the idea of the irreducibility of a differential equation, which he introduced in his next paper [1873b], published simultaneously with the one just described. By analogy with polynomial equations he said a differential equation was irreducible if it had no solutions in common with a differential equation of lower order or one of the same order but lower degree. This implies, for instance, that a differential equation all of whose solutions are algebraic is reducible, for its solutions are the roots of a polynomial equation, which is a differential equation of order zero. More substantially, Frobenius showed (§5) that if the hypergeometric equation is reducible then the hypergeometric series entering in to one of its solutions must be a polynomial. Since this result displays nearly all the features of Frobenius's theory of irreducibility, and since that theory is strikingly similar to the theory of irreducible algebraic equations, the exposition may reasonably be confined to this example.

The hypergeometric equation $x(1 - x)\frac{d^2y}{dx^2} + (\gamma - (\alpha+\beta+1)x)\frac{dy}{dx} - \alpha\beta y = 0$ is reducible if one of its solutions is the solution of a first order equation. Frobenius took as a basis of solutions two which are defined on a neighbourhood of the singular point $x = 1$:

$w = F(\alpha,\beta,\alpha + \beta\ -\gamma + 1, 1 - x),$

$w' = (1 - x)^{\gamma-\alpha-\beta} F(\gamma - \alpha,\ \gamma - \beta,\ \gamma - \alpha - \beta + 1,\ 1 - x)$

because, as he said, each has part of its domain of convergence in common with the series solutions valid near $x = 0$ and $x = \infty$. Indeed, if u and u' are a basis of solutions near $x = 0$ and v and v' are a basis of solutions near $x = \infty$, then Kummer's relations imply that

$$w = au + a'u', \quad w = cv + c'v'$$

$$w' = bu + b'u', \quad w' = dv + d'v'$$

for precise constants a, a', ..., d'. Accordingly, if the equation is reducible then either one coefficient vanishes in each expression for w or in each expression for w', else analytic continuation of w (or w') would yield two linearly independent branches, which must satisfy a differential equation of order two. Frobenius considered each possibility in turn. For example,

$$a = \frac{\Pi(\alpha+\beta-\gamma)\Pi(-\gamma)}{\Pi(\alpha-\gamma)\Pi(\beta-\gamma)}, \quad c' = \frac{\Pi(\alpha+\beta-\gamma)\Pi(\alpha-\beta-1)}{\Pi(\alpha-\gamma)\Pi(\alpha-1)}$$

The function Π never vanishes, but it is infinite at the negative integers, so a and c' vanish if $\gamma - \alpha = \nu + 1$, for some non-negative integer ν. But then in the solution $u' = x^{1-\gamma}F(\alpha-\gamma+1, \beta-\gamma+1, 2-\gamma, x)$ the F term is a polynomial of degree ν. The other cases proceeded similarly.

To put this conclusion in a clearer light, as he said, he rederived it starting from the observation that the hypergeometric equation is completely determined by its exponents at its singular points. Now, if analytically continuing two solutions round a singular point only multiplies them by constants, then the exponent differences must all be integers without the solution containing a logarithmic term. The general first order equation of the Fuchsian type has the form $\frac{dy}{dx} - (\frac{\alpha_1}{x-a_1} + \frac{\alpha_2}{x-a_2} + \ldots + \frac{\alpha_\mu}{x - a_\mu})y = 0$ with solution

$y = C(x-a_1)^{\alpha_1}(x-a_2)^{\alpha_2} \ldots (x-a_\mu)^{\alpha_\mu}$. Its singular points are

$a_1, a_2, \ldots, a_\mu$ and ∞, at which the exponents are $\alpha_1, \alpha_2, \ldots, \alpha_\mu$, and

β respectively, and Fuchs's condition in this case is

$\alpha_1 + \alpha + \ldots + \alpha_\mu + \beta = 0$. Frobenius had shown earlier in this paper

(§4, Theorem I) that the singular points of a differential equation

of lower order are either the singular points of the equation of

highest order with which it has solutions in common, or they are

accidental singular points of the lower-order equation for which

the roots of the corresponding indicial equation are less than the

order of the higher order equation. So here, $\alpha_3 = \alpha_4 = \ldots = \alpha_\mu = 1$.

Furthermore, the common solution of the two differential equations

may be taken to have its singular points at 0, 1, and ∞ with

exponents α, γ, β, so it is

$$y = Cx^\alpha(1 - x)^\gamma(x - a_1)(x - a_2) \ldots (x - a_\mu).$$

In this way Frobenius obtained the necessary conditions $\alpha + \beta + \gamma + \nu = 0$

$\alpha' + \beta' + \gamma' = \nu + 1$. These conditions are also sufficient, as can be

seen from consideration of the function $P \begin{pmatrix} 0 & \infty & 1 & \\ 0 & -\nu & 0 & x \\ \alpha' - \alpha & \beta' - \beta - \nu & \gamma' - \gamma & \end{pmatrix}$

for which the corresponding hypergeometric series is a polynomial.

Frobenius used his theory of irreducibility to clarify three

aspects of the theory of linear differential equations: the behaviour

of the solutions under analytic continuation; the nature of accidental

singular points; and the occurrence of solutions with essential

singularities when the equation is not of the Fuchsian type.

Typical of his results under the first heading is this ([1873b, §3,

Theorem IV]): if a differential equation is reducible there is an

equation of lower order with which it has all its solutions in common.

Indeed, if an equation of order λ has all of its solutions in common

with the equation $Q(y) = 0$ of order $\mu(\lambda > \mu)$ then each of its solutions satisfy an equation of the form $Q(y) = w$ in which w is a solution of a certain equation of order $\lambda - \mu$.

His analysis of the singular points has already been described. To consider his treatment of equations not of the Fuchsian type it is necessary to look also at the work of his contemporary at Berlin, L. W. Thomé. Thomé was the first to consider how Fuchs's theory might be extended to such equations, and starting in 1872 he published a series of increasingly long papers on this topic in the <u>Journal für Mathematik</u>[15]. In the second of these [1873, 266] he introduced the convenient word 'regular' to describe solutions of the kind Fuchs had sought (<u>reguläre Integrale</u>); in this terminology he sought properties of the coefficients of a differential equation which indicated how many linearly independent regular integrals it would have. His first main theorem in this direction [1872, §5, Theorem 2] applied to the equation $\frac{d^m y}{dx^m} + p_1 \frac{d^{m-1} y}{dx^{m-1}} + \ldots + p_m y = 0$, whose coefficients $p_i = p_i(x)$ were single-valued in some neighbourhood of $x = a$ and may be written $\frac{f(x)}{(x-a)^{\pi_i}}$ in lowest terms. He showed that, if in the sequence $\pi_1 + m-1, \pi_2 + m-2, \ldots, \pi_m$, the greatest has value $g > m$, then not all solutions of the equation are regular. Furthermore, if $\pi_h + m-h$ is the first to equal g, then at most $m - h$ solutions are regular. As he observed, in Fuchs's study, $g \leq m$. In [1873] Thomé sought to characterize those equations for which precisely $m - 1$ linearly independent solutions are regular, and obtained an answer which I shall mention below. In subsequent papers Thomé [1874, 1876] looked at the question of reducing differential equations to those of lower order, and this work caught the attention of Frobenius.

Frobenius [1875b] considered the differential equation

$$A(y) = p_m \frac{d^m y}{dx^m} + p_{m-1} \frac{d^{m-1} y}{dx^{m-1}} + \ldots + p_0 y = 0 \text{ and its adjoint}$$

$$\hat{A}(y) = (-1)^m \frac{d^m}{dx^m} (p_m y) + \ldots - \frac{d}{dx} (p_1 y) + p_0 = 0, \text{ an expression}$$

introduced independently by Fuchs, Thomé, and himself earlier (but

known already to Lagrange). He established that, if $B(y) = 0$ denotes

another differential equation, then the equation $A(B(y)) = 0$ has no

fewer regular solutions than $B = 0$ and no more than $A = 0$.

(Solutions of $A(B(y)) = 0$ are either solutions of $B(y) = 0$, or, if

$A(u) = 0$, of $B(y) = u$.) Now, if the equation $C = 0$ has a regular

solution, say $x^\rho v$, where $v(x) = \sum_0^\infty a_n x^n$, then y satisfies the first

order differential equation $B_1(y) = 0$, where

$$B_1(y) : \quad = (\rho v + x v')y - x v y' = 0.$$

Frobenius redefined reducibility here so that, in the new sense of the

word, $C(y) = 0$ is reducible. His definition applied only to equations

whose coefficients near $x = 0$ had the character of rational functions,

and said that such a linear differential equation was reducible if it

had a solution in common with a linear differential equation of lower

order having the same property near $x = 0$. To avoid confusion I shall

refer to this property as local reducibility. In this terminology

Frobenius showed that an m-th order equation of the Fuchsian type is

locally reducible to

$$B_m(B_{m-1}(\ldots(B_1(y))\ldots)) = 0,$$

where each B_i is of the first order and has a regular solution, and

that the converse also holds. So in particular the regular solutions of

a differential equation $C(y) = 0$ all satisfy a differential equation

$B(y) = 0$ of the kind Frobenius considered, and if $C = A(B(y)) = 0$, then

$A(y) = 0$ has no regular solutions. Turning to the adjoint equation

$\hat{A}(y) = 0$ of $A(y) = 0$, Frobenius showed that if $A(y) = 0$ has only

regular solutions then $\hat{A}(y) = 0$ likewise has only regular solutions.

This followed from the result just cited and the reciprocity theorem due to

him and Thomé independently: that $\widehat{A(B(y))} = \hat{B}(\hat{A}(y))$. It was connected

with the number of regular solutions by the observation that the

indicial equation at $x = 0$ has as many roots (counted according to

multiplicity) as the differential equation has regular solutions

(§6 Theorem 3). So Frobenius obtained Thomé's result that $A(y) = 0$ of

order γ has exactly β regular integrals if and only if $\hat{A}(y) = 0$ is

reducible (in the earlier sense) to an equation of order $\gamma - \beta$ whose

indicial equation is a constant. In particular, this implies Thomé's

theorem mentioned, above, that $A(y) = 0$ has all but one of its

solutions regular if and only if its indicial equation is of order

$\gamma - 1$, and the adjoint equation $\hat{A}(y) = 0$ has a solution of the form

$$\exp\{(\frac{c_n}{x^n} + \ldots + \frac{c_1}{x})\}\sum_{\nu}^{\infty} a_{\nu} x^{\rho+\nu}.$$

These results conclude our discussion of Frobenius's work on

differential equations, which is complete except for his short paper

on the existence of algebraic solutions which will be described in its

proper place below (Chapter III). It remains to look briefly at one

further simplification of Fuchs's theory, that of Jordan and Hamburger,

and one addition to it, that of Jules Tannery.

The form of the solution to a differential equation in the neigh-

bourhood of a singular point $x = 0$ was discussed by Fuchs, who, as has

been seen, presented it in this form when the monodromy matrix had a k

times repeated eigenvalue w and $e^{2\pi i r} = w$:

$$u_1 = x^r \phi_{11}$$

$$u_2 = x^r \phi_{21} + x^r \phi_{22} \log x$$

. . . .

$$u_k = x^r \phi_{k1} + x^r \phi_{k2} \log x + \ldots + x^r \phi_{kk}(\log x)^{k-1} \quad [1866, 136, \text{eq. } 11$$

The ϕ_{ij} are functions of x, single-valued near x = 0, and each is a linear combination of ϕ_{11}, ϕ_{21}, ..., ϕ_{k1}. So, in the particular case when k = 2, the solutions in Fuchs's form are

$$u_1 = x^r \phi_{11}$$

$$u_2 = x^r \phi_{21} + x^r \phi_{22} \log x.$$

These arise, as Fuchs had shown, when the analytic continuation of solutions u_1 and u_2 around the singular point produces

$$\tilde{u}_1 = w\, u_1$$

$$\tilde{u}_2 = w_{21}u_1 + w\, u_2 \quad [1866, 135, \text{eq. } 10 \text{ abbreviated}].$$

It is easy to see, by writing $xe^{2\pi i}$ for x, that a suitable basis of solutions in this case is $u_1 = x^r \phi$, $u_2 = \dfrac{x^r \phi \log x}{2\pi i\, w}$ for which $w_{21} = 1$. In other words, the monodromy matrix with respect to this basis has the form $\left(\begin{smallmatrix} w & 0 \\ 1 & w \end{smallmatrix}\right)$, and is in Jordan canonical form.

The Jordan canonical form of a matrix had been introduced by Jordan in his Traité des substitutions et des équations algébriques [1870, Section II, Chapter 2] to simplify the discussion of linear substitutions. That same year Yvon Villarceau drew the attention of the Paris Académie des Sciences to a gap in the general theory of linear differential equations with constant coefficients: the method used to resolve even a second order equation into two first order ones breaks down when the characteristic equation has equal roots. Villarceau wished to resolve $\dot{Y} = AY$, where Y was a vector and A a matrix of constants, into $\dot{y}_1 = \lambda_1 y_1$, $\dot{y}_2 = \lambda_2 y_2$, but of course this is impossible in the case just mentioned.

Jordan observed [1871] that the gap is readily filled by his theory of canonical forms for matrices, which he proceeded to do, so simplifying Fuchs's form of the solution in the case of an equation with constant coefficients (which is, however, not an equation of the Fuchsian type). The introduction of Jordan canonical form into the Fuchsian theory of differential equations was carried out by Hamburger [1873]. He concluded (p.121) that, if w is a k-fold repeated eigenvalue of a monodromy matrix then a basis of solutions $y_1, \ldots, y_k$ can be found for which the transforms under analytic continuation are:

$$\tilde{y}_1 = wy_1$$

$$\tilde{y}_2 = y_1 + wy_2$$

$$\cdots \cdots$$

$$\tilde{y}_m = y_{m-1} + w\, y_m$$

Jordan's theory of canonical forms had been preceeded, in 1858 and 1868, by Weierstrass's theory of elementary divisors which he developed in connection with his study of the problem of simultaneously diagonalizing two bilinear or quadratic forms. Hamburger also sketched the correspondence between Jordan's analysis, which he had presented in quite an elementary form, and that of Weierstrass, which makes considerable use of the theory of determinants. However, the theory of matrices, especially in Germany, continued to rely on Weierstrassian ideas at least until the turn of the century, as Hawkins [1977] has established. The reader is referred to Hawkins's paper for a full discussion of Weierstrass's theory and its influence.

In 1875 Tannery published a paper on linear differential equations which closely resembled those of Fuchs. As observed above, it corrected the mistake in Fuchs's [1866], and it also contained a pleasing converse to Fuchs's main theorem. Tannery showed [1875, 130] that if y_1, y_2, ..., y_m were functions of x continuous except at points a_1, ..., a_ρ around which each became, on analytic continuation, a linear combination of y_1, y_2, ..., y_m, then they were solutions of a linear differential equation of order m with single-valued coefficients:

$$\frac{d^m y}{dx^m} = p_1 \frac{d^{m-1} y}{dx^{m-1}} + \ldots + p_m y, \text{ where } p_i = D_i/D'$$

$$D = \begin{vmatrix} y_1 & \dfrac{dy_1}{dx} & \cdots & \dfrac{d^{m-1} y_1}{dx^{m-1}} \\ y_2 & \dfrac{dy_2}{dx} & \cdots & \dfrac{d^{m-1} y_2}{dx^{m-1}} \\ \cdot & \cdot & \cdots & \cdot \\ y_m & \dfrac{dy_m}{dx} & \cdots & \dfrac{d^{m-1} y_m}{dx^{m-1}} \end{vmatrix}$$

and D_i is obtained from D by replacing the ith column by the transpose of $(\frac{d^m y_1}{dx^m}, \frac{d^m y_2}{dx^m}, \ldots, \frac{d^m y_m}{dx^m})$. So in particular every algebraic function of order m satisfies such a differential equation. The proof is straightforward, and I omit it.

Exercises on Chapter II

1. (a) Show that a necessary condition for the second order linear differential equation

$$\frac{d^2w}{dz^2} + p(z)\frac{dw}{dz} + q(z)w = 0$$

to have z_0 as a regular singular point is that $(z - z_0)p(z)$ and $(z - z_0)^2 q(z)$ are analytic near z_0.

(b) Show, by writing $\tilde{z} = \frac{1}{z}$, that the equation becomes

$$\tilde{z}^4 \frac{d^2w}{d\tilde{z}^2} + \tilde{z}^2(2\tilde{z} - p(\tilde{z}^{-1}))\frac{dw}{d\tilde{z}} + q(\tilde{z}^{-1})w = 0,$$

and deduce that a necessary condition for $z = \infty$ to be a regular singular point is that $zp(z)$ and $z^2 q(z)$ be analytic at infinity.

(c) Show that these conditions are also sufficient.

2. Show that the point $z = 0$ can be a singular point of the coefficients of a differential equation although the solutions are analytic there (such points are called accidental singular points) by carrying through the following argument. Let the equation be

$$w'' + \frac{p(z)}{z} w' + \frac{q(z)}{z^2} w = 0, \text{ where } w' = \frac{dw}{dz}, \text{ etc,}$$

and suppose $w_1(z)$ and $w_2(z)$ are independent solutions analytic near $z = 0$. Form

$$\Delta = \begin{vmatrix} w_1' & w_1 \\ w_2' & w_2 \end{vmatrix}, \quad \Delta_1 = \begin{vmatrix} w_1'' & w_1 \\ w_2'' & w_2 \end{vmatrix}, \quad \Delta_2 = \begin{vmatrix} w_1' & w_1'' \\ w_2' & w_2'' \end{vmatrix}.$$

(a) Show $z^{-1}p(z) = -\Delta_1/\Delta$ and $z^{-2}q(z) = -\Delta_2/\Delta$.

(b) Show $\Delta(0) = 0$.

(c) Show $\frac{d}{dz} \log \Delta = -z^{-1} p(z)$

$$= -z^{-1} p(0) + \frac{dG}{dz}$$

where G is analytic near z = 0, and deduce that

$$\Delta = c \, z^{-p(0)} \, e^{G(z)}$$

(d) Deduce that $p(0)$ must be a negative integer, and that the roots of the indicial equation must be negative integers.

(e) For the singularity to be accidental, no logarithmic terms may appear in the solution. Show that if they are excluded then the roots r_1 and r_2 of the indicial equation are distinct.

(f) Deduce from the fact that Δ has a pole of order $r_1 + r_2 - 1$ at 0 (why?), that z = 0 will be a regular if $r_1 = 1$ and $r_2 = 0$ and not otherwise. When z = 0 is an accidental singular point $p(0) = 1 - (r_1 + r_2) < 0$.

3. Show that the following equations all have accidental singular points at the origin:

(a) $z^2 \frac{dw}{dz} - (4z - z^2)\frac{dw}{dz} + (4 - z)w = 0.$

(b) $z^2 \frac{d^2 w}{dz^2} - 2(4 - \lambda z^2)\frac{dw}{dz} + (6 + \mu z^2)w = 0$

(for all values of the constants λ and μ.)

4. Show that the following third order equation, due to Hurwitz (see Chapter V , n.13), has an accidental singular point at z = 1:

$$z^2 (z - 1)^2 \frac{d^3 w}{dz^3} + (7z - 4)z(z - 1)\frac{d^2 w}{dz^2}$$

$$+ (\frac{72}{7} z(z - 1) - \frac{20}{9}(z - 1) + \frac{3z}{4})\frac{dw}{dz}$$

$$+ (\frac{72 \cdot 11}{73}(z - 1) + \frac{5}{8} + \frac{2}{23})w = 0.$$

5. Show that none of the following equations are of the Fuchsian type.

(a) $z^2 \dfrac{d^2 w}{dz^2} + z \dfrac{dw}{dz} + (z^2 - a^2)w = 0$ (Bessel's equation)

(b) $z \dfrac{d^2 w}{dz^2} + (z_0 + b_0 z)\dfrac{dw}{dz} + (a_1 + b_1 z)w = 0$ (Laplace's equation)

(c) $\dfrac{d^2 w}{dz^2} + (c - z^2)w = 0$ (the Hermite-Weber equation)

(d) $\dfrac{d^2 w}{dz^2} + (a + b \cos 2z)w = 0$ (Mathieu's equation)

Question 5 shows that several of the better known differential equations of mathematical physics are not of the Fuchsian type, and so the study of equations of the Fuchsian type could not be directly connected to physical problems. However, Klein and Bôcher in 1894 showed that it was possible to obtain physically significant equations from those of Fuchsian type by a process of amalgamating singularities in one particular equation. This process is well described in Whittaker and Watson, 203-204, as follows.

6. Show that the general equation of the Fuchsian type with 5 singular points can be written as

$$\frac{d^2 w}{dz^2} + \left(\sum_1^4 \frac{1 - \alpha_r - \beta_r}{z - a_r}\right)\frac{dw}{dz} +$$

$$\left(\sum_1^4 \frac{\alpha_r \beta_r}{(z - a_r)^2} + \frac{Az^2 + 2Bz + C}{(z - a_1) \cdots (z - a_4)}\right)w = 0$$

where the exponents at a_r are α_r and β_r, and at ∞ are μ_1 and μ_2, provided A is chosen so that μ_1 and μ_2 are the roots of

$\mu^2 + \mu\left(\sum_1^4 (\alpha_r + \beta_r) - 3\right) + \sum \alpha_r \beta_r + A = 0$, and B and C are constants.

What in particular does this equation look like when

$\beta_r - \alpha_r = \dfrac{1}{2}$, $\mu_2 - \mu_1 = \dfrac{1}{2}$? (The equation in this form is called the generalized Lamé equation.)

7. Let $a_1 \to a_2$ in the generalized Lamé equation, and show that the new exponents at a_2 are α and β, where

$$\alpha \cdot + \beta = 2(\alpha_1 + \beta_2)$$

$$\alpha\beta = \alpha_1(\alpha_1 + \tfrac{1}{2}) + \alpha_2(\alpha_2 + \tfrac{1}{2}) + \frac{Aa_1^2 + 2Ba_1 + C}{(a_1 - a_3)(a_1 - a_4)}$$

The new exponents depend on B and C, so they may be chosen arbitrarily by making a suitable choice for B and C. This process, called amalgamation, has produced an equation of the Fuchsian type in which three singular points have exponent difference $\frac{1}{2}$ and one (a_2) has arbitrary exponent difference.

8. Show that amalgamating a_3 with a_2 can produce an equation not of the Fuchsian type.

9. Show that equations can be produced in this way having $\underline{\ell}$ regular singular points with exponent difference $\frac{1}{2}$, $\underline{m}$ with arbitrary exponent difference, and $\underline{n}$ irregular singular points, where

$\underline{\ell} = 3,\ \underline{m} = 1,\ \underline{n} = 0,$ (Lamé)

$\underline{\ell} = 2,\ \underline{m} = 0,\ \underline{n} = 1,$ (Mathieu)

$\underline{\ell} = 1,\ \underline{m} = 2,\ \underline{n} = 0,$ (Legendre)

$\underline{\ell} = 0,\ \underline{m} = 1,\ \underline{n} = 1,$ (Bessel)

$\underline{\ell} = 1,\ \underline{m} = 0,\ \underline{n} = 1,$ (Hermite–Weber)

$\underline{\ell} = 0,\ \underline{m} = 0,\ \underline{n} = 1,$ (Stokes).

10. Show that $\frac{dy}{dx} = -y \log^2 y$ has the solution $y = \exp((x - c)^{-1})$ which has an essential singular point at $x = c$ which is independent of the equation.

11. Show that the substitution $y = ue^{-\frac{1}{2}\int p\,dx}$ reduces

$$\frac{d^2y}{dx^2} + p\frac{dy}{dx} + qy = 0 \text{ to}$$

$$\frac{d^2u}{dx^2} + Pu = 0$$

where $P = -\frac{1}{4}p^2 - \frac{1}{2}\frac{dp}{dx} + q$ (p and q being rational functions of x).

Show that if f and g are two linearly independent solutions of

$$\frac{d^2u}{dx^2} + Pu = 0 \text{ and } \eta = \frac{f}{g}, \text{ then } \frac{\eta'''}{\eta'} - \frac{3}{2}\left(\frac{\eta''}{\eta'}\right)^2 = 2P.$$

12. Show that general equation of the Fuchsian type and of the second order, with m finite singularities a_1, a_2, ..., a_m and a singular point at ∞, which has exponents α_1, β_1, α_2, β_2, ..., α_m, β_m, and α, β at those points (where $\alpha_1 + \beta_1 + ... + \alpha_m + \beta_m + \alpha + \beta = m - 1$) is of the form

$$y'' + \left(\sum_1^n \frac{A_i}{z - a_i}\right)y' + \sum\left(\frac{B_i}{(z - a_i)^2} + \frac{C_i}{z - a_i}\right) y = 0$$

where $\sum_1^m C_i = 0$.

Show moreover that $A_i = 1 - \alpha_i - \beta_i$, and $B_i = \alpha_i\beta_i$.

Deduce that the equation can be written as

$$y'' + \left(\sum_1^m \frac{1 - \alpha_i - \beta_i}{z - a_i}\right)y' + \left(F_{m-2}(z) + \sum_1^m \frac{\alpha_i\beta_i \, \psi'(a_i)}{z - a_i}\right)\frac{y}{\psi} = 0 \quad (*)$$

where F_{m-2} is a polynomial of degree $m - 2$ with leading coefficient $\alpha\beta$, and $\psi = (x - a_1) \ldots (x - a_m)$.

Show that the substitution $w = (z - a_1)^{\alpha_1}(z - a_2)^{\alpha_2} \ldots (z - a_m)^{\alpha_m} y$ produces an equation whose exponents are 0 and $\beta_i - \alpha_i$, and (at ∞) $\alpha + \Sigma\alpha_i$, $\beta + \Sigma\alpha_i$. Write down the form of this equation, substituting $\lambda_i = \beta_i - \alpha_i$, $\sigma = \alpha + \Sigma\alpha_i$, $\tau = \beta + \Sigma\alpha_i$.

Show that the substitution of Q.11 here takes the form

$$u = (z-a)^{\frac{1}{2}(1-\lambda_m)}, \text{ and converts (*) into an equation of the form}$$

$$u'' + \frac{u}{4\psi} \left[F_{m-2}(z) + \sum \frac{(1-\lambda_i)^2 \psi'(a_i)}{z-a_i} \right] = 0,$$

where $F_{m-2}(z)$ is a polynomial of degree $m-2$ with leading term $(1 - (\sigma - \tau)^2) z^{m-2}$. Show that the exponents are now $\frac{1}{2}(1-\lambda_i)$, $\frac{1}{2}(1+\lambda_i)$, and $\frac{1}{2}(-1+\sigma-\tau)$, $\frac{1}{2}(-1-\sigma+\tau)$.

13. Verify Fuchs's observation that when $m > 2$ the position and nature of the singular points does not determine the differential equation completely.

14. Show that the general equation of the Fuchsian class with three finite singular points (Heun's equation) is of the form

$$y'' + \left(\frac{\lambda_0}{z} + \frac{\lambda_1}{z-1} + \frac{\lambda_2}{z-a}\right)y' + \frac{\sigma z(z-q)}{z(z-1)(z-a)} y = 0$$

where q is an arbitrary (accessory) parameter.

CHAPTER III   ALGEBRAIC SOLUTIONS TO A DIFFERENTIAL EQUATION

This chapter considers how Fuchs's problem: when are all solutions to a linear ordinary differential equation algebraic? was approached, and solved, in the 1870's and 1880's. First, Schwarz solved the problem for the hypergeometric equation. Then Fuchs solved it for the general second-order equation by reducing it to a problem in invariant theory and solving that problem by ad hoc means. Gordan later solved the invariant theory problem directly. But Fuchs's solution was imperfect, and Klein simplified and corrected it by a mixture of geometric and group-theoretic techniques which established the central role played by the regular solids already highlighted by Schwarz. Simultaneously Jordan showed how the problem could be solved by purely group-theoretic means, by reducing it to a search for all finite monodromy groups of 2×2 matrices with complex entries and determinant 1. He was also able to solve it for 3rd and 4th order equations, thus providing the first successful treatment of the higher order cases, and to prove a general finiteness theorem for the nth order case (Jordan's finiteness theorem). Later Fuchs and Halphen were able to treat some of these cases invariant-theoretically.

The problem occupied the attention of many leading mathematicians in this period, and provided an interesting test of the relative powers of the older methods of invariant theory and the new group-theoretic ones, which favoured the new techniques. It also led to Schwarz's discovery of a new class of transcendental functions associated with the hypergeometric equation which, although not appreciated at the time, were to be of vital importance in the theory of automorphic functions (discussed in Chapter VI).

3.1 Schwarz

On 22nd August 1871, at a meeting of the mathematical section of the Swiss <u>Naturforschenden Gesellschaft</u>, H.A. Schwarz announced the solution to the problem: "when is the Gaussian hypergeometric series $F(\alpha, \beta, \gamma, x)$ an algebraic function of its fourth element?" His paper on this question appeared in the <u>Journal für Mathematik</u> for 1872 (Schwarz [1872]) and his arguments have been popular ever since.

Schwarz wrote the equation for complex x and y as

$$\frac{d^2 y}{dx^2} + \frac{\gamma - (\alpha + \beta + 1)x}{x(1 - x)} \frac{dy}{dx} - \frac{\alpha\beta}{x(1 - x)} y = 0 \qquad (3.1.1)$$

and considered two different cases: where the equation has only one algebraic solution, and where it has two linearly independent algebraic solutions. These cases must be treated separately, and Schwarz first considered the simpler case where one particular integral is algebraic; but either this algebraic function or its logarithmic derivative is a rational function, so the quotient of two branches of the function is constant, and the existence of a second algebraic solution cannot be inferred.

From the work of Fuchs, he said, it is clear that the general solution of (3.1.1) has one of the following forms as a convergent power series:

$$x^a (1 + c_1 x + c_2 x^2 + \dots),$$

$$(\tfrac{1}{x})^b (1 + \frac{c_1}{x} + \frac{c_2}{x^2} + \dots), \text{ or}$$

$$(1 - x)^c (1 + c_1(1 - x) + c_2(1 - x)^2 + \dots)$$

where

$a = 0$ or $1 - \gamma$

$b = \alpha$ or β

$c = 0$ or $\gamma - \alpha - \beta$.

If (3.1.1) has a particular integral, y_1, which is algebraic and whose logarithmic derivative is a rational function of x, then it must be of the form

$$y_1 = x^a (1 - x)^c g(x)$$

where a and c are rational numbers and g is a polynomial function of x of degree n, say. For simplicity one may assume $b = \alpha$, when there are four sub-cases to consider, according as $a = 0$ or $1 - \gamma$ and $c = 0$ or $\gamma - \alpha - \beta$. Schwarz showed that each of them is possible and that the corresponding $F(\alpha, \beta, \gamma, x)$ is algebraic.

The rest of the paper was devoted to the case of two linearly independent algebraic solutions. In this case every solution is algebraic, and the quotient of any two solutions is also algebraic. The equation

$$\frac{d^2 y}{dx^2} + p \frac{dy}{dx} + qy = 0 \qquad\qquad (3.1.2)$$

may be supposed to have two linearly independent solutions y_1 and y_2. They satisfy

$$y_2 \frac{dy_1}{dx} - y_1 \frac{dy_2}{dx} = Ce^{-\int p dx},$$

a result which Schwarz took from Abel [1827], a paper in which Abel derived differential equations, notably Legendre's, for functions defined by definite integrals. In this case

$$y_2 \frac{dy_1}{dx} - y_1 \frac{dy_2}{dx} = Cx^{-\gamma}(1 - x)^{\gamma - \alpha - \beta - 1}$$

so γ and $\alpha + \beta$ must be rational numbers. Indeed, said Schwarz, one sees by consulting entries 9 and 10 in Kummer's table of 24 solutions that α and β must themselves be rational, and for the rest of the paper be therefore assumed α, β, and γ were rational numbers.

Schwarz proposed to consider the quotient of y_1 and y_2, and the related quotients $s = \dfrac{C_1 y_1 + C_2 y_2}{C_3 y_1 + C_4 y_4}$ where $C_1, \ldots, C_4$ are constants. These quotients all satisfy a differential equation obtained by eliminating the three ratios $C_1 : C_2 : C_3 : C_3$ by successive differentiation:

$$\Psi(s,x) = \frac{2\,\dfrac{ds}{dx} \cdot \dfrac{d^3 s}{dx^3} - 3\left(\dfrac{d^2 s}{dx^2}\right)^2}{2\left(\dfrac{ds}{dx}\right)^2} = 2p - \frac{1}{2}p^2 - \frac{dp}{dx} = F(x). \qquad (3.1.3)$$

Following the usage later established by Cayley ([1883, 3]) $\Psi(s,x)$ will be called the Schwarzian of s with respect to x or, more briefly, the Schwarzian derivative[1]. Schwarz regarded the equation $\Psi(s,x) = F(x)$ as a special case of Kummer's equation in [1834] discussed above.

In the present case

$$\Psi(x) = 2p - \frac{1}{2}p^2 - \frac{dp}{dx}$$

$$= \frac{1-(1-\gamma)^2}{2x^2} + \frac{1-(\gamma-\alpha-\beta)^2}{2(1-x)^2} - \frac{(1-\gamma)^2 - (\alpha-\beta)^2 + (\gamma-\alpha-\beta)^2 - 1}{2x(1-x)}$$

or, on setting

$$(1 - \gamma)^2 = \lambda^2, \quad (\alpha - \beta)^2 = \mu^2, \quad (\gamma - \alpha - \beta) = \nu^2,$$

(3.1.3) becomes

$$\Psi(s,x) = \frac{1 - \lambda^2}{2x^2} + \frac{1 - v^2}{2(1-x)^2} - \frac{\lambda^2 - \mu^2 + v^2 - 1}{2x(1-x)} . \qquad (3.1.4)$$

λ, μ, v will be taken to be the positive roots of λ^2, μ^2, v^2 respectively. The advantage of (3.1.3) over (3.1.2) is that, as Heine pointed out to Schwarz, if y_1/y_2 and $e^{-\int pdx}$ are both algebraic then y_1 and y_2 are algebraic, for $y_2^2 \frac{d}{dx}(\frac{y_1}{y_2}) = Ce^{-\int pdx}$. So Schwarz needed only to consider the algebraic nature of one function, y_1/y_2, not two.

The power series solutions of (3.1.1) involve x, 1-x, or $\frac{1}{x}$, so the effect of replacing x by $z = \frac{c_1 x + c_2}{c_3 x + c_4}$ is therefore to be considered:

$$\Psi(s,x) = (\frac{dz}{dx})^2 \Psi(s,z).$$

The effect of replacing x by 1-x, or $\frac{1}{x}$ or compositions thereof on the solutions to (3.1.3) is then readily seen to be, if $s(\lambda, \mu, v, x)$ is one solution:

$$
\begin{aligned}
s(\lambda, \mu, v, z) &= s(\lambda, \mu, v, x) && \text{if } z = x, \\
&= s(v, \mu, \lambda, x) && \text{if } z = 1 - x, \\
&= s(\mu, \lambda, v, x) && \text{if } z = 1/x, \\
&= s(v, \lambda, \mu, x) && \text{if } z = 1/(1-x), \\
&= s(\lambda, v, \mu, x) && \text{if } z = x/1-x, \\
&= s(\mu, v, \lambda, x) && \text{if } z = \frac{x - 1}{x},
\end{aligned}
$$

which agrees with Riemann's theorem concerning his P-function, so

$s(\lambda, \mu, \nu, x)$ is a quotient of two linearly independent branches of the P-function $P(\lambda, \mu, \nu, x)$.

To solve (3.1.3), Schwarz first considered the solutions near a point $x_0 \neq 0$, 1, or ∞. Standard power-series methods, together with Kummer's solutions to (3.1.4) enabled him to establish this theorem:

<u>Theorem 3.1</u>   The map $s(\lambda, \mu, \nu, x)$ from the complex x-plane to the complex s-plane maps each simply-connected region X not containing 0,1 or ∞ onto a simply connected region S containing ∞ once or several times in its interior but having no branch point in its interior.

If in particular x_0 is real and neither 0, 1, nor ∞ then, since λ^2, μ^2, and ν^2 are real, s is real when x is and S is marked out by circular arcs.

Next, he considered the solutions to (3.1.4) in the neighbourhood of the singular points x = 0, 1, ∞. Now he could show

<u>Theorem 3.2</u>   The upper half x-plane E is mapped conformally by s, a particular integral of (3.1.3), onto a simply connected domain S, having no winding point in its interior, which is, in general, a circular arc triangle. The angles at the vertices corresponding to 0, 1, ∞ are $\lambda\pi$, $\mu\pi$ and $\nu\pi$ respectively.

How are these circular-arc triangles connected? To avoid
irksome special cases, Schwarz first assumed that none of the
λ, μ, ν are integers and, to avoid overlapping the triangles
unnecessarily, he reduced λ, μ and ν mod 2. It then turns out by
the reflection principle that each domain S is a circular arc
triangle for which

$$\lambda + \mu + \nu > 1 \qquad\qquad \text{and}$$

$$-\lambda + \mu + \nu < 1, \qquad\qquad \lambda - \mu + \nu < 1, \qquad \lambda + \mu - \nu < 1.$$

In precisely this case S can be taken to be bounded by great circles.
An adjoining triangle S_1 comes from a second copy of the half plane
E, say E_1, and corresponding points in E and E_1 are mapped onto
reciprocal points in S and S_1, that is, to points which are images
under the Möbius transformation of inversion in the common side.
For reasons of symmetry each triangular region S is the image of its
neighbour, and the question reduces to finding all circular-arc
triangles which, when so transformed, give only a finite covering
of the sphere. This is equivalent to finding such triangles as only
occupy a finite number of positions upon successive reflections in
their sides, and in this case s is an algebraic function of x $\in$ E,
for, said Schwarz, the Riemann surfaces of x and s are then closed
surfaces with finitely many leaves. Schwarz observed that this
problem had already been discussed to some extent by Riemann himself,
in a paper published just after his death ([Riemann 1867]) where,
in §12, Riemann considered the case when $\frac{du}{d \log \eta}$ is an algebraic
function of η, and in §18 alluded to the conformal representations
of regular solids onto the sphere.

Indeed, Schwarz noted, the solution of his (Schwarz's) problem is precisely that the triangles must either fit together to form a regular double pyramid (angles $\frac{\pi}{2}$, $\frac{\pi}{2}$, $\nu\pi$, $\nu = \frac{1}{n}$) or a regular solid. This gave him a table of 15 cases (up to an ordering of λ, μ, ν) in s was algebraic, given in Table 3.1, and in all other cases (when $\lambda + \mu + \nu \leq 1$ or $\lambda + \mu + \nu > 1$ but λ, μ, or ν not as tabled) s was transcendental. In each case it is possible to write down the associated Gaussian hypergeometric series and exhibit it directly as an algebraic function. Schwarz gave as an explicit example the case $\lambda = \frac{1}{3} = \mu$, $\nu = \frac{1}{2}$, for which the regular solid is a tetrahedron, divided by its symmetry planes into 24 triangles with angles $\frac{\pi}{3}$, $\frac{\pi}{3}$, $\frac{\pi}{2}$ and, in slightly less detail, the cases of the octahedron, icosahedron, and dodecahedron. His analysis was extended by Brioschi [1877a, b] and completed by Klein [1877], see Section 3.3.

It remained for Schwarz to consider the special cases where some of λ, μ, ν are integers. If $\lambda = 0$ the function s is necessarily transcendental. When $\lambda = m \neq 0$ he showed s can only be algebraic if $B_m = 0$ and $x = 0$ is an accidental singularity. This, it turned out, would happen if and only if one of $|\lambda| - |\mu + \nu|$ or $|\lambda| - |\mu - \nu|$ was an odd positive integer. Further conditions are necessary for s to be algebraic, namely, that all of λ, μ, ν are non-zero integers, their sum is odd, and the sum of the absolute value of any two exceeds the absolute value of the third.

Schwarz did more than solve the problem he set himself of finding algebraic solutions to the hypergeometric equation. His thorough treatment of the tetrahedral case revealed elegant connections with elliptic function theory, as one might expect, but even more important for the direction of future work was his

investigation of the simplest transcendental cases when

$\lambda + \mu + \nu < 1$. This occupied §5 of his paper.

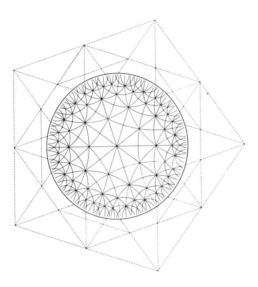

Figure 3.1

No.	λ''	μ''	ν''	area π	
I.	$\frac{1}{2}$	$\frac{1}{2}$	ν	ν	dihedron
II.	$\frac{1}{2}$	$\frac{1}{3}$	$\frac{1}{3}$	$\frac{1}{6} = A$	tetrahedron
III.	$\frac{2}{3}$	$\frac{1}{3}$	$\frac{1}{3}$	$\frac{1}{3} = 2A$	
IV.	$\frac{1}{2}$	$\frac{1}{3}$	$\frac{1}{4}$	$\frac{1}{12} = B$	cube or
V.	$\frac{2}{3}$	$\frac{1}{4}$	$\frac{1}{4}$	$\frac{1}{6} = 2B$	octahedron
VI.	$\frac{1}{2}$	$\frac{1}{3}$	$\frac{1}{5}$	$\frac{1}{30} = C$	dodecahedron
VII.	$\frac{2}{5}$	$\frac{1}{3}$	$\frac{1}{3}$	$\frac{1}{15} = 2C$	or
VIII.	$\frac{2}{3}$	$\frac{1}{5}$	$\frac{1}{5}$	$\frac{1}{15} = 2C$	icosahedron
IX.	$\frac{1}{2}$	$\frac{2}{5}$	$\frac{1}{5}$	$\frac{1}{10} = 3C$	
X.	$\frac{3}{5}$	$\frac{1}{3}$	$\frac{1}{5}$	$\frac{2}{15} = 4C$	
XI.	$\frac{2}{5}$	$\frac{2}{5}$	$\frac{2}{5}$	$\frac{1}{5} = 6C$	
XII.	$\frac{2}{3}$	$\frac{1}{3}$	$\frac{1}{5}$	$\frac{1}{5} = 6C$	
XIII.	$\frac{4}{5}$	$\frac{1}{5}$	$\frac{1}{5}$	$\frac{1}{5} = 6C$	
XIV.	$\frac{1}{2}$	$\frac{2}{5}$	$\frac{1}{3}$	$\frac{7}{30} = 7C$	
XV.	$\frac{3}{5}$	$\frac{2}{5}$	$\frac{1}{3}$	$\frac{1}{3} = 10C$	

Table 3.1

Here he showed that a circular-arc triangle with angles $\lambda\pi$, $\mu\pi$, $\nu\pi$ can be formed with sides perpendicular to a fixed boundary circle, and that successive reciprocation can then fill out the interior of this fixed circle with copies of the original triangle. A picture of the case $\lambda = \frac{1}{5}$, $\mu = \frac{1}{4}$, $\nu = \frac{1}{2}$ was supplied. The function s is necessarily transcendental, but it is single valued whenever $\frac{1}{\lambda}$, $\frac{1}{\mu}$, $\frac{1}{\nu}$ are integers. In such a case the fixed circle is a natural boundary of s, and s cannot be analytically continued onto the boundary. This phenomenon had been noticed earlier by Weierstrass in 1863, indeed, Schwarz went on, Kronecker had pointed out that the θ-series for $\sqrt{\frac{2K}{\pi}}$

$$\sqrt{\frac{2K}{\pi}} = 1 + 2q + 2q^4 + 2q^9 + \cdots$$

gives an example where q cannot be taken on or outside the unit circle, nor can the function be analytically continued past $|q| = 1$ in any other way[3]. Other examples pertained to the case $\lambda + \mu + \nu = 0$, and had been discussed by Weierstrass in 1866. Thus, if in the usual notation:

$$K = \int_0^1 \frac{dx}{\sqrt{[(1-x^2)(1-k^2x^2)]}} \ , \quad K' = \int_0^1 \frac{dx}{\sqrt{[(1-x^2)(1-k'^2x^2)]}}$$

where $k^2 + k'^2 = 1$ then

$$s = \frac{a'K + b'K'}{aK + bK}$$

is a function of k^2 when a, b, a', b' are given, and s is bounded but k^2 unbounded.

Yet another Weierstrassian example given by Schwarz pertained to the case $\lambda + \mu + \nu = 1$, when the triangles tessellate the plane and s is a single valued function provided, up to order, either $\lambda = \frac{1}{3} = \mu = \nu$, or $\lambda = \frac{1}{2}$, $\mu = \nu = \frac{1}{4}$, or $\lambda = \frac{1}{2}$, $\mu = \frac{1}{3}$, $\nu = \frac{1}{6}$. In these cases x is a single-valued function of s as s runs over the plane.

Schwarz's elegant solution of the all-solutions algebraic problem for the case of the hypergeometric equation indicated the important role to be played by geometrical reasoning in analysing such problems. But it was an open question for his contemporaries as to how to proceed with the attractive problem of dealing either with the second order equation when more than three singular points are present, or with equations of higher order. The challenges were to be met both with the then-traditional methods of invariant-theory, and, more directly, with the then-novel methods of group theory. I shall suggest that the relative success of the new group-theoretic methods, coupled with their triumphant extension into Schwarz's transcendental case $(\lambda + \mu + \nu < 1)$, did much to advance group theory at the expense of invariant theory.

3.2 Generalizations

Schwarz's solution to the algebraic-solution problem for the hypergeometric equation had been simple, because the hypergeometric equation is unique amongst second order equations of the Fuchsian class in that the exponents at the singular points uniquely determine the nature of the solutions, as Fuchs had shown. So the question extending his results was a difficult one. It was to prove much easier to treat the second order case with n singular points and several attempts were made upon this problem. These may be divided into two kinds, and briefly characterized as invariant-theoretic and group-theoretic.

In the early 1870's invariant theory was a central domain of mathematics. It is the study of forms (homogeneous polynomials of some degree in various indeterminates) enjoying special properties under linear changes of the indeterminates. Expressions, in the coefficients of a form, which were unchanged by all such transformations were called absolute invariants, or else, if they altered by a constant which depended on the coefficients of the form but was independent of the linear transformation, they were called relative invariants. Analogous expressions involving the indeterminates were called covariants. A computational procedure was developed for producing new covariants from old, and the central question in any given problem in invariant theory was always to find a complete basis of covariants in terms of which all other covariants could be expressed as sums and products. Gordan, the leading invariant theorist of his day, had established the existence of a finite basis for binary forms in his important [1868].
The well-developed theory of binary forms, however, stood in

marked contrast to the difficulties encountered in the extending of
the study of invariants to forms in three or more variables, and so
it is not surprising to find that Fuchs and Gordan, who sought to use
invariant theory to solve the algebraic-solution problem, first
considered the second-order differential equation.

Group theory on the other hand was a new subject in the 1870's
although implicitly group-theoretic ideas had been around for some
time (see Wussing [1969]). Crucial papers stressing the importance
of the group idea were Jordan: Traité des substitutions et des
équations algébriques (1870), Sylow [1872], and, if only for what it
tells us of Klein himself, Klein's Erlanger Programm [1872]. Klein
drew his inspiration from Jordan, whose Traité he was later to
describe as a book with seven seals (Klein [1922]) and he was at pains
in the Erlanger Programm to stress the value of relating group theory
to the invariants as a classificatory principle. (He had gone to
Paris in 1870 with his friend Sophus Lie to learn the new ideas,
although his studies were interrupted by the Franco-Prussian war.)
Jordan's Traité contains a range of applications of group theory to
mathematics, and so it was natural for him and Klein to take up the
algebraic-solutions problem in differential equations from that point
of view, but the Traité is even more remarkable for its
reformulation of the ideas of Galois theory[4]. Lagrange's work on the
theory of equations emphasized polynomials called resolvents with
certain invariance properties under permutations of the roots.
Replacing the indeterminate, say x, in these polynomials by a
homogenous coordinate $x_1 : x_2$ produces a binary form, and indeed
invariant theory derives in part from this kind of study. Jordan's
Traité is the first book to place the underlying groups of

permutations in the foreground and to diminish the importance of
specific polynomials. Jordan took the burden of Galois' ideas to lie
in a permutation-theoretic form of the theory of groups and their
normal subgroups which Galois had begun, and he developed this theory
of such groups without regard for the hierarchies of invariant forms
which correspond to the subgroups of a given Galois group. German
mathematicians on the other hand, notably Kronecker and Dedekind,
sought to explore this correspondence between groups and families of
invariant functions, as is described in Chapter IV. This is also the
approach Klein envisaged for the algebraic-solutions problems, as is
particularly clear in his treatment of the transcendental functions
discovered by Schwarz. Klein was also extremely interested in the
resolvents of equations and their geometric representation. This
particular emphasis, present in his work and Gordan's, but absent from
Jordan's, led Klein and Gordan in conversation to call their study
'hyper-Galois' theory, Klein [1922, 261; 1967, 90]. So one may say
the schools of thought brought to bear on the algebraic-solutions
problem derived from the traditional approach to theory of
polynomials, as it had developed into invariant-theory; the modern
group theoretical approach, pioneered by Jordan, of Galois theory;
and a geometric blending of the two, preferred by Klein.

The first person to follow up Schwarz's paper was Lazarus Fuchs,
returning to the problem he had first stated in 1865. His methods were
traditional, and, as will be seen, by no means completely successful.
Later work by Gordan resolved the matter fully in invariant-theoretic
terms, but by then Klein and Jordan had also solved the problem group
theoretically, and Jordan had gone beyond it to a study of the third
order case. It seems that on this question the new methods surpassed
the old.

Fuchs' solutions

Fuchs published four accounts of his work: a summary [1875], a complete account [1876a], a short note for French readers in the form of a letter to Hermite [1876b], and more definitively, [1878]. I shall proceed to give a summary of the main paper, [1876a].

If a second order differential equation with rational coefficients

$$\frac{d^2u}{dz^2} + p\,\frac{du}{dz} + qu = 0 \tag{3.2.1}$$

has an algebraic solution u_1, then u_1 is a root of some irreducible polynomial equation

$$A_m u^m + A_{m-1} u^{m-1} + \ldots + A_0 = 0 \tag{3.2.2}$$

where the A_i, $i = 0, 1, \ldots, m-1$, are rational functions of z. Any other root of (3.2.2) is also a solution to (3.2.1), because (3.2.2) is irreducible, so the question arises: how are the roots of (3.2.2) related? Essentially two cases can arise, which are the cases considered by Schwarz. Either there are two roots, say u_1 and u_2, which do not have a quotient which is a rational function of z, in which case u_1 and u_2 can be taken as a basis of solutions to (3.2.1), or there are not. This latter case is simpler and will be dispensed with at once. If u_k is another root of (3.2.2) and $u_k = ju_1$, say, then one sees at once on substituting ju_1 into (3.2.2), that j must be a constant and indeed a primitive root of unity[5]. Now this is true for every root of (3.2.2) so it reduces to

$$A_m u^m + A_0 = 0$$

and u_1 is a root of a rational function. Accordingly, in this case every solution to (3.2.1) is a root of a rational function. But, said Fuchs, this case can be detected in advance by purely algebraic means which are adequate in fact to deal with the nth order differential equation, and need be discussed no further[6]. There remains the more interesting case when (3.2.1) has two independent algebraic solutions, and hence all its solutions are algebraic.

Fuchs reduced (3.2.1) in this case to

$$\frac{d^2y}{dz^2} + Py = 0 \tag{3.2.3}$$

by means of the substitution $u = \mu y$, $\mu = e^{-\frac{1}{2}\int p\,dz}$, where

$$P = q - \frac{1}{4} p^2 - \frac{1}{2} \frac{dp}{dz} .$$

Since the general theory of equations of the Fuchsian class implies that

$$p(z) = \sum_{i=1}^{\rho} \frac{\alpha_i}{z-a_i} , \qquad q(z) = \sum_{i=1}^{\rho} \frac{\beta_i + (z-a_i)\gamma_i}{(z-a_i)^2} , \tag{3.2.4}$$

where $a_1,\ldots,a_\rho$ are the singular points of (3.2.1) and the α_i, β_i, and γ_i, $i = 1,\ldots,\rho$ are constants such that $\Sigma\gamma_i = 0$,

$$\mu = (z-a)^{-\frac{1}{2}\alpha} \ldots (z-a_\rho)^{-\frac{1}{2}\alpha}.$$

So all the solutions to (3.2.3) are algebraic if all the solutions to (3.2.1) are, and conversely all the solutions to (3.2.1) are algebraic if all solutions to (3.2.3) are and, in addition, $\alpha_1, \ldots, \alpha_\rho$ are rational (i.e. ρ is the logarithm of a rational function). Henceforth Fuchs worked with (3.2.3).

The general solution to (3.2.3) has the form

$$\alpha y + \beta \phi (y) \qquad\qquad (3.2.5)$$

where α, β are constants, y is an algebraic solution to (3.2.3), and $\phi(y)$ a rational function of y and z, say

$$\phi(y) = c_0 + c_1 y + \ldots + c_m y^m, \qquad\qquad (3.2.6)$$

where $c_0, \ldots, c_m$ are rational functions of z. Taking y and $\phi(y)$ as a basis of solutions to (3.2.3), Fuchs next considered the monodromy of the differential equation, and obtained various constraints upon the nature of the solutions that can arise which will be discussed below.

If y is any algebraic solution to (3.2.3) then it satisfies an irreducible equation of degree m in y, and Fuchs showed that the number m does not depend on the choice of the solution of the differential equation [1876a, §8]. Suppose the equation for y is

$$A_m y^m + A_{m-1} y^{m-1} + \ldots + A_0 = 0, \qquad\qquad (3.2.7)$$

and y_1, ..., y_{m-1} are its other roots. These roots may be divided up into equivalence classes, where one is equivalent to another if and only if their quotient is constant. A family of pair-wise inequivalent roots was said by Fuchs to form a <u>reduced root system</u> y, y_1, ..., y_{n-1} say, and y_i, y_{ij}, ..., $y_i j^{\ell_i - 1}$ to be the roots corresponding to y_i, j being a primitive ℓ_i^{th} root of unity. Fuchs called the least common multiple of the ℓ's as y_i runs through a reduced root system the <u>index</u> of the equation, and he showed [§9 Theorem 2]:

$$\ell n = m, \tag{3.2.8}$$

since (3.2.7) is irreducible. For that reason, too, the analytic continuation of any two distinct roots of (3.2.7) along a closed path produces distinct roots [§9 Theorem 3].

Fuchs was now ready to translate his problem into invariant-theoretic terms. He let η, η_1, ..., η_{n-1} be a reduced root system of index ℓ, and y_1 and y_2 any basis of solutions of the differential equation (3.2.4). Then $\eta_i = A_{i1}y_1 + A_{i2}y_2$, and a form of the nth degree is constructed:

$$f(y_1, y_2) = (A_{01}y_1 + A_{02}y_2) \cdots (A_{n-1,1}\, y_1 + A_{n-1,2}y_2) \tag{3.2.9}$$

Any circuit in z changes f by at most a multiple of a primitive ℓ^{th} root of unity, so f is a root of a rational function of z, with highest exponent ℓ. Conversely, any form which is a root of a rational function of z and has $\eta = A_{01}y_1 + A_{02}y_2$ as a factor also has factors $\eta_1 = A_{i1}y_1 + A_{i2}y_2$, where η, η_1, ..., η_{n-1} are a reduced root system for the irreducible equation which η satisfies. Fuchs [§10] called any form which only has as factors the members of a reduced root system and in which every factor has degree one a ground form (Primform). Accordingly, every ground form is the root of a rational function, and if two ground forms have a common factor they are identical up to a common factor. Conversely, any form which is the root of a rational function can be written as a product of ground forms and contains each member of a reduced root system equally often. Fuchs's purpose in introducing these forms was to replace the solutions to the original differential equation, which are algebraic functions, with related quantities which are simpler, being roots of rational functions.

Although m depends only on the differential equation, the index ℓ, and the order of the reduced root system, n, may depend upon the choice of solution y_1. Fuchs therefore chose y_1 so that n was as small as possible, say N, and let the corresponding index be L, when m = NL. N is then the least degree of a form which is a root of a rational function, and any such form is a ground form. Fuchs's approach was to calculate the possible values of N, using the machinery of invariant theory whenever possible. As it stands the problem would be solved if one could characterize these binary forms of degree N all of whose covariants of lower degree vanish identically, but this Fuchs confessed himself unable to do, in letter to Hermite, January 1876 (= Fuchs [1876b]). He therefore resorted to less direct methods.

The Hessian covariant of a form $\Phi(y_1, y_2)$ is the determinant

$$\begin{vmatrix} \dfrac{\partial^2 \Phi}{\partial y_1^{\,2}} & \dfrac{\partial^2 \Phi}{\partial y_1 \, \partial y_2} \\[4mm] \dfrac{\partial^2 \Phi}{\partial y_2 \, \partial y_1} & \dfrac{\partial^2 \Phi}{\partial y_2^{\,2}} \end{vmatrix}$$

. If Φ is a ground form of degree N its

Hessian, which Fuchs denoted Ψ, is again a root of a rational

function, and is of degree 2N-4. It is therefore a ground form, since

it cannot be a product of ground forms. It cannot vanish identically

unless it is a power of linear function, in which case N = 1, so

Fuchs henceforth assumed N > 1.

The case N = 2 was disposed of simply enough [§13]. Fuchs was

able to show that in this case the solutions can be written in the

form

$$y_1 = \psi(z)^{-\frac{1}{4}} [p + q \ (z)^{\frac{1}{2}}]^{\alpha}$$

$$y_2 = \psi(z)^{-\frac{1}{4}} [p + q \ (z)^{\frac{1}{2}}]^{-\alpha} \qquad (3.2.10)$$

for p, q rational functions of z and α a rational number. They are

algebraic if and only if $\int \psi(z)^{\frac{1}{2}} dz$ is the logarithm of an algebraic

function.

The remaining possible values of N are associated with the more complicated forms of the monodromy relations, and required more ad hoc methods. After four paragraphs and seven pages Fuchs finally produced the following table of possibilities:

N	L	$\Phi(y_1, y_2)$
N>2,	L>2	
4	3 or 6	$a_0 y_1^4 + a_3 y_1 y_2^3$
6	4	$a_1 y_1^5 y_2 + a_3 y_1^3 y_2^3 + a_5 y_1 y_2^5$
6	8	$a_1 y_1^5 y_2 + a_5 y_1 y_2^5$
8	3 or 6	$a_1 y_1^7 y_2 + a_4 y_1^4 y_2^4 + a_7 y_1 y_2^7$
10	8	$a_1 y_1^9 y_2 + a_5 y_1^5 y_2^5 + a_9 y_1 y_2^9$
12	5 or 10	$a_1 y_1^{11} y_2 + a_6 y_1^6 y_2^6 + a_{11} y_1 y_2^{11}$

(3.2.11)

together with the cases already discussed: N=4, L=2, N=4, L=1; N=2; and N=1 (§19).

The most interesting fact in the table is the upper bound on $N \leq 12$. It is testimony to the intricacies of Fuchs's method that it contains incorrect cases (N=6 and L=4; N=8, L=3 or 6; and N=10) which are not possible values, as Klein was quick to notice (Klein [1876, = 1922, 305]). Klein also commented on these cases in Fortschritte, VII, 1877, 172-3.

Fuchs summarized his finding in two theorems [§20 Theorems 1 and 2]: if a differential equation has algebraic solutions then either the general solution or one of the tabulated forms in y_1 and y_2 (a basis of arbitrary solutions) is a root of a rational function. Conversely, if a fundamental system of solutions can be made to yield a form (of degree >2 and not a power of a form of degree 2) which is a root of rational function then the original differential equation has algebraic solutions.

It can hardly be said that this conclusion, although very interesting, amounts to a characterization of the differential equations having algebraic solutions. It does not yield for instance, a simple test which can be applied to a given equation. Fuchs spent the last fifteen pages trying to improve matters, as follows.

Setting $v = y^\mu$ turns the original equation (3.2.3) into

$$\frac{d^{\mu+1}v}{dz^{\mu+1}} + P_{\mu-1}\frac{d^{\mu-1}v}{dz^{\mu-1}} + \ldots + P_0 v = 0 \qquad (3.2.12)$$

with the same singular points as the original equation, and Fuchs showed the necessary condition for (3.2.3) to have algebraic solutions is that for some value of $\mu = 1, 2, 4, 6, 8, 10, 12$ (3.2.12) is satisfied by the root of a rational function. If this is the case for $\mu \neq 2$ then conversely, if (3.2.12) has algebraic solutions so does (3.2.3). The case $\mu = 2$ reduces to the case $N = 2$.

He made various attempts to avoid the passage to (3.2.12) but without much success. Two theorems appear that are worthy of notice (§24). They concern the solutions with exponent r = p/q in a neighbourhood of a singular point z = a. A circuit of that singular point sends y to yj^p, j a primitive q^{th} root of 1, and q|m, the degree of the irreducible equation satisfied by y. So, if ν is the number of members of the reduced root system containing y, $\nu \geq N$, so $\nu \leq m/q$, i.e $\nu \leq NL/q$ from which it follows that $q \leq L$. If N > 2 then from the table above $q \leq 10$. Fuchs stated the theorem [§24]: if any denominator, in lowest terms, of the exponents is greater than 10 then (3.2.3) has no algebraic solution unless it, or (3.2.10) with $\mu = 2$ is satisfied by a root of a rational function. He went on to prove that if (3.2.12) is satisfied by a root of a rational function and the denominators of the exponents are not 1, 2 or 3 then (3.2.3) is satisfied either by no algebraic functions or by roots of rational functions.

He compared this result with the result of Schwarz on the hypergeometric equation where unless two of λ, μ and ν equal 2, no value of them in Schwarz's table exceeds 5. Fuchs noted in the case (2, 2, n) that if all the denominators of the exponents, q_i say, are equal to 2, and the necessary subsidiary condition for the avoidance of logarithmic terms is satisfied, then (3.2.3) is satisfied by a square root of a rational function.

3.3 Klein and Gordan

Klein first heard of Schwarz's work in the late Autumn of 1874 or
Spring of 1875, after Gordan came to join him at Erlangen. Klein
wrote, in the introduction to the Vorlesungen über das Ikosaeder
[1884] "I had at that time already commenced the study of the
Icosahedron for myself (without then knowing of Professor Schwarz's
earlier works, ...) but I considered my whole manner of attacking
the question rather in the light of preliminary training". He always
spoke warmly of the semester he spent with Gordan, his last semester
at Erlangen before going to Munich, and it seems to have been a
rewarding clash of styles and approaches to mathematics for both men.
Klein, only 25 in 1874, had already published over thirty papers on
line geometry, on surfaces of the third and fourth degrees, on
non-Euclidean geometry, and on the connection between Riemann surfaces
and algebraic curves. He was an enthusiast for anschauliche geometry,
stressing the importance of a visual and tangible presentation of
mathematical ideas, and preferring the conceptual framework of group
theory to the more algebraic and computational study of explicit
invariants. Gordan, then 37, had spent a year in Göttingen with
Riemann, who, however, had been very ill. Together with Clebsch,
Gordan had attempted to put Riemann's function-theoretic ideas on a
sound algebraic footing in their Theorie der Abelschen Funktionen
[1866], but Gordan soon turned to what was to be his great love:
the formal side of the theory of invariants. In his [1868] he proved
the important theorem that for any binary form there is
always a finite basis for its rational invariants and covariants.
He became the acknowledged master of the theory of invariants, greatly
preferring the mechanisms of the algebra to the more suggestive domain
of geometry, quarrying deeply where Klein chose to soar aloft, seeking

detailed and difficult results where Klein sought to unify the
disparate parts of mathematics.

Their work from 1875 to 1877 reveals many reciprocal influences
and contrasts. Klein wrote several papers on invariants, chiefly on
the icosahedral invariants which he was beginning to connect with the
unsolvability of the quintic equation. In 1875 his work received a
special impulse, as he later said (Klein [1922, 257]) from Fuchs's
work on the algebraic solutions problem. It seemed to Klein that, with
his understanding of the role of the regular solids, he could complete
Fuchs's treatment and, indeed, simplify it. His approach, as he was
later to stress to Poincaré, was much closer to that of Schwarz than
that of Fuchs, and he seems to have had little respect for Fuchs's
achievements, which he found ungeometric (a letter from Klein to
Poincaré, 19 June 1881 = Klein [1923, 592]). Klein placed the
regular solids and their groups at the centre of his study of the
algebraic-solution problem, and when he produced the appropriate forms
it was done *"without any complicated calculation only with the ideas
of invariant theory"*. Klein [1875, = 1922, 276] (italics in original).
Gordan for his part found Klein's geometric considerations "very
abstract and not bound up of necessity with the question in any way"
and proposed to show how the finite groups of linear transformations
in one variable could be found algebraically (Gordan [1877a, 23]).
Klein replaced Fuchs's indirect resolution of his form problem with an
ingenious reduction of the whole question to Schwarz's five cases.
Gordan accepted Fuchs's terms and solved the form-problem directly.
Klein's method will now be discussed.

Klein's solution

Klein began his [1875/76] with the problem of finding all finite groups of motions of the sphere or, equivalently, of finding all finite groups of linear transformations of x + iy. This connection between metric geometry and invariant theory was one he had been pleased to make in the Erlanger Programm (see particularly §6). The root of the connection is that, in the case of the icosahedron, say, there is a group of order 60, the symmetry group of proper motions of the solid. A typical point of the sphere can be moved to a total of 60 different points under the action of this group, but certain points have smaller orbits. There is an orbit of order 12 corresponding to the vertices of the icosahedron, an orbit of order 20 corresponding to the midpoints of the faces, and an orbit of order 30 corresponding to the midpoints of the edges. An orbit of each of these four kinds can be specified as the zeros of a form of the appropriate degree. Klein denoted the forms of order 12, 20 and 30 by f, H and T respectively, and calculated each explicitly. But between three forms of order 60 there must be a linear relationship (since a form is known up to multiplication by a constant once its zeros are given), and in this case the relationship is $T^2 + H^3 - 1728f^5 = 0$. Furthermore, as one would expect, H and T are known when f is known (the vertices specify the icosahedron completely) so it is enough to determine f and to explain its relationship to H and T. Accordingly the invariant theory can be easily developed, once it has been established what the possible finite groups are, and Klein accomplished this classical task anew in §2 of the paper.

Klein regarded the general motion of the sphere as screw-like or loxodromic about an axis. If a finite group is to be constructed each element in it must have finite order, and so be of the form $z' = \varepsilon z$, ε a rational root of unity. Furthermore, the axes of any two motions must meet inside the sphere, and indeed all the axes must meet in the same point. But this reduces the finite group to one of the five classical examples: the cyclic, dihedral, tetrahedral, octahedral, and icosahedral groups. As Klein remarked in §3, this argument simultaneously determines all the finite groups of non-Euclidean motions.

The groups having been found, the forms in each case can be written down and related to one another by what Klein called a general principle (§5). This asserted that, if $\Pi = 0$ and $\Pi' = 0$ are the equations of two sets of points which arise from two given points by the action of a finite group and they are of the proper degree (say, the order of the group) then $\kappa\Pi + \kappa'\Pi' = 0$ represents any other such system of points, for some choice of parameter κ/κ'. Otherwise put, if Π and Π' are two G-invariant polynomials of the same degree any third G-invariant polynomial is linearly dependent on them. This is true, Klein observed, because there is only a single infinity of orbits of the same degree, parameterised, one might add, by the sphere.[7]

To give the forms explicitly in the case of the octahedron Klein took the canonical form obtained by Schwarz and wrote it in homogeneous co-ordinates as

$$f = x_1 x_2 (x_1^4 - x_2^4) \quad (§6).$$

The general procedures of invariant theory then suggested the following manoeuvres: $f = a_x^6$, which is the symbolic notation for f as a binary form of degree 6; the 6th, 4th, and 2nd transvectants[8] of f with itself are: $(ab)^6 = A$, a constant;

$$(ab)^4 a_x^2 b_x^2, \text{ of degree 4; and}$$

$$(ab)^2 a_x^4 b_x^4 = H, \text{ of degree 8, the Hessian of } f.$$

Any form of the sixth degree with isolated roots can be reduced to the form $x_1 x_2 \phi_4$, where ϕ_4 is of degree four and the coefficients of x_1^4 and x_2^4 are, respectively, 1 and −1. Accordingly if its fourth transvectant (Überschiebung) $(ab)^4 a_x^2 b_x^2$ is made to vanish identically it corresponds to the canonical form of the octahedron, and the converse is also true. H itself cannot vanish identically (else all the roots of f would coincide).

The eight roots of $H = 0$ give the centres of the eight faces of the octahedron. For the same reason the functional determinant of f and H, T, is of the twelfth degree and equated to zero locates the twelve mid-edge points of the octahedron. Together with the invariant A these forms satisfy

$$\frac{Af^4}{36} + \tfrac{1}{2}H^3 + T^2 = 0$$

as is easily seen in the canonical case.

Analogous reasoning dispatched the icosahedral equation, and enabled Klein to discuss the irrational covariants in each case. Klein concluded the paper with a brief reference to the Galois group of the icosahedral equation. There are, he said, 60 proper motions of the icosahedron itself, and for each motion four elements of the Galois group corresponding to the map $\varepsilon \rightarrow \varepsilon^{\nu}$, $\varepsilon = e^{2\pi i/5}$, $\nu = 1, 2, 3, 4$. The Galois group, then, has 240 elements, but adjoining ε reduces it to a group of 60 elements. These sixty motions permute the five octahedrons inscribed in the icosahedron, which establishes an isomorphism between the Galois group and the even permutations on five elements.

Klein's next paper [1876] described the implications of this work for the algebraic-solutions problem. Suppose the equation is $y'' + py' + qy = 0$. It is algebraically integrable if and only if p is the logarithm of an algebraic (indeed, rational) function, so Klein has assumed this was the case. Then y, a quotient of linearly independent solutions to the second order equation, is a solution of the Schwarzian equation

$$[n]_z = f(z) = 2q - \frac{1}{2} p^2 - p', \tag{3.3.1}$$

where $[n]_z$ is Klein's notation for the Schwarzian derivative, nowadays written $\{n, z\}$, and Klein sought to characterize the rational function $f(z)$.

The monodromy group of the original equation is known because it is finite and so a specific one of the known possibilities. To that group G, say, is associated a canonical G-invariant rational function $Z = Z(Y)$ which is such that the inverse function $Y = Y(Z)$ satisfies a canonical Schwarzian equation

$$[n]_Z = R(Z). \tag{3.3.2}$$

But y is also such that its inverse function $z = \zeta(y)$ is G-invariant, and rational, and so ζ and Z are rational functions of each other. So, finally, the substitution $Z = \phi(z)$ where ϕ is a rational function, converts $Y = Y(Z)$ into $Y(\phi(z)) = y(z)$ and the canonical equation (3.3.2) into the given (3.3.1), and conversely the inverse rational substitution $\zeta = \psi(Z)$ converts (3.3.1) into (3.3.2). The functions ϕ, ψ are subject to no other constraint, since any rational function of a G-invariant rational function is G-invariant, and so Klein deduced the elegant result that those second order linear homogeneous differential equations which are algebraically integrable are precisely those which can be obtained from the canonical ones by an arbitrary rational change of variable, so the problem is solved in principle.

Furthermore, their form is known, for if $Z = Z(Z_1)$ is any function

$$\{\eta, Z_1\} = \left(\frac{dZ}{dZ_1}\right)^2 \{\eta, Z\} + \{Z, Z_1\},$$

and so the conversion of $\{\eta, Z\} = R(Z)$ by $Z = \phi(z)$ into $\{\eta, z\} = f(z)$ converts the canonical equation into $\{\eta, z\} = \left(\frac{dZ}{dz}\right)^2 R(Z) + \{Z, z\} = f(z)$. However, it must be said that the appearance of $f(z)$ in any given case is not such as to suggest immediately what substitution ϕ should be made. Klein gave a suggestive method in his [1877 = 1922, 307-320], which is briefly discussed in Forsyth, [1900 Vol. 4, 184-187] but it must be said that neither Klein nor Forsyth is explicit about how such a

calculation might actually be carried out[9]; one notes Katz'
despairing remark [1976, 556]"...but even in cases when one knows
the answer ahead of time, it seems hopeless ever to carry out
Forsyth's test procedure."

It may be of interest to give Klein's list of the ground forms
which arise in each of the five cases as they appear in the terms
of the original differential equations. Starting from the
Schwarzian form

$$[\eta] = P(x),$$

where $P(x) = y_1/y_2$ is a quotient of two arbitrary particular solutions
of the hypergeometric equation

$$\frac{d^2 y}{dx^2} + p \frac{dy}{dx} + qy = 0,$$

he differentiated $\eta = y_1/y_2$ and wrote η' in the form

$$\frac{C \, e^{-\int pdx}}{y_2^2},$$

C an arbitrary constant. He took for definiteness the icosahedral
equation

$$1728 \, \frac{H^3(\eta)}{f^5(\eta)} = R(x)$$

and, observing that $T^2 = 12f^5 - 12^4 H^4$ implies

$$T = \overset{\bullet}{C}(3H'f - 5f'H),$$

deduced $y_2^2 = C \dfrac{H^2(\eta)T(\eta)}{f^6(\eta)\ R'(x)}\ e^{-\int pdx}$.

So, finally, $f(y_1, y_2) = C^6\ \dfrac{R^4(1 - R)^3}{R'^6}\ e^{-6\int pdx}$,

and, if $p = 0$, f is rational.

Klein listed the ground forms in the table reproduced on p.130 , where C denotes an arbitrary constant.

Simultaneously with Klein, the Italian mathematician Brioschi was also looking at algebraic solutions to the hypergeometric equation, and in 1877 the Mathematische Annalen carried two contributions from him. The first, [1877a], was in the form of a letter to Klein, and in it he showed how the study of forms whose fourth transvectant vanishes can be connected with a second order differential equation. In his [1877b] Brioschi obtained the precise form of the Schwarzian equation for eleven of the fifteen cases in Schwarz's list. Since Schwarz had done one case (no. I), this left three (XII, XIV, XV) outstanding for which Brioschi's method was inadequate. These were tackled by Klein. Brioschi found that the equations $[t]_x = -2p$, where t was (in the order II, III, ... X, XI, XIII):

$$x,\ \frac{-4x}{1-x^2}\ ,\ x,\ \frac{-(1-x)^2}{4x}\ ,\ x,\ \frac{-4x}{1-x^2}\ ,\ \frac{-(1-x)^2}{4x}\ ,\ \frac{-(x-4)^3}{27x^2}\ ,$$

$$-\frac{1}{4^3}, \quad \frac{x(x+8)^3}{(1-x)^3}, \quad \frac{4}{27}\cdot\frac{(x^2-x+1)^3}{x^2(1-x)^2}, \quad \frac{1}{4.27}\cdot\frac{(x^2+14x+1)^3}{x(1-x)^4}.$$

Brioschi's approach was more invariant-theoretic than Klein's, as befits a man of the previous generation. It was also more modest, for it accepted the solution to Fuchs's problem already proposed. The only man to offer a solution to that problem entirely in the terms of invariant theory was the acknowledged master of the subject, Paul Gordan.

(I)
$$y_1 = Ce^{-\frac{1}{2}\int p\,dx}\cdot\frac{R^{\frac{n+1}{2n}}}{R'},$$

$$y_2 = Ce^{-\frac{1}{2}\int p\,dx}\cdot\frac{R^{\frac{n-1}{2n}}}{R'}.$$

(II)
$$y_1 y_2 = Ce^{-\int p\,dx}\cdot\frac{R^{\frac{1}{2}}(R-1)^{\frac{1}{2}}}{R'},$$

$$y_1^n + y_2^n = 2C^{\frac{n}{2}}e^{-\frac{n}{2}\int p\,dx}\cdot\frac{R^{\frac{n+2}{4}}\cdot(R-1)^{\frac{n}{4}}}{R'^{\frac{n}{2}}},$$

$$y_1^n - y_2^n = 2C^{\frac{n}{2}}e^{-\frac{n}{2}\int p\,dx}\cdot\frac{R^{\frac{n}{4}}(R-1)^{\frac{n+2}{4}}}{R'^{\frac{n}{2}}}.$$

(III)
$$y_1^4 - 2\sqrt{-3}\,y_1^2 y_2^2 - y_2^4 = C^2\,e^{-2\int p\,dx}\cdot\frac{R^{\frac{5}{3}}\cdot(R-1)}{R'^2},$$

$$y_1^4 + 2\sqrt{-3}\,y_1^2 y_2^2 - y_2^4 = C^2\,e^{-2\int p\,dx}\cdot\frac{R^{\frac{4}{3}}\cdot(R-1)}{R'^2},$$

$$2\sqrt[4]{-27}\cdot y_1 y_2(y_1^4 - y_2^4) = C^3\,e^{-3\int p\,dx}\cdot\frac{R^2(R-1)^2}{R'^2}.$$

(IV)
$$\sqrt[4]{108}\cdot y_1 y_2(y_1^4 - y_2^4) = C^3\,e^{-3\int p\,dx}\cdot\frac{R^2(R-1)^{\frac{3}{2}}}{R'^3},$$

$$y_1^8 + 14 y_1^4 y_2^4 + y_2^8 = C^4\cdot e^{-4\int p\,dx}\cdot\frac{R^3(R-1)^2}{R'^4},$$

$$y_1^{12} - 33 y_1^8 y_2^4 - 33 y_1^4 y_2^8 + y_2^{12} = C^6\cdot e^{-6\int p\,dx}\cdot\frac{R^4(R-1)^{\frac{1}{2}}}{R'^6}.$$

(V)
$$\sqrt[5]{12^3}\,y_1 y_2(y_1^{10} + 11 y_1^5 y_2^5 - y_2^{10}) = C^6\,e^{-6\int p\,dx}\cdot\frac{R^4(R-1)^3}{R'^6},$$

$$-(y_1^{20} + y_2^{20}) + 228(y_1^{15} y_2^5 - y_1^5 y_2^{15}) - 494 y_1^{10} y_2^{10}$$
$$= C^{10}\cdot e^{-10\int p\,dx}\cdot\frac{R^7(R-1)^5}{R'^{10}},$$

$$(y_1^{30} + y_2^{30}) - 522(y_1^{25} y_2^5 - y_1^5 y_2^{25}) - 10005(y_1^{20} y_2^{10} + y_1^{10} y_2^{20})$$
$$= C^{15}\,e^{-15\int p\,dx}\cdot\frac{R^{10}(R-1)^8}{R'^{15}}.$$

TABLE   3.2

3.4 The Solutions of Gordan and Fuchs

Gordan considered the same geometric problem as Klein from two aspects: as a question involving finite groups, and following Fuchs, as a question about binary forms with vanishing covariants, and dealt with them in two papers in the Mathematische Annalen of 1877.

In the first paper [1877a] Gordan considered transformations of the form $\eta' = \frac{\alpha\eta + \beta}{\gamma\eta + \delta}$, $\alpha\delta - \beta\gamma \neq 0$, essentially as rotations, which he treated in the language of invariant theory. If, he said, $\eta = \frac{x_1}{x_2}$ and $\eta' = \frac{y_1}{y_2}$, then these transformations can be thought of as the vanishing of a bilinear form. As such it has two invariants, its determinant and another which he denoted $- 2 \cos \phi$, where he called ϕ the argument. He showed that, if S is a transformation with argument ϕ, T has argument ψ, and ST has an argument θ, then $S^{-1}T$ has argument H, where $\cos \theta + \cos H = \cos(\theta + \psi) + \cos(\phi - \psi)$.

If the transformations form a finite group then each transformation has finite period, and is equivalent to one of the form $\eta' = \rho\eta$, so each ϕ is some submultiple of π, and the transformations can be grouped into families according to the basic transformations of which they are powers. He reduced the problem to establishing the maximal periods, and then supposed S and T were each of this maximal period, and so had the same argument, ϕ. Then the arguments of ST and $S^{-1}T$ would be θ and H, say, where $\cos \theta + \cos H = \cos 2\phi + 1$. θ and H must also be rational multiples of π, so, setting $2\phi = \phi_1$, $\pi - \theta = \phi_2$ and $\pi = H - \phi_3$ he got the equation

$$1 + \cos \phi_1 + \cos \phi_2 + \cos \phi_3 = 0,$$

which is to be solved in angles which are rational multiples of π.
He found, quoting [Kronecker 1854], that there were very few
solutions, indeed just the known cases. The periods could be
2, 3, 4, or 5, giving rise to the various groups of the regular solids,
whose elements be presented explicitly up to conjugacy in each case.

In his second paper [1877b] Gordan solved Fuchs's ground form
problem directly, by showing, as he put it, that "the result follows
immediately from the general rules which I have developed in my text
Ueber das Formensystem binärer formen...". The problem Fuchs had
raised was to characterize those forms, f, of least degree all of
whose covariants of lower degree vanish identically and all of whose
covariants of higher order which are powers of forms of lower order
also vanish. Gordan's approach was to consider a second form, P,
of degree $\mu + 1$, with the property that its final transvectant with
f, $(f, P)^{\mu+1}$, vanished. He was able to show [1877b, 147] that P
enabled him to detect when covariants of f were powers of linear forms,
in which case f could not be a ground form. He was able to show that
the only forms of order greater than 4 satisfying Fuchs's criteria
belonged to the octahedral or icosahedral system, but his argument
proceeded through seven subcases and cannot be summarized here.

The work of Gordan and Klein enabled Fuchs to return to his
original list of ground forms and prune it of its spurious members.
In his paper [1878a] he analyzed each item in his list, indicating
new properties of the genuine ones and providing valid, ad hoc,
invariant-theoretic reasons for deleting those to which Klein had
objected. For the two members that remained with N > 4 he
characterized the ground forms of each degree that can arise. For

example, he showed [1878a §7 Theorems III-VI = 1906, 128] that for
N = 6, L = 8 every ground form of degree 24 is of the form
$v\bar{H}(f_6)^3 + \lambda f_6{}^4$, where f_6 is the unique ground form of degree
6, $\bar{H}(f_6)$ its Hessian[10] (of degree 8), and v and λ are constants not
both zero and such that $\frac{\lambda}{v} \neq 108$. Similarly for N = 12, L = 10, every
ground form of degree 60 is of the form $\sigma\bar{H}(f_{12})^3 + \lambda f_{12}{}^5$, where f_{12}
is the unique ground form of degree 12, and λ, v are constants not
both zero and such that $\frac{\lambda}{v} \neq 1728$. The exceptional cases $\bar{H}(f_6)^3 +$
$108 f_6{}^4$ and $\bar{H}(f_{12})^3 + 1728 f_{12}{}^5$ are squares of rational functions.

Fuchs obtained these theorems by considering the highest degree,
μ, a ground form could have, and comparing it with the lowest degree,
N. If f is a form of degree N > 4 then $\phi: = H(f)^3 - \lambda f^2 H^2(f)$ cannot
vanish, for ground forms only have a common factor if they are
identical, and $H(f)^3 = \lambda f^2 H^2(f)$ would imply $H^2(f)$ divides $H(f)^3$
which is impossible (since the degree of $H(f)^3$ is greater than the
degree of $H^2(f)$). However, $\frac{f^2 H^2(f)}{H(f)^3}$ is not altered on substituting
jf for f, where j is a root of unity, and so it is not altered by any
circuit of z. So ϕ must be a root of a rational function, and is
therefore a product of ground forms. However, if F is a ground form
of degree μ having no factors in common with f, H(f), or $H^2(f)$, then
for some suitable λ, F divides ϕ, and so $\mu \leq 6N - 12$. Consequently
when N = 6 the maximal degree of a ground form is 24, and when N = 12
the maximal degree is 60. These bounds are attained, as we have seen.
Furthermore, μ is related to m, the degree of the algebraic equation
satisfied by a solution of the differential equation, for Fuchs
showed [1878a, §2 Thm. III = Werke II, 120] that $\mu = m$ or m/2, and if
L is 8 or 10 then in fact $\mu = m/2$. This result was implicit in one of
his earlier tabulations of L and N, [1876, §17 = Werke II, 39].

Fuchs also deduced that any ground form of maximal degree μ is a rational function, and that between any three there is a linear relation with constant coefficients. But this relationship was, for him, a consequence of the theorems III-VI just quoted, and not, as it was for Klein, a means to understanding the forms.

3.5 *Jordan's solution*

Camille Jordan's solution to the algebraic solutions problem, couched in terms of the finiteness of the related group of linear transformations, was put forward in a note in the Comptes Rendus of March 1876, [1876a]. It erred in omitting one of the possible cases, the icosahedral one, and in November of the same year he restored the missing case, [1876b], remarking:

> "..a calculating error, which in no other way invalidates
> the principles of our reasoning, caused us to omit one of
> these groups...",

and he acknowledged that the first solution of the group-theoretic problem was due to Klein.

In a third short paper, [1877], Jordan sketched, for the first time, the answer to the problem for a third-order differential equation, and in June 1877 he submitted his long paper [1878] on the question to the Journal für Mathematik. It gives not only a treatment of the nth order differential equation, but a full account of Jordan's methods. It is this paper which will now be discussed.

Jordan observed that the solutions to a given linear differential equation are all algebraic if and only if the corresponding monodromy group is finite. So, in order to enumerate the different

types of mth order equation with that property, it is enough to
construct the different groups of finite order which can be
represented as linear groups in m variables. He found five such
groups when m = 2, eleven when m = 3, and was able to show in the
general case that the finite groups which can arise (and hence the
solutions of the differential equation) satisfied certain additional
conditions. His methods, as befits the leading group theorist of the
day, were those of the newly-discovered Sylow theory[11]. Jordan first
employed them in the case of the second-order differential equation,
to which he devoted Chapter I of the paper.

He denoted the typical linear substitution[12] in two variables

$$S = |u_1, u_2 \quad \alpha u_1 + \beta u_2, \gamma u_1 + \delta u_2|.$$

After a linear change of variable S can be written either in the
canonical form

$$|x, y \quad ax, by|,$$

when he said it was of the first kind, or in the form

$$|x, y \quad ax, a(y + \lambda x)|.$$

He said it was of the second kind if $\lambda = 0$, and otherwise of the third
kind. In either case the roots of the characteristic equation for S
coincide. If G is to be a finite group all of its elements S must be
of finite order, so no element can have infinite order. G cannot then
contain any element of the third kind.

Jordan next showed that the elements T which commute with an element of S of the first kind are precisely those which are also in canonical form when a basis is chosen with respect to which S is diagonalized (i.e. S and T have the same eigenspaces). Consequently the elements T which commute with an S of the first kind commute with each other (indeed, form a commutative subgroup of G). The finite groups G can now be divided into two types. Those of the first type contain sets of elements of the form

$$S = |x, y \qquad ax, by|$$

with respect to a fixed basis $\{x, y\}$, where a and b are roots of unity. Those of the second type are formed from the first by adjoining to a single set of the first type an element of the form

$$T = |x, y \qquad y, kx|$$

where k is a root of unity. For, every finite group H all of whose elements are conjugate to those of a group of the first type, and which contains an element of the first kind, belongs to the first or second type.

Jordan's problem was now to determine all finite groups of either type. He let G be such a group, Ω its order, and g the subgroup of elements of the second kind, with order ω, and sought a formula for Ω analogous to Schwarz's formula for the λ, μ, ν above (p. 106). He let $S \epsilon G$ be an arbitrary element of the first kind, and defined F_S to be the elements of G which commute with S. Evidently $F_S \supset g$ and F_S has $\mu\omega$ elements, say, with $\mu > 1$ since $S\epsilon F_S$, Sg. So G is a union of sets F_S, and no S appears in two different sets F_S, $F_{S'}$, as can be easily seen. If S has the form

$$S = |x, y \quad ax, by|$$

and E is the subgroup of elements, g, of G such that $g^{-1}F_S g = F_S$ then the elements of E have one of the forms

$$|x, y \quad \alpha x, \delta y|, \text{ in which case they lie in } F_S,$$
$$\text{or } |x, y \quad \beta y, \gamma x|.$$

Therefore E either has order $K \mu \omega = 2 \mu \omega$ or $\mu \omega$, depending on whether or not it has an element of the form $|x, y \quad \beta y, \gamma x|$. The number of sets F_S conjugate to F_S in G is then $\Omega/k\mu\omega$. Each F_S contains the ω elements of g and $(\mu - 1) \omega$ elements of the first kind. The total number of elements of the first kind in the totality of sets conjugate to F_S is therefore $\frac{\Omega}{k\mu\omega} (\mu - 1) \omega = \frac{\Omega(\mu - 1)}{k\mu}$. If this argument is repeated for each $F_{S'}$ not conjugate to F_S until the group is exhausted, the following formula for the order of the group G is obtained:

$$\Omega = \omega(1 - \frac{\mu - 1}{k\mu} - \frac{\mu' - 1}{k' \mu'} - \dots) \geq 1. \qquad (3.5.1)$$

This is the necessary formula. It can only contain two or three terms $\frac{\mu - 1}{k\mu}$. If it has two terms $\frac{\mu - 1}{k\mu}$ and $\frac{\mu' - 1}{k' \mu'}$ then either $k = 2$, $k' = 1$, $\mu' = 2$ and $\Omega = 2\mu\omega$, G is of the second type, or $k = 2$, $\mu = 2$, $k' = 1$, $\mu' = 3$, and $\Omega = 12\omega$. If there is also a term $\frac{\mu'' - 1}{k'' \mu''}$ then necessarily $k = k' = k'' = 2 = \mu''$ and either $\mu' = 2$ and $\Omega = 2\mu\omega$, or $\mu' = 3$ and $\mu = 3, 4,$ or 5, in which cases

$\Omega = 12\omega$, 24ω, or 60ω respectively. In short, every G of the required kind and not of the second type has order $r\omega$ where r is either 12, 24, or 60.

There is an evident representation of G as 2×2 matrices, by means of the function $z = \dfrac{mx + ny}{px + qy}$, which annihilates the ω elements of g. The image of G under this representation is a group Γ of order r, and Jordan studied Γ using Sylow theory. His treatment of the case $r = 60$ is typical. Γ in this case contains 6 groups of order 5 and indeed is the homomorphic image of a group of permutations of 6 letters of order 60. But then it must also be isomorphic to the alternating group on five letters, A_5. Its generators can be written down, as permutations they are

$$A^1 = (\alpha \ \beta \ \gamma \ \delta \ \varepsilon), \ B^1 = (\beta \ \varepsilon)(\gamma \ \delta), \ C^1 = (\beta \ \delta)(\gamma \ \varepsilon)$$

and as linear substitutions they are

$$A = |x, y \qquad \theta x, \ \theta^{-1} \ y|, \qquad \theta^{10} = 1,$$

$$B = |x, y \qquad y, \ -x \ |,$$

$$C = |x, y \qquad \lambda x + \mu y, \ \mu x - \lambda y| \text{ where } \mu^2 + \lambda^2 + 1 = 0,$$

$$\lambda = \frac{1}{\theta^2 - \theta^{-2}} \ .$$

G is obtained from Γ by adjoining elements $|x, y \qquad ax, \ ay|$ where a is a primitive ωth root of unity.

The five types of groups which Jordan found are, as one would expect, the cyclic and dihedral groups of arbitrary order, and the tetrahedral, octahedral, and icosahedral groups. Corresponding to each group is a particular type of solution function to the appropriate differential equation. For instance, in the case of the icosahedral group A_5, let z be an arbitrary solution to a differential equation whose monodromy group is A_5. Then z^ω is a rational function of the roots of an equation of the fifth degree whose discriminant is a perfect square. If u is another solution then u^ω is a rational function of z^ω and the independent variable t.

In the third and final chapter of his [1878] Jordan sought to give a complete analysis of the third order linear differential equation by extending the methods used to discuss the second order case. There are trivial extensions of the two dimensional groups to three dimensions whereby a finite cyclic group is added on as a direct summand and alone affects the z-variable [no. 62]. Jordan looked for non-trivial three dimensional representations as groups of matrices with determinant 1. If one such group is H his first task was to find an equation for the order Ω of H, analogous to (3.5.1). H may well have a subgroup, K, consisting of the direct sum of a two-dimensional group, K^1, and a one-dimensional group. A lengthy consideration of the various cases that can arise, depending in part on the choice of K^1 [nos. 71-96], finally yielded the sought-for equality [no. 96, equation 63]

$$\Omega = \frac{\Phi}{1 - \Sigma}$$

where Σ is a sum of terms from the following list:

$$1 - \frac{1}{120\lambda} \; , \; 1 - \frac{1}{48\lambda}, \; \frac{65}{96}, \; \frac{33}{48}, \; 1 - \frac{1}{24\lambda}, \; \frac{15}{24}, \; \frac{m-1}{km}, \; \frac{1}{2}, \; \frac{1}{4}, \; \frac{1}{8}.$$

Here k = 1, 2, 3 or 6, m is subject to certain restrictions, and λ is arbitrary.

Jordan's next task was to extract from this formula a list of the possible groups it could refer to. He considered the fourteen different summations that Σ could be, and came up with a table of 47 associated orders for groups [no. 124], with, in each case, an indication of the kinds of subgroups that would be present. To get some control over the proliferating chaos Jordan next observed that the groups he sought were either simple groups from the table, which could be added to the list in no. 62, or had a group in the new list as a normal subgroup - let them be added to the list - or had a group in the extended list as a normal subgroup, and so on [no. 125].

It turned out that there were very few simple groups in the table. Relatively simple considerations involving Sylow theory eliminated all but seven of them [nos. 127-149], and six of those in fact correspond to no group at all [nos. 150-186]. The outstanding case, XXXII of order 60 Φ, corresponds to the icosahedral group, and an extension of it to a group of order 60.3 = 180 [nos. 187-193].

The construction of a group whose elements normalized a simple group in the list could be carried out in three distinct ways, yielding a group of order 27.2.12 and two of its subgroups, of orders 27.2.4, 27.2.2 [no. 202]. No new group had the icosahedral group as a normal subgroup. So finally Jordan produced the six groups (in addition to the give trivial types of no. 62) listed in footnote 13.

As has been remarked, this list is incomplete, since it lacks the simple group of order 168. When Jordan revised his Journal für Mathematik paper for the Atti della Reale Accademia in 1880, he rectified this omission, which had been brought to light by Klein[14]. It had derived from a too-hasty interpretation of his equation 63 [no. 96]. Correctly interpreted it led directly to a group of order 24.7 containing 8 cyclic groups of order 7, which necessarily must be isomorphic to the group of transformations

$$\left| t \quad \frac{\alpha t + \beta}{\gamma t + \delta} \right|$$

$t = \infty, 0, 1, \ldots, 6 \pmod 7$ and $\alpha\delta - \beta\gamma$ a quadratic residue mod 7, i.e. the simple group G_{168}. Jordan showed this group had generators

$$A = \begin{pmatrix} \tau & 0 & 0 \\ 0 & \tau^2 & 0 \\ 0 & 0 & \tau^4 \end{pmatrix} \qquad \tau^7 = 1, \text{ of order } 7,$$

$$B = \begin{pmatrix} 0 & 1 & 0 \\ 0 & 0 & 1 \\ 1 & 0 & 0 \end{pmatrix} \qquad \text{of order 3, and}$$

$$C = \begin{pmatrix} a & c'' & b' \\ c'' & b' & a \\ b' & a & c'' \end{pmatrix}, \qquad \text{of order 2, where}$$

$$a\tau + b'\tau^2 + c'' \tau^4 = 0$$
$$a\tau^{-1} + b'\tau^{-2} + c''\tau^{-4} = 0$$

in which form it precisely matched Klein's description of it in Klein [1878/79, §4], discussed in Chapter V. Jordan was lazy about

such matters and mentioned Klein but did not give the reference[15].
Jordan also showed that G_{168} was the only group which had been
omitted of order 24.7.q, the case under discussion.

Jordan had also shown in his [1878] that in some sense only
finitely many linear differential equations of order n have finite
monodromy groups[16]. He showed that the finite subgroups of
$G\ell(n ; ¢)$ could be classified into types, in such a way that the
situation for arbitrary n resembled that for n = 2, when there are
five types: two infinite families and three other groups. This
result, for general n, is of independent interest in the study of
groups, and is known as Jordan's finiteness theorem. He found his
proof of 1878 imperfect and re-worked it for a second publication
in 1880. This proof, which amplifies and clarifies the earlier one,
will only be discussed briefly here. [1880, = Oeuvres, II,
177-217]. References are to his numbered paragraphs.

He took G to be a fixed finite subgroup of the linear group in
n variables (7) and F to be an abelian subgroup (faisceau). F can
be simultaneously diagonalized, and when this is done its elements
may be written in the form

$$(a_{1i}, a_{2i}, \ldots, a_{ni}) \text{ for some index i, } 1 \le i \le |F|$$

(modifying his notation slightly). Jordan defined F to be
irreducible with respect to an integer λ if each ratio

$$a_{\ell i}/a_{ki} \qquad 1 \le k < \ell \le |F|$$

took more than λ values as i went from 1 to $|F|$, provided
$a_{\ell i} \neq a_{ki}$, (4). He said F was maximal (complet, 7) if it was the
largest abelian subgroup in G whose elements had the same form, say,
for example

$$(a_{1i}, a_{2i}, \ldots, a_{ni}), \qquad a_{1i} = a_{2i} \, .$$

His precise result is (8):

<u>Theorem</u>   There are integers λ_n and μ_n depending only on n, such that
G has a maximal abelian normal subgroup F irreducible with respect
to λ_n and of index $\leq \mu_n$ in G, and F contains every other abelian
subgroup irreducible with respect to λ_n.

His proof was by induction on n ; n = 1 is trivial, and
$\lambda_1 = \mu_1 = 1$. To prove the theorem for n when it was known for
1, 2, ..., n - 1 he considered two cases. He said G was decomposable
(9) if $\mathbb{C}^n$ can be decomposed into proper subspaces such that each
element of G either preserves the subspaces or permutes them. In this
case, Jordan showed that G contained a subgroup G' which preserved the
subspaces, and the inductive hypotheses readily supplied a subgroup
F of G' and integers λ_n and μ_n satisfying the theorem with respect to
G'(14). Jordan then showed that F also satisfied the theorem with
respect to G (15-20). If a group is not decomposable Jordan called it
indecomposable; the modern word is 'primitive'. This paper is one of
many displaying a high level of abstraction in Jordan's work, which
goes far beyond the theory of permutations in which it is couched.
In this paper the theory of 'systems of imprimitivity' is thoroughly
outlined.

The preface to his [1880] is interesting for the light it sheds on Jordan's approach to mathematics. Speaking of the problem of determining the finite groups of linear transformations in two variables, he remarked that it was first solved by Klein [1875/76] and then confirmed by Fuchs [1876] and Gordan [1877a, b] using entirely different methods. Then he went on:

"In spite of the considerable interest which attaches to the work of these eminent geometers, one could want a more direct method for solving this question. The determination of the sought-for groups is only in effect a problem of substitutions, which must be capable of being treated by the sole resources of that theory without recoursing as M. Klein, to non-Euclidean geometry or, as MM Fuchs and Gordan, to the theory of Forms. Besides, the new method which is to be found must, in order to be entirely satisfactory, be capable of being extended to groups in more than 2 variables."

Jordan as a mathematician believed in propaganda by deeds rather than words, but he was here asserting that the proper approach to the question originally raised by Fuchs is group theory, and proposing that the test for all methods must be their capability to deal with the higher order cases, for which methods were currently the only ones. By his example Jordan established group theory as a subject in its own right, and as one capable of many applications. It became increasingly regarded as the 'natural' abstract structure underlying many mathematical problems and, following Jordan's example, French mathematicians came to prefer group theory to the theory of invariants. So, although Hermite had been strongly attracted to invariant theory, the next generation in France were not, and the subject developed much more strongly in Germany. On the other hand, Halphen's successful treatment of the algebraic solutions problem for differential equations of higher order did depend essentially on invariant theory, as will be seen. But it was eclipsed almost at once by the group-theoretic methods of Poincaré, so once again invariants seemed to be less powerful than the newer techniques.

3.6 *Equations of higher order*

Frobenius had observed [1875a] that if $P = 0$ is a homogeneous linear differential equation of order λ all of whose solution are algebraic, then the solutions satisfy an irreducible algebraic equation of order $\nu \geq \lambda$, all the roots of which, $y_1,\ldots,y_\nu$ also satisfy the differential equation. Only λ of these roots will be linearly independent, so constants $c_1,\ldots,c_\nu$ can be found such that

$$y = c_1 y_1 + \ldots + c_\nu y_\nu$$

takes $\nu!$ distinct values as the ν roots are permuted in all possible ways. By theorems of Abel and Galois, $y_1,\ldots,y_\nu$ are therefore expressible as rational functions of y and x. Frobenius investigated the converse, and found that if an irreducible linear differential equation of order greater than 2 has a solution in terms of which all the other solutions can be written rationally, then all the solutions are algebraic functions. He argued as follows. If the differential equation has a solution y and all the other solutions can be expressed rationally in terms of y but y is transcendental, then under continuation y can only transform as $y \mapsto \dfrac{ay+b}{cy+d}$, where a, b, c, and d are rational functions. Straightforward monodromy considerations

produce a transformation y ↦ ky + r, where k is a constant, so the
rational function r satisfies the differential equation. Any rational
function satisfies a first order linear differential equation, and so
the original equation is reducible. It might happen that r was zero,
but then the original equation can only be of first or second order.
The reverse of this conclusion is that, if the given differential
equation is irreducible and of order greater than 2 then y, if it
exists, is algebraic, and Frobenius's conclusion is established.

Not much else was done with the differential equations of order
greater than 2 all of whose solutions were algebraic, for Jordan's
work indicated how technically complicated it could become. In
[1882a, b] Fuchs showed that if y_1, y_2, and y_3 are a basis of solutions
to a third order differential equation of the Fuchsian class which
furthermore satisfy a homogeneous polynomial $f(y_1, y_2, y_3) = 0$, of
order m > 2, then the equation is algebraically integrable. When
m = 2 the equation is satisfied by the square of the solutions of a
certain second-order differential equation. Such equations had been
studied earlier and in a different way by Brioschi [1879]. The
polynomial f is a projective embedding of a Riemann surface, and Fuchs
showed that the genus, p, of the surface has a strong effect on the
number, n, of reduced roots of any algebraic equation satisfied by any
solution of the differential equation: p > 1 implied n ≤ 4; p = 1
implied n = 2, 3, 4 or 6; and p = 0 reduced to the case of an
algebraically integrable second order differential equation. Fuchs's
method involved the pth order differential equation satisfied by the
φ's which appear in the homogeneous form of the integrands of the
first kind on the Riemann surface

$$\frac{\phi(y_1, y_2, y_3) \ \Sigma \pm c_1 \ y_2 \ dy_3}{\Sigma \ c_1 \ \frac{\partial f}{\partial y_1}}$$

(see Chapter V), and, when p = 1, the earlier results of Briot and Bouquet [1856a]. It would be too long an excursion to show more precisely how Fuchs obtained this result, but the simpler result that the differential equation is integrable algebraically if the curve $f(y_1, y_2, y_3)$ is algebraic, was somewhat simplified by Forsyth [1902, IV, 214-216], and can be presented. It rests on the observation that, n being greater than 2, the Hessian of f is a single-valued non-constant function of z, and in fact a rational function (since the differential equation is of the Fuchsian class). Consequently every other covariant of f is a rational function of z (or a constant). Let $k = \psi$ be a non-constant covariant other than H, then $f = 0$, $H = \phi$ (a rational function) and $K = \psi$ provide three algebraic equations from which y_1, y_2, and y_3 can be found, and the result is proved.

Halphen devoted most of his paper [1884] to the relationships which exist between a linear differential equation of order q and the curve defined projectively by a basis of solutions $(y_1 : \ldots : y_q)$. He was particularly interested in the differential invariants which survive the transition from the equation

$$\frac{d^q Y}{dX^q} + qP_1 \frac{d^{q-1} Y}{dX^{q-1}} + \frac{q(q-1)}{2!} P_2 \frac{d^{q-2} Y}{dX^{q-2}} + \ldots qP_{q-1} \frac{dY}{dX} + P_q Y = 0$$

to the equation

$$\frac{d^q y}{dx^q} + qP_1 \frac{d^{q-1} y}{dx^{q-1}} + \frac{q(q-1)}{2!} P_2 \frac{d^{q-2}}{dx^{q-2}} + \ldots + qP_{q-1} \frac{dy}{dx} + P_q y = 0 \quad (3.6.1)$$

under the arbitrary changes of variable $\frac{dx}{dX} = \mu(X)$, $Y = y\, u(x)$, where $\mu(X)$ and $u(X)$ are indeterminate functions. He defined an absolute invariant as a function ϕ of the P_i and their derivatives such that

$$\phi(P_1, P_2, \ldots, \frac{dP_1}{dX}, \ldots) = \phi(P_1, P_2, \ldots, \frac{dp_1}{dx}, \ldots).$$

Such an invariant is

$$V = -P'' + 3(P'_2 - 2P_1 P'_1) - 2(P_3 - 3P_1 P_2 + 2P^3_1),$$

where $p'_i = \frac{dp_i}{dX}$, etc. [1884, 112] and he noted that, surprisingly, this invariant does not depend on the order of the differential equation. He found a sequence of q-1 invariants for equations of order more than three which enabled him to prove quite general results about differential equations of arbitrary order $q > 3$. He observed that if the coefficients are algebraic they have a genus, which he termed the genus of the equation. A reduction of (3.6.1) to an equation with constant coefficients, or rational coefficients (genus zero), or doubly periodic coefficients (genus 1) or to an equation of genus p being sought, Halphen could express necessary and sufficient conditions for this to be possible. The first task is possible if and only if the absolute invariants are all constants, the others if and only if the q - 2 relations between q - 1 absolute invariants are of genus p [1884, 126-130]. For a third-order equation he showed that V vanished identically if and only if the curve defined by $(Y_1 : Y_2 : Y_3)$ was a conic.

In awarding this essay the <u>Grand Prix</u> in 1881, Hermite said that
it "...showed a talent of the highest order. Nothing is more
interesting than to see the introduction of the algebraic notions of
invariants into this research into the integral calculus, which have
originated in the theory of forms, and these new combinations make the
hidden elements appear on which, in its various analytic guises, the
integration of a given equation depends... They are here joined to
a consideration which equally plays an essential role in these
researches: that of the genus of an algebraic equation between two
variables, introduced into analysis by Riemann and which is so often
employed in the works of our time."

Exercises on Chapter III

1.　　　　Verify Schwarz's derivation of (3.1.4) from (3.1.1).

2.　　　　Verify that, if $z = \dfrac{c_1 x + c_2}{c_3 x + c_4}$,

$$\Psi(s,x) = \frac{d^2 z}{dx^2} \Psi(s,z).$$

3.　　　　Show that the only rational values of λ, μ and ν satisfying
Schwarz's conditions on p.103　are those given in the
accompanying table.

4.　　　　Show that if $u_k = j u_i$ (as on p.112) then j is a primitive
root of unity.

5.　　　　Verify Fuchs's Theorem 3 (p.115).

6.   Verify the claim made about the functions defined in (3.2.10) on p. 117 .

7.   Verify Klein's claim (p. 124) that the axes of any two motions R_1 and R_2 of the sphere (which lie in the same finite group) must intersect. [Consider the plane which perpendicularly bisects p and $R_1(p)$, where p is a fixed point of $R_2 R_1$.]

8.   Let G be a finite group of 2×2 matrices, and $\{n, Z\} = R$ be the canonical Schwarzian equation associated to G, with algebraic solution $z = z(Z)$ as described on p. 127 . Let $\{n, \zeta\} = P$ also have monodromy group G, and suppose $y(\zeta)$ is an algebraic solution of it. Suppose $Z = Z(y)$ is a G-invariant rational function inverse to $z = z(Z)$. Show ζ is a rational function of Z by showing $\zeta = \zeta(z(Z))$ is algebraic and single-valued.

9.   Verify Klein's table of solutions on p 130.

10.   Verify Gordan's result (p.131) that if S has argument ϕ, T argument ψ, and ST argument θ, then $S^{-1}T$ has argument H, where $\cos \theta + \cos H = \cos(\theta + \psi) + \cos(\phi - \psi)$. [Gordan's quantity $- 2 \cos \phi$ is the negative of the trace of the rotation matrix considered.]

11.   Verify Kronecker's result quoted on p. 132 .

12.   Show that the symmetry group of the projective self-transformations of Desargues' configuration is the full

permutation group on five elements by finding five objects
in the configuration which are permuted.

13.   Find the symmetry group of Pappus' configuration and
relate it to Hesse's group.

14.   To construct a circular-arc triangle with angles α, β
and γ (where $\alpha + \beta + \gamma < \pi$) draw a circle centre O and mark off
two radii OA and OB enclosing an angle of 2β, then draw OC where
$\hat{BOC} = \pi - (\alpha + \beta + \gamma)$, then draw OD where $\hat{COD} = 2\gamma$. Let AB
and DC meet at E.   Show that the circular-arc triangle BCE has
angles β at B, γ at C, and α at E.   Show how to draw a circle
centre E which is orthogonal to the arc CE.   How is the
construction to be modified if $\alpha + \beta + \gamma > \pi$?

        Construct Schwarz's tessellation.   You will find it helpful
to draw all arcs in the figure out to the boundary circle, C.
Verify the following simple construction for inverting an arc
a in a circle b, where both arcs are orthogonal to the boundary
circle C.   Join the extremities A_1 and A_2 of a to the centre of
b and let these lines meet C at A'_1 and A'_2.   Then the image
of a is the arc joining A'_1 and A'_2 which is orthogonal to C.

        These constructions can be interpreted as isometries for the
non-Euclidean geometry of this disc, see Appendix 4.   Verify
what Schwarz's tessellation suggests but Poincaré was the first
to prove rigorously, that the tessellation reaches arbitrarily
close to the boundary circle.

CHAPTER IV MODULAR EQUATIONS

The study of the algebraic-solutions problem for a second-order
linear ordinary differential equation had brought to light the conceptual importance of considering groups of motions of the sphere, and,
in particular, finite groups. Klein connected this study with that of
the quintic equation, and so with the theory of transformations of
elliptic functions and modular equations as considered by Hermite,
Brioschi, and Kronecker around 1858. Klein's approach to the modular
equations was first to obtain a better understanding of the moduli,
and this led him to the study of the upper half plane under the action
of the group of two by two matrices with integer entries and determinant
one; his great achievement was the production of a unified theory of
modular functions. Independently of him, Dedekind also investigated
these questions from the same standpoint, in response to a paper of
Fuchs. So this chapter looks first at Fuchs's study of elliptic
integrals as a function of a parameter, and then at the work of Dedekind.
The algebraic study of the modular equation is then discussed, and
the chapter concludes with Klein's unification of these ideas.

4.1 Fuchs and Hermite.

Fuchs's interest in the elliptic integrals K and K', J and J', had
been reawakened by a letter he received from Hermite written on 1st
July 1876 (Fuchs, [1906, 113]). Hermite wrote: "You should without
doubt be able to show, by means of the principles at your command, that
on setting $\frac{K'}{K}$ = ω and k = f(ω), k is a single-valued function of
ω = x + yi for all positive x, but what I cannot work out, and it
interests me very much, is how to see clearly that on setting $\frac{J'}{J}$ =

x + yi one ceases to have a single-valued function. Your methods, I don't doubt, should immediately give the reason for the difference in nature of the functions defined by the two equations".

Fuchs replied in November 1876 and a lengthy extract was published [1877a = Fuchs 1906, 85-114]. As before (see Ch.II.2) Fuchs studied K and K' as functions of k^2, the modulus, by means of the differential equation which they satisfy. Analytic continuation around closed circuits in the k^2-plane transform K and K' to $a_1K + b_1K'$, $a_2K + b_2K'$ where a_1, b_1, a_2, b_2 are independent of the start and finish point. Fuchs found these numbers and so was able to show that the real part of

$$H = \frac{a_2K + b_2K'}{a_1K + b_1K'} \qquad (4.1.1)$$

is always positive or zero. When $q = e^{-\pi H}$ is considered as the independent variable, k^2 as a function of q is holomorphic inside the unit circle in the q-plane, but cannot be continued analytically onto or beyond the circle. On the other hand, whilst J and J', the integrals of a second kind, permit the definition of a function

$$Z = \frac{\alpha_1J + \beta_1J'}{\alpha_2J + \beta_2J'} \qquad (4.1.2)$$

and the introduction of a new independent variable $s = e^{\pi Z}$, when $\frac{1-k^2}{k^2}$ is considered as a function of s it can be extended analytically to a holomorphic function in the whole finite s-plane.

In more detail, Fuchs argued that K and K' are functions of k^2, which, on setting $k^2 = 1/u$, yield two functions η_1 and η_2 of u, $K = \frac{1}{2}\sqrt{u}.\eta_1$, $K' = \frac{1}{2}\sqrt{u}.\eta_2$. These satisfy Legendre's equation,

$$2u(u-1)\frac{d^2\eta}{du^2} + 2(2u-1)\frac{d\eta}{du} + \frac{1}{2}\eta = 0. \tag{4.1.3}$$

η_1 and η_2 can be given power series expansions in a neighbourhood of the singular points 0, 1 and ∞, of this equation, and they transform upon analytic continuation around closed circuits of each singular point according to the following monodromy matrices:

around $u = 1$: $\quad S_1 = \begin{pmatrix} 1 & -2i \\ 0 & 1 \end{pmatrix}$,

around $u = \infty$: $\quad S_\infty = \begin{pmatrix} -1 & 0 \\ 2i & -1 \end{pmatrix}$,

and, since $S_0 = S_1^{-1}S_\infty^{-1}$,

around $u = 0$ $\quad S_0 = \begin{pmatrix} 3 & -2i \\ -2i & -1 \end{pmatrix}$.

The monodromy relation for any closed circuit is therefore given by some product of the form $S_0^{k} S_1^{\ell} S_0^{k_1} S_1^{\ell_1} \ldots$, and this, like all the separate powers of S_0 and S_1, has the form

$$\sigma = \begin{pmatrix} \lambda & \mu i \\ \nu i & \rho \end{pmatrix}$$

where λ, μ, ν, ρ are real integers, and $\lambda\rho + \mu\nu = 1$.

To specify all the values of $H = \eta_2/\eta_1 = K'/K$ at a given point u, he let H_0 be an arbitrary value, then all values are then necessarily of the form

$$\frac{\nu i + \rho H_0}{\lambda + \mu i H_0} = \frac{i}{\mu}\frac{1}{\lambda + \mu i H_0} - \frac{\rho i}{\mu}.$$

He specified a single branch of H by the conditions

$$H_0 = \begin{cases} -i & \text{at } u = 0 \\ 0 & \text{at } u = 1 \\ \dfrac{4 \log 2}{\pi} + \dfrac{1}{\pi} \lim_{u \to \infty} \log u & \text{at } u = \infty. \end{cases}$$

Accordingly H could take the values

$$\frac{\nu - \rho}{\lambda + \mu} i \qquad \text{at } u = 0$$

$$\nu i/\lambda \qquad \text{at } u = 1 \qquad \text{or, as } u \to \infty, \ H \to \rho i/\mu \text{ or to a number}$$

with real part equal to $+\infty$.

Near $u = 0, 1, \infty$, H is likewise a transform of H_0 in that neighbourhood, and H_0 can be written down explicitly from the solutions to (4.1.3). But now H_0 and H always have real part greater than or equal to zero, and so $|q| < 1$, as was to be shown. It remained to show that u is holomorphic inside the unit q-disc, but cannot be analytically continued beyond it. The representation of η_1 and η_2 as solutions to (4.1.3) establishes that u is holomorphic inside the q-disc, but in establishing that the circle $|q| = 1$ is a natural boundary for u, Fuchs made a slight mistake. Although the boundary values are u = 0, 1, and ∞ distributed in everywhere dense sets along the circle, Fuchs failed to notice the value ∞, perhaps sharing the widespread contemporary lack of awareness of 'bad' point sets, as Schlesinger suggests [Fuchs, 1908, 113]. His mistake was detected by Dedekind and will be discussed more fully below (p 168). In any case the conclusion that $|q| = 1$ is a natural boundary for u remains valid.

As for J and J' as functions of the modulus k^2, Fuchs again preferred to work with $u = 1/k^2$, and so he introduced $\zeta_1 = 2\sqrt{u}J$, and $\zeta_2 = 2\sqrt{u}J'$, which satisfy the differential equation

$$2u(u-1) \frac{d^2\zeta}{du^2} + 2u \frac{d\zeta}{du} - \frac{1}{2} \zeta = 0.$$

Fuchs found that $Z = \zeta_1/\zeta_2$ transformed under analytic continuation around different circuits in the n-plane into expressions of the form $\frac{\lambda + \mu i Z_0}{\nu i + \rho Z_0}$, and introducing $s = e^{-\pi Z}$ he found u as a function of s was holomorphic inside the unit s-disc. But now s as a function of u is well behaved near $|s| = 1$ and so he showed that the inverse function $u = u(s)$ can be extended analytically to the whole s-plane.

Hermite replied to Fuchs's letter on November 27th 1876. He was, he said, delighted with it. Not only had it explained the difference between $\frac{K'}{K}$ and $\frac{J'}{J}$ but it had done so in a way "which I judge to be of the greatest importance for the theory of elliptic functions. The truly fundamental point that the real part of H is essentially positive I had sought in vain to establish by elementary methods, in order not to be obliged to turn to the new method discovered by Riemann." He commented particularly on one aspect of Fuchs's work, the elementary derivation of famous equation of Jacobi

$$4\sqrt{k} = 2q^{1/8} \frac{(1 + q^2) \, (1 + q^4) \, \cdots}{(1 + q) \, (1 + q^3) \, \cdots}$$

as follows: Fuchs, [1906, 103]. "Is there not some point in observing that on setting $\chi^2 = f(H)$, it follows from your analysis that all solutions of $f(H) = f(H_0)$ are given by the formula

$$H = \frac{\nu i + \rho H_0}{\lambda + \mu i H_0} ,$$

and insisting on the extreme importance of this result for the determination of the singular moduli[1] of M. Kronecker, and on remarking that the beautiful discoveries of that illustrious geometer concerning the applications of the theory of elliptic functions to arithmetic seem to rest essentially on this proposition, of which a proof has not been given before?" It is interesting to observe that it was Hermite, and not Fuchs, who preferred to emphasize the inverse function to the quotient of the solutions of the differential equation which has the more readily comprehensible property

$$f(H_0) = f(\frac{\nu i + \rho H_0}{\lambda + \mu i H_0}).$$

Two profound ideas emerged during this exchange between Hermite and Fuchs:

(i)   the study of the function inverse to the quotient of two independent solutions to a differential equation, which had earlier been broached by Schwarz [Schwarz, 1872] and

(ii) the invariance of such functions under a certain group of transformations, although the group concept was not yet made explicit in this context.

The transformation of modular functions

Hermite himself had published a short but crucial paper [1858] on the transformation of modular functions. The paper was chiefly devoted to the solution of quintic equations by modular functions, and for that reason it is described in section 5.4, but it contained a mysterious table of transformations[2]. He defined $q := e^{-\pi K'/K} = e^{i\pi\omega}$, denoted $k^{\frac{1}{4}}$ by $\phi(\omega)$ and the complementary modulus $k'^{\frac{1}{4}}$ by $\psi(\omega)$, so

$$\phi^8(\omega) + \psi^8(\omega) = 1,$$

$$\phi(-1/\omega) = \psi(\omega),$$

$$\phi(\omega + 1) = e^{-i\pi/8} \frac{\phi(\omega)}{\psi(\omega)},$$

$$\psi(\omega + 1) = \frac{1}{\psi(\omega)}.$$

The values of $\phi(\frac{c+d\omega}{a+b\omega})$ and $\psi(\frac{c+d\omega}{a+b\omega})$ can be deduced easily from these formulae, where a, b, c, and d are integers and ad−bc = 1. Hermite found the value depended only on the residue class of $\begin{pmatrix} a & b \\ c & d \end{pmatrix}$ mod 2. There are six such classes, and the corresponding transformations were, he said:

$$\text{I} \qquad \begin{pmatrix} a & b \\ c & d \end{pmatrix} \equiv \begin{pmatrix} 1 & 0 \\ 0 & 1 \end{pmatrix} \quad \phi(\frac{c+d\omega}{a+b\omega}) = \phi(\omega)\, e^{\frac{i\pi}{8}(d(c+d)-1)}$$

$$\text{II} \qquad \equiv \begin{pmatrix} 0 & 1 \\ 1 & 0 \end{pmatrix} \qquad = \psi(\omega)\, e^{\frac{i\pi}{8}(c(c-d)-1)}$$

$$\text{III} \qquad \equiv \begin{pmatrix} 1 & 1 \\ 0 & 1 \end{pmatrix} \qquad = \frac{1}{\phi(\omega)}\, e^{\frac{i\pi}{8}(d(d-c)-1)}$$

$$\text{IV} \qquad \equiv \begin{pmatrix} 1 & 1 \\ 1 & 0 \end{pmatrix} \qquad = \frac{1}{\psi(\omega)}\, e^{\frac{i\pi}{8}(c(c+d)-1)}$$

$$\text{V} \qquad \equiv \begin{pmatrix} 1 & 0 \\ 1 & 1 \end{pmatrix} \qquad = \frac{\phi(\omega)}{\psi(\omega)}\, e^{\frac{i\pi}{8}cd}$$

$$\text{VI} \qquad \equiv \begin{pmatrix} 0 & 1 \\ 1 & 1 \end{pmatrix} \qquad = \frac{\psi(\omega)}{\phi(\omega)}\, e^{-\frac{i\pi}{8}cd} \;.$$

Many years later, in 1900 Hermite [1908, 13-21] wrote to J. Tannery to explain how he had come by such formulae. His method rested on a formula for transforming θ-functions due to Jacobi, but, he said (p.20), it "... does not please me at all: it is long, above all, indirect; it rests entirely on the accident of a formula of Jacobi, forgotten and as lost among all the discoveries due to that genius".

By then the transformations had been derived several times. The first to do so was Schläfli [1870] who used the approach later employed by Fuchs (above) and considered the monodromy relations of K and iK' under analytic continuation around their poles at 0 and 1. The monodromy matrix at 0 is $\begin{pmatrix} 1 & 0 \\ 2 & 1 \end{pmatrix} = \begin{pmatrix} 0 & 1 \\ 1 & 2 \end{pmatrix}\begin{pmatrix} 0 & 1 \\ 1 & 0 \end{pmatrix}$, and at 1 the monodromy matrix is $\begin{pmatrix} 1 & -2 \\ 0 & 1 \end{pmatrix} = \begin{pmatrix} 0 & 1 \\ 1 & 0 \end{pmatrix}\begin{pmatrix} 0 & 1 \\ 1 & -2 \end{pmatrix}$. Schläfli observed that a method due to Gauss [Disq. Arith. §27] enables one to write any matrix $\begin{pmatrix} a & b \\ c & d \end{pmatrix}$, in which $ad-bc = 1$, $a \equiv d \equiv 1 \pmod 4$ and b and c are even, as a product

$$\begin{pmatrix} a & b \\ c & d \end{pmatrix} = \begin{pmatrix} 0 & 1 \\ 1 & a_1 \end{pmatrix}\begin{pmatrix} 0 & 1 \\ 1 & b_1 \end{pmatrix} \cdots \begin{pmatrix} 0 & 1 \\ 1 & a_n \end{pmatrix}\begin{pmatrix} 0 & 1 \\ 1 & b_n \end{pmatrix} ,$$

where each a_i and b_i is even. The expressions for a, b, c, and d involve continued fractions. When the calculations are done,

$$\phi(\frac{c+dz}{a+bz}) = e^{i\frac{\pi}{8}\sum a_i}\phi(z) \text{ and } \psi(\frac{c+dz}{a+bz}) = e^{i\frac{\pi}{8}-\sum b_i}\psi(z), \text{ where}$$

$\sum a_i \equiv ac + a^2 - 1 \pmod{16}$ and $-\sum b_i \equiv -ab+a^2-1 \pmod{16}$, so Hermite's results are obtained. Schläfli also gave transformations for the function $\tilde{\psi}(z)$ defined by $\tilde{\psi}^{24}(z) = \phi\psi$, for which ([Jacobi, Fund Nova p.89]) $\tilde{\psi}(z) = \dfrac{2^{1/6} e^{i\pi z/24}}{\Pi(1 + q^n)}$, and for K independently in the 6 cases distinguished by Hermite. Independently of Schläfli, Koenigsberger also solved the problem (Koenigsberger [1871]). He analyzed the matrices $\begin{pmatrix} a & b \\ c & d \end{pmatrix}$ as Schläfli had done, using continued fractions, but found the transformations

$$\phi(\frac{-1}{\tau}) = \psi(\tau), \ \phi(\tau + 1) = e^{i\frac{\pi}{8}}\frac{\phi(\tau)}{\psi(\tau)} , \text{ and } \psi(\tau + 1) = \frac{1}{\psi(\tau)}$$

via the expressions for ϕ and ψ as quotients of infinite products. He made no mention of monodromy.

Schläfli's and Koenigsberger's works suggest that the transformations which must be understood are $z \to z+1$ and $z \to -1/z$. Although they did not say so explicitly, when taken mod 2 these transformations generate the six element group of integer matrices $\begin{pmatrix} a & b \\ c & d \end{pmatrix}$ for which $ad - bc \equiv 1 \pmod 2$ and each entry is 0 or 1, which, as a set, had been distinguished by Hermite. The passage from integer elements to integers modulo 2 is not exactly explained in either case, but, for example, $z = \dfrac{K(x)}{iK'(x)}$ is an infinitely many-valued function of x for which the values for a common x are of the form z and its transforms $\dfrac{cz+d}{az+b}$, where $\begin{pmatrix} a & b \\ c & d \end{pmatrix} \equiv \begin{pmatrix} 1 & 0 \\ 0 & 1 \end{pmatrix} \pmod 2$, so x is invariant as a function of z under such transformations $z \to \dfrac{cz+d}{az+b}$. That the transformations $z \to z + 1$ and $z \to -1/z$ generate a group was a commonplace by that time, for it occurs in the calculation of cross-ratios. However, the connection between that group and the group $\{ \begin{pmatrix} a & b \\ c & d \end{pmatrix} \equiv \begin{pmatrix} 1 & 0 \\ 0 & 1 \end{pmatrix} \bmod 2 \}$ seems not to have called for any explanation in the minds of Schläfli or Koenigsberger, nor need it have done so unless it would be worthwhile to place this reasoning in a larger context. The man who saw the need for that was Dedekind.

4.2 *Dedekind.*

The first person to emphasize the 'invariance' interpretation of this analysis of functions like the modulus, k^2, was Richard Dedekind, in his justly celebrated paper of 1877. It was published in the same issue of Borchardt's Journal für Mathematik as Fuchs's paper, and like it is a letter, addressed in this case to Borchardt. Since Dedekind's reputation does not even now reflect his true worth it may not be out of place to comment on his career at this point. Dedekind contributed much more to mathematics than his constructive definition of the real numbers ('Dedekind cuts', discovered in 1858 but only published in Stetigkeit und irrationale Zahlen [1872]). The modern esteem in which this work is held is entirely justified, but Dedekind's other achievements are generally known only to specialists, not just because of their difficulty but, I fear, from an exaggerated attention paid by historians and popularizers to the foundations of mathematics. Dedekind did much more for mathematics than just arithmetizing elementary analysis. He was a profound unifier of mathematics and one of the creators of modern algebraic number theory; the concepts of ring, module, ideal, field, and vector space are as much his contributions as anyone else's. He was a great problem solver, particularly in his favourite subject of algebraic number theory. He was a gifted expositor and dedicated editor of the work of others, chiefly his mentors Gauss, Dirichlet, and Riemann, all of whom he had known personally. The sensitive and illuminating article on him by K.R. Biermann [1972] will perhaps restore the balance and help to make his reputation more truly reflect his many achievements[3].

Dedekind's first work, done under Gauss, had been on Eulerian integrals. In the 1860's he edited Gauss's manuscripts on number theory Gauss, [Werke, 1st ed., vol. II, 1863] and in the same year brought out the first of four influential editions of Dirichlet's Vorlesungen über Zahlentheorie. His interest in algebraic number theory led him to most of his own original work in this period, but in 1876 he edited Riemann's papers with his lifelong friend Heinrich Weber. This paid a debt Dedekind felt to Riemann, and marked an intellectual return to Göttingen where he had been a student and instructor in the 1850's. In 1880, in a joint paper with Weber, he laid the foundations of the arithmetic theory of algebraic functions and Riemann surfaces. Biermann [op cit, 2] tells us that Dedekind was perhaps the first person ever to lecture on Galois theory, (these lectures are discussed in Section 3) but only two students were present to hear him replace the permutation group concept by that of the abstract group (see Purkert, [1976] and below, p. 177). As we shall see, this immersion in the work of another man had its customary effect of deepening and broadening Dedekind's own understanding of mathematics.

Transformations of modular functions

Dedekind began his letter to Borchardt more or less where
Hermite's comments ended. He observed that he had already been
engaged for several years on the determination of the ideal-class
number of cubic fields, which was connected with Kronecker's work on
singular moduli and complex multiplication. In seeking a simpler
route to the 'exceptionally beautiful' but difficult results of
Kronecker he found he had been led to appreciate the fundamental
importance of Hermite's point (quoted above), that $k = k(\omega)$ is
invariant under transformations $\omega_1 = \dfrac{\gamma + \delta\omega}{\alpha + \beta\omega}$, α, β, γ, δ rational
integers satisfying $\alpha\delta - \beta\gamma = 1$ and β, γ even. He remarked that this
observation enables one to derive the usual theory of elliptic functions
without difficulty, but set as his main task the elaboration of the
theory of modular functions independently of elliptic functions.

Almost all of the usual introductory material on elliptic modular
functions appears for the first time in this paper[4]. The upper half
of the complex plane, which Dedekind denoted S, is divided into
equivalence classes by the action of the group of matrices of the
form $\left(\begin{smallmatrix} \alpha & \beta \\ \gamma & \delta \end{smallmatrix}\right)$, $\alpha\delta - \beta\gamma = 1$, α, β, γ, δ integers. Those equivalent to
0 are the rationals on the real line, together with the point $i\infty$;
Dedekind denoted this set R and called it the boundary of S. As a
complete system of representatives (of the equivalence classes) he
chose the domain defined by the three conditions:

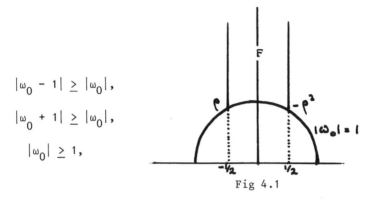

$$|\omega_0 - 1| \geq |\omega_0|,$$

$$|\omega_0 + 1| \geq |\omega_0|,$$

$$|\omega_0| \geq 1,$$

Fig 4.1

which is symmetrically situated with respect to the y-axis, and remarked that the proof that this is a complete system follows the lines of the analogous theorem for binary quadratic forms of negative determinant. This domain he called the principal field (Hauptfeld, denoted here by F), and he added that it is bounded by the lines $x_0 = \pm \frac{1}{2}$ and the arc $x_o^2 + y_0^2 = 1$ ($\omega_0 = x_0 + y_0 i$), each point on the boundary except i being equivalent to one other point. The effect of $\begin{pmatrix} \alpha & -\gamma \\ -\beta & \delta \end{pmatrix}$ is to move the circular-arc triangle bounded by

$$\rho = \frac{-1 + i\sqrt{3}}{2} \, , \, -\rho^2, \, \infty, \text{ to the triangle with vertices}$$

$$\frac{-\gamma + \alpha\rho}{\delta - \beta\rho} \, , \, \frac{-\gamma - \alpha\rho^2}{\delta + \beta\rho^2} \, , \, \frac{-\alpha}{\beta} \, .$$

Dedekind then sought a complex-valued function v on S which took the same value at all equivalent points in such a way that, conversely, to each value of the unbounded variable v there corresponded a unique equivalence class. To obtain such a function he followed the principles laid down by Riemann in his inaugural dissertation (§21). Each half of F can be mapped onto one half of the complex plane by the Riemann mapping principle, so that the y-axis is mapped onto the real axis, and corresponding points on the boundary are given conjugate complex values under the map. The map is extended to the whole of S by the reflection principle, and made unique by insisting that $v(\rho) = 0$, $v(i) = 1$, $v(\infty) = \infty$. This function v Dedekind denoted val(ω)

and called the valency (<u>Valenz</u>); it is nowadays known as Klein's J-function. Klein studied it the next year, making due acknowledgement to Dedekind. The inverse function to the valency function Dedekind described as a covering of the complex sphere, branched over 0,1, and ∞. The point 0 is of order 2, 1 is of order 1, and ∞ is a logarithmic branch point.

To study functions like val(ω) Dedekind introduced the differential expression

$$[v,u]: \quad = \frac{-4}{\sqrt{(\frac{dv}{du})}} \frac{d^2}{dv^2} (\frac{dv}{du})^{\frac{1}{2}},$$

which is twice the Schwarzian derivative, and therefore satisfies such identities as $[v,u] = [v, \frac{C + Du}{A + Bu}]$ A, B, C, D constants. The function $v = val(\omega)$ itself has the Schwarzian derivative

$$[v,\omega] = f(\omega)$$

which has a single valued inverse $[v,\omega] = F(v)$, and $F(v)$ is finite except at $v = 1,0,\infty$, where, respectively, the products

$$(1-v)^{-\frac{1}{2}} \frac{dv}{d\omega} , \quad v^{-\frac{2}{3}} \frac{dv}{d\omega} , \quad v^{-1} \frac{dv}{d\omega}$$ are finite and non-zero, and it is

soon clear   that

$$F(v) = \frac{36v^2 - 41v + 32}{36v^2(1 - v^2)} .$$

Otherwise put, $v = val(\omega)$ is a solution of the third order differential equation $[v',\omega] = F(v')$ and the general solution to that equation has the form val $(\frac{C + D\omega}{A + B\omega})$.

Dedekind was now ready (§6) to introduce the elliptic modular functions themselves. $(\frac{dv}{d\omega})^{1/2}$ satisfies a second order linear differential equation with respect to v, and u = const. $v^{-1/3}(1-v)^{-1/4}$ $(\frac{dv}{d\omega})^{1/2}$ satisfies the hypergeometric equation

$$v(1-v) \frac{d^2u}{dv^2} + (\frac{2}{3} - \frac{7v}{6}) \frac{du}{dv} - \frac{u}{144} = 0,$$

whose general solution in terms of the Gaussian hypergeometric series F is const. $F(\frac{1}{12}, \frac{1}{12}, \frac{2}{3}, v)$ + const. $F(\frac{1}{12}, \frac{1}{12}, \frac{1}{2}, 1 - v)$, which may also be expressed as a Riemannian P-function. Dedekind found the square root of u of special interest; it is the function

$\eta(\omega)$ = const. $v^{-1/6} (1-v)^{-1/8} \frac{dv}{d\omega}$ nowadays called the Dedekind η-function. It is a single valued function on S, finite and non-zero on the interior of S, as $\omega \to \infty$, $\eta \to 0$ like $v^{-1/24}$ and thus like $1^{\omega/24}$: $= e^{2\pi i \omega/24}$. He chose the constant so that $\eta(\omega) = 1^{\omega/24}$ at $\omega = \infty$.

$\eta(\frac{\gamma + \delta\omega}{\alpha + \beta\omega}) = c.(\alpha + \beta\omega)^{1/2} \eta(\omega)$ where $c^{24} = 1$, in particular

$\eta(1 + \omega) = 1^{1/24}\eta(\omega)$, $\eta(\frac{-1}{\omega}) = 1^{-1/8} \omega^{1/2} \eta(\omega)$ where $\omega^{1/2} = 1^{1/8}$ when $\omega = 1^{1/4} = i$. Now η is completely determined, for if $f(\omega)$ is another function with these properties then $f(\omega)/\eta(\omega)$ is an everywhere finite single-valued function of v = val(ω) i.e. a constant, which takes the value 1 at v = ∞, and so the constant necessarily equals 1.

Dedekind could now make clear the relationship between $\eta(\omega)$ and the modulus of an elliptic integral or its square root k. He introduced three auxiliary functions corresponding to the three transformations of order 2

$$\eta_1(\omega) = \eta(2\omega), \ \eta_2(\omega) = \eta(\frac{\omega}{2}), \ \eta_3(\omega) = \eta(\frac{1+\omega}{2}),$$

for which the following identity holds:

$$\eta_1(\omega) \ \eta_2(\omega) \ \eta_3(\omega) = 1^{1/48} \ \eta(\omega)^3.$$

In terms of these functions it turns out that

$$k^{1/8} = 1^{1/48} \sqrt{2} \cdot \frac{\eta_1(\omega)}{\eta_3(\omega)} = \phi(\omega), \text{ say}$$

$$k'^{1/8} = 1^{1/48} \eta_2(\omega)/\eta_3(\omega) = \psi(\omega), \text{ say}$$

where ϕ and ψ are the functions introduced by Hermite [1858] (see above, p158), and K and K' can be defined by the equations

$$\sqrt{\frac{2K}{\pi}} = 1^{-1/24} \eta_3(\omega)^2/\eta(\omega) \text{ and } K'i = K\omega.$$

Finally, calculating $\phi(1 + \omega)^8 = \frac{k}{k-1}$ and $\phi(\frac{-1}{\omega}) = 1 - k$ led Dedekind to the conclusion

$$v = \mathrm{val}(\omega) = \frac{4}{27} \frac{(k+\rho)^3 (k+\rho^2)^3}{k^2(1-k)^2}.$$

The function $k = \phi(\omega)^8$ can be completely determined from this information, just as η was. It satisfies $[k,\omega] = \frac{(k+\rho)(k+\rho^2)}{k^2(1-k)^2}$, which reduces to $\frac{d}{dk}(k(1-k)\frac{dK}{dk}) = \frac{1}{4}K$, which, Dedekind noted, was Fuchs's starting point, and k has therefore the same properties as the quotient of theta-functions

$$\frac{\theta_2(0,\omega)^4}{\theta_3(0,\omega)^4}, \text{ from which it follows that}$$

$$\eta(\omega) = 1^{\omega/24} \prod_{n=1}^{\infty} (1 - 1^{\omega n}) = q^{1/12} \prod_{n=1}^{\infty} (1 - q^{2n}),$$

where $q = 1^{\omega/2}$. Dedekind regretted that he could not represent η explicitly as a function of ω without invoking the theory of elliptic functions, a task later to be accomplished by Hurwitz using Eisenstein series. Dedekind had defined η as an infinite product in his earlier discussion of a fragment of Riemann's (see Riemann Werke, 466–478 and Dedekind's Werke XIII) where he also introduced the Dedekind symbol to elucidate the transformation properties of $\eta(\frac{\gamma + \delta\omega}{\alpha + \beta\omega})$.

At this point Dedekind drew upon this work in editing Riemann's papers to correct the mistake of Fuchs mentioned earlier. Riemann had considered the boundary values of elliptic modular functions [Riemann <u>Werke</u> XXVIII] and found that as $x + yi \to \frac{m}{n}$ along the vertical line $x = \frac{m}{n}$:

$$k \to \infty \quad \text{if } m \equiv n \equiv 1 \qquad \text{mod } 2,$$
$$k \to 1 \quad \text{if } m \equiv 0 \quad n \equiv 1 \quad \text{mod } 2,$$
$$k \to 0 \quad \text{if } m \equiv 1 \quad n \equiv 0 \quad \text{mod } 2.$$

Fuchs's mistake had been, as Schlesinger observed, not to distinguish sharply between a set which continuously fills out a curve from one which is merely everywhere dense in a given curve. Fuchs's method of considering k^2 via its inverse function is, of course, less direct and informative than the Riemann-Dedekind one.

Dedekind concluded his remarkable paper with a thorough discussion of the modular equation from his point of view. For an integer matrix $\begin{pmatrix} A & B \\ C & D \end{pmatrix}$ of determinant n, he set $v_n = \text{val} \frac{C + D\omega}{A + B\omega}$ and asked how many different functions v_n there can be. There are, he showed, with the now customary proof,

$$\psi(n) = n \prod_{p \mid n} (1 + \frac{1}{p})$$

such functions $v_1, \ldots, v_{\psi(n)}$, say. Then the function

$$F: \quad = \prod_{r=1}^{\psi(n)} (\sigma - v_r) = f(\sigma, v)$$

is a single-valued function of σ and ω which is a polynomial in σ of degree $\psi(n)$ with coefficients single-valued functions of $v = \text{val}(\omega)$. The $\psi(n)$ roots of this equation are the v_n's, which are therefore algebraic functions of v. He also showed that the polynomial F is irreducible and symmetric, and conjectured correctly that its coefficients are rational integers. In conclusion, he pointed out that a study of F would illuminate the theory of singular moduli and complex multiplication, but that further developments in which the composition of quadratic forms would play an essential role would have to wait for another opportunity[5].

Comments

One is struck by modernity of this paper. The concept of a function invariant under a certain group is clearly grasped, although to be sure the matrices $\begin{pmatrix} \alpha & \beta \\ \gamma & \delta \end{pmatrix}$ are not explicitly said to form a group. Even so Dedekind casually uses the fact which he stated (§6), that the matrices $\begin{pmatrix} \alpha & \beta \\ \gamma & \delta \end{pmatrix}$ are all expressible as products of $\begin{pmatrix} 1 & 1 \\ 0 & 1 \end{pmatrix}$ and $\begin{pmatrix} 0 & -1 \\ 1 & 0 \end{pmatrix}$. The idea that any single-valued function defined on F or, equivalently, automorphic under the matrices $\begin{pmatrix} \alpha & \beta \\ \gamma & \delta \end{pmatrix}$ and single-valued on S, is a rational function of $val(\omega)$ is clearly stated in the paper (§6) and underlies the discussion of $val(\omega)$ in §3 and of $\eta(\omega)$ in §6. Many of the arguments given by Dedekind occur virtually unchanged in modern presentations of this material, all his arguments have a conceptual clarity and depth seldom found in Fuchs, for example. To what extent are they novel?

The most crucial novelty is the presentation of the theory of modular functions and the relationship between different moduli almost entirely divorced from the theory of elliptic functions. What had previously been derived by astute use of the multitude of equations concerning infinite sums and products in Jacobi's theory was here rederived with a new lucidity and economy of ideas. Only the formula for η as an infinite product eluded Dedekind's reformulation. The theory of modular functions could now be studied independently of elliptic functions, a natural task which had been in the air for some time. In a passage quoted by Dedekind at the end of §1 Hermite had observed [1862 = 1908 II, 163n] ".... no other way for reaching the modular equations has yet offered itself than that which has been given by the founders of the theory of elliptic functions" and Klein in his [1878/79] sought to emphasize the implications these new ideas might have for the study of elliptic functions.

The new theory gave a geometric interpretation of modular functions. They were now obtained by an invariance principle from their definition on a fundamental region, and could be seen to be constrained naturally to the upper half-plane. Dedekind's study of Legendre's equations and the transformations of order two ($\eta_1 := \eta(2\omega)$, $\eta_2(\omega) := \eta(\frac{\omega}{2})$, $\eta_3 := (\frac{1+\omega}{2})$) and of order N, suggested that the existence of a rich theory of modular transformations could be obtained by extending the invariance idea, and that this should shed light on the hard-won but obscure results in the theory of moduli. Dedekind did not explicitly stress the concept of a group – that was Klein's decisive contribution – so a certain vagueness of terminology is in order.

The point of departure for Dedekind was the lattice of periods of an elliptic function. Parallelogram lattices had been studied in connection with quadratic forms by Gauss [1840 = Werke 1863, II, 194] and Dirichlet [1850 = Werke II, 1863, 194] whence Dedekind surely came to hear of them. He may well not have known of Kronecker's study "Über bilineare Formen mit vier Variabeln" presented to the Berlin Academy in 1866 but not published until 1883 [= Werke II, 425-495]. In §2 of that work Kronecker discussed the six cosets (as they would be called today) of SL(2; Z) when the entries are reduced mod 2, and applied his results to the reduction of quadratic forms $ax^2 + 2bxy + cy^2$. The stripping-down of elliptic function theory to the lattice idea allowed Dedekind (and Klein) to introduce modular functions as functions on lattices. A lattice has a basis ω_1, ω_2 such that $\omega_1/\omega_2 = \tau \in H$, the upper half plane, and any other basis ω_1, ω_2 is obtained from ω_1 and ω_2 by an element of $SL(2; Z)$. The ratio of the periods, τ, is thus determined only up to the action of this group, and the inverse function (eg. k^2) is invariant under the action. It is this realization of the importance for modular functions of a well-known result about lattices and quadratic forms that was Dedekind's starting point. He cited Dirichlet Zahlentheorie, 2nd ed., §65, which discusses quadratic forms of negative determinant[6].

The invariance of elliptic functions under transformations of the variable, e.g. $\wp(\omega) = \wp(\omega + \mu_1\omega_1 + \mu_2\omega_2) = z$ was, of course, a very well-known idea. But it had always been regarded as a generalization of the periodicity of the trigonometric functions; ω_1 and ω_2 were periods, corresponding to closed loops in the z-domain. Fuchs's presentation of the invariance of k^2 followed this approach exactly as did Weierstrass in his presentation of the theory of elliptic functions in the 1870's (subsequently published as Werke, V). Dedekind broke with it and took invariance under a group or family of transformations as his starting point. Ultimately the two approaches are not all that different; their unification via the theory of Riemann surfaces was the achievement of Klein, but the group-theoretic approach is at least as natural as the topological one, and was to prove to be the way historically towards the 'right' generalization of elliptic functions. In this work of Dedekind one can detect also

an attempt to prefigure what a Riemannian theory of functions for
arbitrary closed domains might be like, and what the role of the
'universal covering surface' might be. The elaboration of these
ideas was due, however, to Klein and his students. To understand it,
we must look first at the Galois theory of the modular equation as
it had been developed in the 1850's; this will be the theme of the
next two sections.

4.3 Galois theory, groups and fields

On 29 May 1832, the evening before his fatal injury in a duel,
Galois wrote the now-famous letter to his friend Ausguste Chevalier.
It begins

"My dear friend,

I have done several new things in analysis.

Some concerning the theory of equations; others integral

functions".   [Oeuvres, p.25].

In the theory of equations he described how the solvability of
an equation by radicals is connected to the solvability of the group
of permutations of its roots. If G is such a group (Galois did not
define a group, but this is the first use of the word in this sense)
and H a subgroup, then one has two (coset) decompositions

$$G = H + HS + HS' + \ldots$$
$$G = H + TH + T'H + \ldots$$

The decompositions were said by Galois to be proper ("propre")
if they coincide, i.e. if H is what is nowadays called a normal subgroup. If the group of the equation is successively decomposed until
no further proper decomposition is possible, then Galois said of the

indecomposable groups which result that;

"If these groups each have a prime number of permutations the equations will be solvable by radicals; otherwise not." (p.26)

Galois' proof of this theorem was found amongst his papers after his death, and published for the first time by Liouville in his Journal de Mathématiques for 1846. It will not be described here. In his letter, Galois went on to apply his theory of equations to the modular equations of elliptic functions, studied earlier by Abel and Jacobi.

"One knows that the group of the equation which has its roots the sine of the amplitude of the p^2-1 divisions of a period is this:

$$x_{k,\ell}, \ x_{ak+b\ell,ck+d\ell};$$

in consequence the corresponding modular equation will have as its group

$$x_{\frac{k}{\ell}}, \ x_{\frac{ak+b\ell}{ck+d\ell}}$$

in which k/ℓ can have the $p+1$ values

$$\infty,0,1,\ldots,p-1" \quad)p. \ 27)$$

He remarked that this group, let us call it G, has $(p+1)\ p(p-1)$ elements, and a normal subgroup, G' of size $\frac{1}{2}(p+1)\ p(p-1)$, consisting of the substitutions x_k, $x_{\frac{ak+b}{ck+d}}$ where $ad - bc$ is a quadratic residue mod p, but that G' has no proper decomposition $p \neq 2,3$. Furthermore, the degree of the modular equation can sometimes be reduced. It cannot be reduced below p, because then the prime p would not appear in the size of the group of the equation, but it can be reduced from p+1 to p. This can occur only if the group G has a (non-normal) subgroup H of size $(p+1) \frac{p-1}{2}$. For example, when p = 7 he paired the symbols ∞ and 0, 1 and 3, 2 and 6, 4 and 5 and denoted the group of substitutions obscurely as

$$x_k, \quad x_{a\frac{k-b}{k-c}}$$

where b and c are paired and a and c are either both residues or both non-residues mod p. He claimed this gives a subgroup of order (p+1) $\frac{p-1}{2}$ = 24; a clearer example will be given below. A similar reduction is possible when p = 5 or 11, and, Galois concluded,

"Thus, for the case of p = 5,7,11, the modular equation reduces to degree p.

"With complete rigour, this reduction is not possible in the higher cases." (p. 29)

The modern notation for these elements $\begin{pmatrix} a & b \\ c & d \end{pmatrix}$ a,b,c,d ∈ Z/pZ = F_p the field of p elements is quite irresistible. The action $k \mapsto \frac{ak+b}{ck+d}$ permutes the lines in the plane over this field, which have the p + 1 possible slopes 0,1,...,p−1,∞. The substitutions of H when p = 7, say, permute the pairs of lines with slopes 0 and ∞, 1 and 3, 2 and 6, 4 and 5 and so fill out an octahedral subgroup of G' of index 7. Because it is not normal in the whole group, conjugation represents the group G as acting on 7 elements[7].

(This was not all; let us note in passing that Galois concluded his letter with a staggering summary of the theory of Abelian integrals which remarkably foreshadows Riemann's work. (p. 30).)

Responses (1) Jordan.

The publication of Galois' work in Liouville's <u>Journal</u>
stimulated the emerging generation of mathematicians to try to
understand and apply it. This process has been described recently
by Wussing, and may be summarized as follows (deferring details of
the modular equation to the next section). There was an initial
period in which several authors, notably Betti, Kronecker, Cayley,
and Serret, filled in holes in Galois' presentation of the idea of a
group. These commentaries on Galois presented the connection between
group theory and the solvability of equations by radicals, and went
on to explore the solution of equations by other means, to be
described below. The group idea was elaborated in terms of permutations
of a finite set of objects, thus following Cauchy's presentation of
the theory of permutation groups in 1844-6, and Wussing refers to the
formulation as that of permutation groups. The elements are often
called substitutions or operations, they come with a set of objects
which they permute. The crucial presentations of permutations groups
were made by Jordan in his "Commentaire sur Galois" [1869] and his
<u>Traité des substitutions et des équations algébriques</u> [1870].

Jordan presented a systematic theory of permutation groups, in
terms of abstract properties such as commutativity, conjugacy,
centralizers, transitivity, normal subgroups and quotient groups,
group homomorphisms and isomorphism. It seems to me that Jordan came
close to possessing the idea of an abstract group. He remarked "One
will say that a system of substitutions forms a group (or a <u>faisceau</u>)
if the product of two arbitrary substitutions of the system belong to
the system itself." (<u>Traité</u>, 22, quoted in Wussing, 104) and he

spoke of isomorphisms (= "isomorphisme holoédrique") between groups
as one to one correspondences between their substitutions which
respect products (Traité, 56, Wussing 105). One might well regard
the use of words like "substitution" and the permutation notation
($|x,x'$ ... $ax + bx' + ...,$ $a'x + b'x' + ...|$, for example) as well-
adapted to their purpose, but not to be taken too literally. On the
one hand Jordan presented an extensive battery of technical concepts
of increasing power - we have already seen his use of Sylow theory
in the treatment of monodromy groups - on the other hand he analysed
in the Traité a wide range of situations in which groups could be
found permuting lines (the 27 lines on a cubic surface, the 28
bitangents to a quartic) and appearing as symmetry groups of the
configuration of the nine inflection points on a cubic and of Kummer's
quartic with sixteen nodal points. Jordan's capacity to articulate
a powerful theory of finite groups and to recognize them 'in nature'
surely argues for an implicit understanding of the group idea presented
for convenience only in the more familiar garb (to his audiences)
of permutation groups. This is not to deny the role of permutation
theoretic ideas in Jordan's work, indicated by the emphasis on
transitivity and degree (= the number of elements in the set being
permuted), but rather to indicate that ideas of composition and action
(as for example change of basis in linear problems) were prominent,
and could be seized upon by other mathematicians.

(2) Kronecker

Nonetheless, Wussing can point to a valid distinction in the
degree of abstraction between Jordan's work and the less well known
treatments of Kronecker and Dedekind. Wussing shows clearly that

Kronecker, who learned the new theory of solvability of polynomial
equations from Hermite and others during his stay in Paris in 1853,
was chiefly concerned to further the study of solvable equations.
He sought to construct all the equations which are solvable by
radicals, as for instance, the cyclotomic equations $x^p - 1 = 0$, p a
prime. Those equations in particular he called Abelian [Werke, IV,
6] and it seems that this is the origin of the designation Abelian
for a commutative group; the Galois group of a cyclotomic equation
permutes the roots cyclically. In a later work [1870], also relating
to number theory, Kronecker gave this abstract definition of a
finite commutative group (quoted in [Wussing, 47]:

"Let θ', θ'', θ''' ... be a finite number of elements so
constituted that from any two of them a third can be derived by means
of a definite operation. By this, if the result of this operation
is denoted by f, for two arbitrary elements θ', and θ'', which can
be identical to one another, a θ''' shall exist which equals $(f(\theta',\theta''))$.
Moreover, it will be the case that

$$f(\theta',\theta'') = f(\theta'',\theta')$$

$$f(\theta',f(\theta'',\theta''')) = f(f(\theta',\theta''),\theta''')$$

and also, whenever θ'' and θ''' are different from one another, $f(\theta',\theta'')$
is not identical with $f(\theta',\theta''')$.

"This assumed, the operation denoted by $f(\theta',\theta'')$ can be replaced
by the multiplication of the elements $\theta'\theta''$, if one thereby introduces
in place of complete equality a pure equivalence. If one makes use
of the usual sign for equivalence: ~, then the equivalence $\theta'.\theta'' \sim \theta'''$
is defined by the equation $f(\theta',\theta'') = \theta'''$."

Here the elements are abstract and not necessarily presented as permutations, but Kronecker was always concerned to use the group-idea to advance other domains of mathematics, chiefly number theory. He concentrated therefore on the study of the roots of equations, regarding them as given by some construction, and so increasingly elaborated his theory of the "Rationalitätsbereich", which may be regarded as a constructive presentation of field extensions of certain ground fields (usually the rationals, $\mathbb{Q}$). In this he resembles his contemporary, Dedekind, and indeed the resemblance seems to have been uncomfortable to Kronecker, who on occasion claimed priority for his theory of algebraic numbers over Dedekind's, and suggested that his ideas may have influenced Dedekind. Kronecker also delayed the publication of the important paper of Dedekind and Weber [1882] in the Journal für Mathematik for well over a year[8]. It seems likely that Dedekind was gradually discovering and publishing ideas which Kronecker had had earlier but had not brought forward to his (Kronecker's) own satisfication. Kronecker often referred to ideas he had had in the 1850's when finally writing some of them down in the 1880's. In view of the probable destruction of the Kronecker Nachlass (Edwards [1979]) the matter of priority is unlikely ever to be settled.

(3)   Dedekind

As Purkert has shown [1976], Dedekind also developed the idea of an abstract group in the context of Galois theory. He lectured on Galois theory at Göttingen in 1856-1858, although he published nothing on it until 1894, in the famous eleventh supplement to Dirichlet's lectures on number theory (4th edition). Purkert presents an updated manuscript of Dedekind, now in the Niedersächsischer Staats und

Universitätsbilbliothek zu Göttingen, which he dates entirely
plausibly at around 1857-1858. It goes far beyond the limited ideas
about groups published in his Werke, II, paper LX1. In brief, the
manuscript describes the following: the idea of permutations of a
finite set of objects is generalized to that of a finite abstract
group. For permutations, Dedekind showed

Theorem 1   If $\theta\theta' = \phi$, $\theta'\theta'' = \psi$, then $\phi\theta'' = \theta\psi$, or, more briefly,

$$(\theta\theta')\theta'' = \theta(\theta'\theta'')$$

Theorem 2   From any two of the three equations $\phi = \theta$, $\phi' = \theta'$,

$\phi\phi' = \theta\theta'$ the third always follows.

The subsequent mathematical arguments, however, "are to be
considered valid for any finite domain of elements, things, ideas
θ, θ', θ'', ..., having a composition $\theta\theta'$ of θ and θ' defined in any
way, so that $\theta\theta'$ is again a member of the domain and the manner of
the composition corresponds to that described in the two fundamental
theorems". [Purkert, 1976, 4].

The ideas of subgroup (Divisor) and coset decomposition of a
group G are defined, and a normal subgroup (eigentlicher Divisor) is
defined as one, K, satisfying

$$K = \theta_1^{-1}K\theta_1 = \theta_2^{-1}K\theta_2 = \ldots = \theta_h^{-1}K\theta_h$$

where the θ's are coset representatives for K (cf. Galois: "propre").
Dedekind showed that the cosets of a normal subgroup themselves
formed a group, in which K played the role of the identity element.
He went on to apply this theory to the study of polynomials; unhappily,
it is clear that the manuscript is incomplete, and a 'field-theoretic'
part is missing. However, a considerable amount of the theory of

polynomial equations survives, and is analysed by Purkert. One may
well suppose, incidentally, that Riemann knew of Dedekind's ideas
on the subject, although he never felt inclined to work in the area
himself. This lends a certain irony to Klein's[9] description of his
own work as a 'Verschmelzung Galois mit Riemann'.

(4) Klein

To return to the theme of the development of group theory in
the period up to the 1870's, there is one final source, the work of
Felix Klein. Klein learned about group theory from Jordan in 1870,
when he and Sophus Lie went jointly to Paris, a sojourn interrupted
in Klein's case by the outbreak of the Franco-Prussian war. It came
to represent the third part of his characteristic style of mathematics
throughout the 1870's, the others being the invariant theory he had
learned the previous year from Clebsch, and the geometric impetus
which had marked him from his earliest studies with Plücker in Bonn.
It is well-known that one of the earliest fruits of Klein's visit to
Paris was the successful study of non-Euclidean geometry from the
projective point of view, described in the two papers called "Über
die sogenannte Nicht-Euklidische Geometrie" [1871, 1873], and the use
of groups to classify geometries as described in the so-called
Erlanger Program. As for Klein's interest in group theory per se,
Freudenthal [1970, 226] has written with his character vigour that

"... Klein's interest in groups was always restricted to those
which came from well-known geometries, regular polyhedra or the non-
Euclidean plane. His point of departure was never a group to which
he associated a geometry - an operation of which the exploitation will

be reserved for E. Cartan... More and more, Klein's activity concerning groups recalls that of a painter of still life."

It is indeed true that Klein never considered group theory abstractly (to have done so would have been to run counter to his geometric and pedagogic inclinations) but Klein would surely have regarded his attitude to group theory as the very opposite of a painter of still life; one thinks of his disagreements with Gordan quoted in Chapter III. As to Freudenthal's passing reference to the non-Euclidean plane, I shall suggest in the final chapter how this might be otherwise expressed, the better to capture a weakness of Klein's thought.

It might be supposed that birational transformations of projective algebraic curves would be another source of group theoretic ideas, as for instance in a study of the group of birational automorphisms of a given curve. Birational transformations had been brought to the fore by Riemann in his study of algebraic functions [1857c], but it seems that they retained their original significance of changes in the equation of a given curve for some time. The study of their group theoretic implications will therefore be deferred until the next chapter on chronological grounds, and I shall turn from this synopsis of the history of group theory 1850-1870 to look at particular problems deriving from Galois' ideas as they were considered before Klein.

4.4 The Galois Theory of modular equations c. 1858.
Betti

The first to consider Galois' work on the reduction of the degree of the modular equation was Betti, [1853]. He considered the

equation for snw/p, where w is a period of the elliptic integral, so

$$\frac{w}{4} = \int_0^1 \frac{dx}{((1-x^2)(1-k^2x^2))^{\frac{1}{2}}} \quad , \text{ and p is a prime. This equation has}$$

p^2-1 roots $x_{m,n} := \text{sn}\left(\frac{mw + n\tilde{w}}{p}\right)$ where $\tilde{w}$ is the other period and m and

n are integers $0 \le m, n \le p$, $(m,n) \ne (0,0)$, and it is invariant under

the substitutions $\left\{ \begin{matrix} x_{m,n} \\ \\ x_{b'm+a'n, \ bm+an} \end{matrix} \right\}$ a, b, a', b' integers between

0 and p - 1, in his notation. These he said formed the group G of

the equation. G was the product of a group H whose elements were

$\left\{ \begin{matrix} y_{q,i} \\ \\ y_{pq,i} \end{matrix} \right\}$ and K, the group of an equation of degree p + 1, consisting

of the substitutions $\left\{ \begin{matrix} y_{q,i} \\ \\ y_{q, \ \frac{ai + b}{a'i + b'}} \end{matrix} \right\}$ or, letting t take the

values $0,1,...p-1, \frac{1}{0}$, K consists of the substitutions $\left\{ \begin{matrix} t \\ \\ \frac{at + b}{a't + b'} \end{matrix} \right\}$

of which there are p(p-1)(p+1). K is a product of a group of order

2 and one of order $\frac{1}{2}$ p(p-1)(p+1), consisting of the elements for

which ab' - ba' is a square mod p, which in turn has p subgroups of

order $\frac{1}{2}$(p-1)(p+1). Betti listed them explicitly when p = 5, defining

them by stating what permutation each element was of the six objects

0, 1, ..., ∞. Betti then showed that when p = 5, 7 or 11 it is

possible to pair off the objects 0, 1, ..., p-1, ∞ as Galois had

done, but that this is not possible for p > 11, and so was led to

claim, as his Theorems I and II: that the modular equation was not solvable by radicals but could be reduced from degree p+1 to p

when p = 5, 7, 11; and that the equation for $\text{sn}\frac{w}{p}$ decomposed into

p+1 factors of degree p-1 which were solvable by radicals when the

root of an equation of degree p+1 was adjoined, but that the

equation was not solvable by radicals. Its degree could come down

from p+1 to p when p = 5, 7, 11.

Hermite

Hermite published a decisive paper [1858], on the connection

between modular functions and quintic equations. He began by

observing that the general cubic equation can be put in the form

$x^3 - 3x + 2a = 0$, $a = \sin \alpha$, when it has the three solutions

$2\sin \alpha/3$, $2\sin \frac{\alpha+2\pi}{3}$, $2\sin \frac{\alpha+4\pi}{3}$. The general quintic equation can

likewise be reduced by root-extraction to

$$x^5 - x - a = 0.$$

a reduction Hermite called "the most important step...since Abel" and

attributed to Jerrard (Klein, in Vorlesungen über das Ikosaeder, II,

1, §2, argues convincingly that this reduction is originally due to

the Swedish mathematician E.S. Bring [1786]). In the reduction form

the quintic equation is readily solvable by modular functions, as

follows. As usual, let

$$K = \int_0^{\pi/2} \frac{d\theta}{(1-k^2\sin^2\theta)^{\frac{1}{2}}} \quad , \quad K' = \int_0^{\pi/2} \frac{d\theta}{(1-k'^2\sin^2\theta)^{\frac{1}{2}}}, k^2 + k'^2 = 1$$

and introduce ω by $e^{i\pi\omega} = e^{-\pi K/K'} = q$. Let λ be a modulus related

to k by a 5th order transformation, then, as Jacobi [1829] and

Sohnke [1834] had shown, $\sqrt[4]{k} = u$ and $\sqrt[4]{\lambda} = v$ are related by the modular equation

$$u^6 - v^6 + 5u^2v^2 (u^2 - v^2) + 4uv (1 - u^4v^4) = 0,$$

which is to be regarded as an equation for v containing a fixed but arbitrary parameter u. If, then, $u = \phi(\omega)$ and $v = \phi(\lambda)$, the solutions of the equation are

$- \phi(5\omega)$, and $\phi(\frac{\omega + 16m}{5})$, $m = 0, 1, 2, 3, 4$, which may be denoted v_∞, v_0, $v_1, \ldots, v_4$.

Hermite then defined

$$\Phi(\omega) := (\phi(5\omega) + \phi(\tfrac{\omega}{5}))(\phi(\tfrac{\omega+16}{5}) - \phi(\tfrac{\omega+4.16}{5}))(\phi(\tfrac{\omega+2.16}{5}) - \phi(\tfrac{\omega+3.16}{5}))$$

$$= - (v_\infty - v_0) (v_1 - v_4) (v_2 - v_3).$$

and found that $\Phi(\omega + r.16)$, $r = 0, 1, \ldots 4$ were the roots of a quintic equation for y:

$$y^5 - 2^4.5^3.y.u^4(1 - u^8)^2 - 2^6\sqrt{5}^5\ u^3(1 - u^8)^2(1 + u^8) = 0,$$

which can be further reduced, by means of the transformation

$$y = 2.4\sqrt{5}^3\ u. (1 - u^8)^{\frac{1}{2}} . t,\ to$$

$$t^5 - t - \frac{2}{\sqrt[4]{5}^5}\ \frac{1 + u^8}{u^2(1-u^8)^{\frac{1}{2}}},$$

thus exhibiting the quintic as a typical one in Bring–Jerrard form. The solution of a given equation $y^5 - y - a = 0$ is then known once u is found such that

$$\frac{2}{\sqrt[4]{5}^5} (\frac{1 + u^8}{u^2(1-u^8)})_{\frac{1}{2}} = a, \text{ which happily reduces to solving a}$$

quartic equation.

In Hermite's paper the crucial idea is to use the possibility
of reducing the modular equation at the prime 5, whose solutions
can be assumed to be known, to a quintic, whose solutions are thus
obtained, and to show that any quintic equation can be obtained in
this way. The impossibility of performing that reduction when $p > 11$
is not explained (Hermite admitted to this lacuna in the argument).
Referring to Betti in a footnote, he said that Betti had published
on the subject after his own first work had been done and the
results announced in Jacobi's Werke (first edition II, 249 [= 2nd
edition, 1969, II, 87-114]), while the present paper remained
unpublished.

Kronecker

Other mathematicians who concerned themselves with the modular
equations were Kronecker and Brioschi. Kronecker's work [1858a] on
the modular equation of order 7 was presented to the Berlin Akademie
der Wissenschaften by his friend Kummer on 22nd April 1858, shortly
after Hermite's work on the quintic was presented to the Paris
Académie. It contains a polynomial function of degree 7 in 7 variables,
which takes only 30 different values under permutations of the
variables, and which is in Kronecker's terminology eight-fold cyclic,
i.e. unaltered by 8 cyclic permutations[10]. It is clearly left
invariant by a group of order $7!/30 = 168$. In general if the seven
variables are the seven roots of a seventh degree polynomial equation
then that equation has a property "which is more general than its
solvability". This property Kronecker called its affect ('Affecte')
but did not define, beyond remarking that one property of an

affected equation is that all of its roots are rational functions of
any three roots. He remarked that all polynomial equations of degree
7 which can be obtained by reducing the modular equation of degree
8 had this affect, and conjectured that conversely all equations of
degree 7 with the affect could be solved in this way by equations
derived from the theory of elliptic functions. But, he went on,
"To prove this last seems indeed to be difficult; at least, I have
not brought my researches, which I started for this purpose two
years ago, and are concerned with the object of the present note,
to conclusion." He had, he said, been able to make a direct connection
between the quintic and its corresponding modular equation just as
had Hermite, but he could not see that if offered any application
to equations of degree 7.

On the 6th June 1858 Kronecker wrote to Hermite [1858b],
enclosing a copy of his note on the equation of degree 7, remarking
that it was already two years old but he had hoped to obtain more
general results before publishing. However, he felt certain only
that methods such as Jerrard's would not lead to the solution of
equation of higher degree, and so he had sought instead to get a
better grasp of the solution of the quintic by means of modular
functions. In this case he felt the crucial element was the
existence of functions of 6 variables which are 6 fold cyclic, of
which he gave examples leading to the solution of the quintic.

Brioschi

Independently of these two, Brioschi had joined in the chase,

also giving in his [1858] a solution of the quintic, derived from
his study of the multiplier equation[11] rather than the modular
equation. So one may say that the quintic equation was by then
solved, but that the generalization to higher degrees remained obscure.
It is likely that Brioschi also wrote to Hermite about this[12], at
all events Hermite's reply of 17 December 1858 was published in the
Annali di Mathematica (1859, vol II = Oeuvres, II, 83-86). Hermite
considered the reduction of the modular equation from the eighth
degree to the seventh, and gave two explicit forms for the solutions
of the seventh degree equation that arise. He expressed the group
of the equation in this way. For x an integer modulo 7, he let
$\theta(x) \equiv 2x^2 - x^5$ (mod 7) and considered the group of substitutions
(generated by)

$$Z_x, \ Z_{ax+b}, \ Z_{a\theta(x+b)+c} \text{ in his notation,}$$

where a is a quadratic residue mod 7 and b and c are arbitrary. The
group has $3.7 + 3.7^2 = 168$ elements. It gave one function of 7
letters having only 30 values, and the other explicit form of the
solution gave another, where in this case $\theta(x) \equiv -2x^2 - x^5$ (mod 7).
He showed the two systems of substitutions were conjugate. The
same issue of the Annali di Matematica carried Kronecker's
observation, also in the form of a letter to Brioschi [Werke IV, 51,
52]; that the two systems of functions of 7 letters were really the
same, being obtainable the one from the other by relabelling the
letters.

Betti also wrote to Hermite (24 March 1859), and Hermite
included the letter in one of his Comptes Rendus notes on modular

equations for that year [Hermite Oeuvres II, 73-75]. He was led to consider the subgroup of substitutions $\theta(K) = \frac{aK+b}{cK+d}$, a, b, c, d integers mod 7, consisting of a $\frac{K-3b}{K-b}$, $-$ a $\frac{K-b}{K-3b}$, aK, $-\frac{a}{K}$ where a and b are residues mod 7 (a group of order 24) with analogous results for the prime 11. Again he claimed that, for number-theoretic reasons, such subgroups could not be found for p > 11. The existence of these subgroups correspond to the reducibility of the modular equation.

So one may conclude by saying that by 1859 Betti, Brioschi, Hermite, Kronecker at least had caught up with Galois. The modular equations at the primes 5, 7 and 11 were connected to groups of order $\frac{1}{2}$ p(p-1)(p+1) = 60, 168, 660 respectively. The reducibility of those equations was connected to the existence of 'large' subgroups of these groups, of index p in each case. Hermite had also shown that the general quintic was solvable by modular functions, Kronecker had conjectured that the general polynomial equation of degree 7 was not, and Betti had a proof that reducibility stopped at p = 11. Finally Jordan in his [1868] and the Traité (348) gave a short proof of this result of Galois and Betti. These results were also published in J. A. Serret, Cours d'Algèbre [4th ed, 1879, 393-412], and most accessibly in Dickson, [1900, esp. p. 286].

4.5. Klein

The previous chapter (§3) discussed how Felix Klein approached the question of solving the algebraic-solutions problem from a group-theoretic point of view. The largest group which presented itself there, the group of the icosahedron, has a significance which derives from a mysterious unity between several problems.

First, there is the striking fact that the modular equation at the prime p = 5, can be reduced from degree six to degree five.

Second, there is Hermite's solution of the general quintic equation by means of modular functions. The occurrence of modular functions, while not unexpected, seemed to Klein to require a more profound explanation than the mere analogy with the trigonometric functions offered by Hermite. Equally, as Klein said, the studies of Kronecker and Brioschi "gave no general ground why the Jacobi resolvents of degree six are the simplest rational resolvents of the equations of the fifth degree" [1879a = 1922, 391]. (Jacobi resolvents are defined in Chapter V, n. 9.)

Third, there is the invariance of certain binary forms under the appropriate groups of linear substitutions discussed in Chapter III.

It seemed to Klein that a deeper study of the icosahedron would not only illuminate the underlying unity of these problems in elliptic function theory, the theory of equations, and invariant theory, but also suggest generalizations to modular equations of higher degree, to ternary and higher forms and to the study of function on general Riemann surfaces. He discussed the connection between the icosahedron and the quintic equation in [1875/76] and again in [1877a]. In the first paper he proceeded from the classification of the finite groups of linear substitutions in two

variables to an analysis of the covariants of the binary form
corresponding to the regular solids. In the second paper he
reversed the process, at Gordan's instigation[13], deriving the theory
of equations of the fifth degree from a study of the icosahedron.
If μ_1 and μ_2 are projective coordinates on the Riemann sphere, then
the twelve vertices of an icosahedron are specified by a binary form,
f, in μ_1 and μ_2 of degree twelve. Klein took as the fundamental
problem: given f and its covariants H (the Hessian of f) and T (the
Jacobian of f and H) as functions of μ_1 and μ_2, find μ_1 and μ_2 as
functions of f, H and T. This problem will not be pursued here,
beyond noticing that an inverse question is raised, and solved by
Klein.

He raised the connection between the icosahedron and transformations of elliptic functions in a third paper [1878/79a] which, as
he admitted overlaps considerably with Dedekind's paper discussed
above. This paper began the series of papers and books on elliptic
modular functions which form Klein's greatest contribution to
mathematics, and in which he and his students pioneered a geometric
approach to function theory allied to a 'field theoretic' treatment
of the classes of analytic functions which were brought to light.
The central element in this work is the geometric role of certain
Galois groups, which generalize the group of the icosahedron, and
are connected to appropriate Riemann surfaces (no longer necessarily
the Riemann sphere). The nature of this connection was to occupy
Klein deeply for several years. Klein's analysis of the role of the
icosahedron in the theory of transformations of elliptic functions

and modular equations will now be presented. The nature of the generalizations he made to further problems described in the next chapter.

In Section I of his [1878/79a] he drew together certain observations on elliptic functions which, he said, were not strictly new but seemed to be little known in their totality. The elliptic integral

$$I = \int \frac{dx}{\sqrt{f(x)}} \, , \quad f(x) = a_0 x^4 + \ldots + a_4$$

possesses two invariants which, following Weierstrass[14], he called

$$g_2 = a_0 a_4 - 4a_1 a_2 + 3a_2^2, \text{ and } g_3 = \begin{vmatrix} a_0 & a_1 & a_2 \\ a_1 & a_2 & a_3 \\ a_2 & a_3 & a_4 \end{vmatrix}$$

The discriminant of g_2 and g_3 he denoted Δ: $= g_2^3 - 27g_3^2$. For the absolute invariant he chose g_2^3/Δ and denoted it J, rather than the more usual g_2^3/g_3^2. The reason for Klein's choice was presumably that J has slightly simpler mapping properties than $g_2^3/g_3^2 = 27(\frac{1}{J-1})$. J can be written in terms of the cross-ratio, σ, of the four roots of $f(x)$, taken in some order; Klein gave

$$J = \frac{4}{27} \left(\frac{(1-\sigma+\sigma^2)^3}{\sigma^2(1-\sigma)^2} \right),$$

observing that it is not altered by the substitution of any other value of the cross-ratio σ, $\frac{1}{\sigma}$, $1-\sigma$, $\frac{1}{1-\sigma}$, $\frac{\sigma-1}{\sigma}$, $\frac{\sigma}{\sigma-1}$. It is therefore, in the terminology of Schwarz and himself, of the double pyramid type, the double pyramid in this case having 6 faces.

Klein also gave expressions for g_2, g_3, Δ, and J in terms of the ratio, ω, of the periods ω_1 and ω_2, of the elliptic integral I, based on some of the formula in Jacobi's Fundamenta Nova. In particular, setting $q = e^{-\pi i \omega}$,

$$\omega_2{}^{12}\sqrt{\Delta} = 2\pi \, q^{1/6} \prod_v (1-q^{2v})^2,$$

and in a footnote (§5n.8) Klein observed that $q^{1/6}\prod_v (1-q^{2v})^2$ was the square of Dedekind's function $\eta(w)$. Furthermore

$$g_2\left(\frac{\omega_2}{2\pi}\right)^4 = \frac{1}{12} + 20 \sum \frac{n^3 q^{2n}}{1-q^{2n}},$$

and so as q tends to zero J behaves like $\dfrac{1}{1728q^2}$.

To describe the mapping properties of J, Klein considered it as a function of ω, itself a function of the modulus k^2 of the integral. Indeed ω maps a half plane of k^2 onto a circular arc triangle and so the whole k^2-plane onto two adjacent triangles with all angles zero in the ω plane as in Figure 2. Klein has here unknowingly rediscovered Riemann's observation on k^2 as a function of ω (Appendix 2   ) but not published until 1902; Klein refers only to Schwarz [1872, 241, 2]. Further analytic continuation of ω extends its image by successive reflections in the sides of the triangles until the whole of the upper half plane is reached. In this way the logarithmic branching of ω as a function of k^2 is displayed, the branch points being $k^2 = 0$, 1, ∞. Now J can be described as a function of ω as follows. Each image of a k^2 half plane in the ω plane can be divided into six congruent triangles having angles $\frac{\pi}{2}$, $\frac{\pi}{3}$, 0 as shown (fig 3) and Klein said, each such small triangle is mapped by J onto a half plane. He called such a triangle

elementary, and the corresponding domain, mapped by J onto the

complex J-plane, an elementary quadrilateral (fig 4) and observed

that exactly this quadrilateral figure had been introduced by

Dedekind on purely arithmetic grounds (as a fundamental domain for

the group SL(2,Z)), although he, Klein, preferred to use the well

known results of elliptic function theory. In terms of the effect

 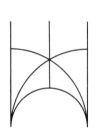 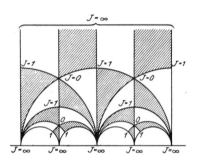

Fig 4.2,3 = [1878/79 figs 5.6]   Fig 4.4 = [1878/79, fig 7]

of substitutions $\omega' = \frac{\alpha\omega+\beta}{\gamma\omega+\delta}$ on the figure, Klein showed there are

elliptic substitutions, which he defined as those having fixed

points. There are fixed points of period 3 at $\rho = \frac{-1+\sqrt{-3}}{2}$ and

all equivalent points in H, where J is zero, and others of period two

with a fixed point at i, and all other equivalent points, where J = 1.

There are parabolic substitutions, whose fixed points are by

definition i∞ and all real, rational points, where J is infinite.

Finally he called all other substitutions hyperbolic (they have two

distinct fixed points on the real axis). It follows from Schwarz's

general considerations of the mapping of triangular regions onto a

half plane or plane that ω, as a function of J, satisfies the

hypergeometric equation with

$$\alpha = \frac{1}{12} = \beta, \ \gamma = \frac{2}{3} :$$

$$J(1-J) \ \frac{d^2 z}{dJ^2} + (\frac{2}{3} - \frac{7J}{6}) \ \frac{dz}{dJ} - \frac{z}{144} = 0, \text{ and Klein gave various forms}$$

for the solution of this equation and for the separate periods ω_1

and ω_2 as functions of J. This concluded the first section of his

paper.

In the second section he investigated the polynomial equations

which arise from transformations of elliptic functions. For a prime

p, a pth order transformation between two elliptic functions is a

transformation between a lattice of periods and one of its p+1

sublattices of order p, and it gives rise accordingly to a polynomial

equation of degree p+1 between the associated absolute invariants (J

and J' in Klein's notation),

$$\phi(J,J') = 0.$$

Klein observed that these equations, first obtained by F. Müller

[1867, 1872], are much simpler than the ones connecting the moduli

directly, and he proposed to study them geometrically, interpreting

J and J' as variables on two Riemann surfaces. He took J' so that

at the value of $\omega = \omega(J)$ the corresponding $\omega' = \omega'(J)$ took the values

$$\omega' = \frac{\omega}{p}, \ \frac{\omega+1}{p}, \ \ldots, \ \frac{\omega+(p-1)}{p}, \ \frac{-1}{p\omega} \ (p > 3).$$

J' is branched over J = 0, 1, and ∞, and Klein showed that

(i)   at J = 0, if p = 6m + 5 the p + 1 leaves are arranged in

           cycles of threes, but if p = 6m + 1, p - 1 leaves are

           arranged in cycles of three with two leaves left isolated;

(ii)   at J = 1, if p = 4m + 3 the leaves are joined in pairs but if p = 4m + 1 there are two isolated leaves;

(iii)   at J = ∞, J' = ∞, p leaves are joined in a cycle but one

leaf is isolated.

The genus of the transformation equation ϕ was calculated from
the formula

$$g = -p + \Sigma \frac{\sigma-1}{2}$$

where σ is the number of leaves cyclically interchanged at a branch
point and the sum is taken over all branch points. It depends
crucially on the value of p mod 12, because of the branching
behaviour, and Klein noted that for p = 5, 7, 13, g = 0;

and for p = 11, 17, 19, g = 1; etc.

He added that separate considerations showed that g was also zero
when p was 2 or 3 (§8–13).

The map from J' to J given by ϕ associates p+1 fundamental
quadrilaterals in the ω plane to each fundamental quadrilateral in
the ω' plane. The boundary of the new fundamental polygon is mapped
onto itself, yielding a closed Riemann surface, by means of the
substitutions $\omega' = \frac{\alpha\omega+\beta}{\gamma\omega+\delta}$ which identify points in the ω-plane having
the same J and J' values. In particular ω and ω/p must be equivalent,
which forces β to be divisible by p, so Klein considered those
substitutions $\begin{pmatrix} \alpha & \beta \\ \gamma & \delta \end{pmatrix}$ $\beta \equiv 0$ mod p. Two are parabolic $\begin{cases} \omega' = \omega + p \\ \omega' = \frac{\omega}{\omega + 1} \end{cases}$

two are elliptic, and the rest hyperbolic. When these identifications
are made, the genus g of the Riemann surface so obtainable can be
calculated from Euler's formula v + f - e = 2-2g, and the results

agree with the earlier calculation. This method also applies directly to the cases p = 2, 3, and even 4, as Klein showed.

When the genus is zero, J can be expressed as a rational function of a parameter τ. Klein discussed the case p = 7 in detail (§14) $J = \frac{\phi(\tau)}{\psi(\tau)}$, where ϕ and ψ are polynomials of degree 8. When J = ∞ 7 leaves are joined in a cycle and one is isolated, so ψ factorises into a simple and a sevenfold term. Let τ be so chosen that the sevenfold term vanishes at $\tau = \infty$, the simple term at $\tau = 0$, i.e. $\psi = c.\tau$ for some constant c. Similarly ϕ has the form

$$(\tau^2 + \alpha\tau + \beta)(\tau^2 + A\tau + B)^3$$

β is an arbitrary non zero constant, Klein chose $\beta = 49$. (β cannot be zero, for then ϕ and ψ would have a common factor.) Now $J - 1 = \frac{\phi-\psi}{\psi}$ should have four double zeros, arising from the nature of the branch point 1, so $\phi-\psi$ is the square of a quartic expression. This quartic must, furthermore, be a simple factor of the functional determinant $\begin{vmatrix} \phi & \psi \\ \frac{d\phi}{d\tau} & \frac{d\psi}{d\tau} \end{vmatrix}$. After a little work this leads to equations for α, A, and B, and Klein concluded that

$$\phi = (\tau^2 + 13\tau + 49)(\tau^2 + 5\tau + 1)^3$$
$$\phi - \psi = (\tau^4 + 14\tau^3 + 63\tau^2 + 70\tau - 7)^2$$
$$\psi = 1728\tau.$$

J' is likewise $\frac{\phi(\tau')}{\psi(\tau')}$ for some τ', and Klein asked for the relation between τ and τ'. It must, he said, be linear, $\tau' = \frac{a\tau+b}{c\tau+d}$, since τ and τ' are single-valued functions on the sphere, and of period 2 since repeating the transformation takes one from J' to J

and τ' to τ. Indeed $\tau\tau' = C$ for some constant C, since $\tau = 0$ or ∞
at the two points where $J = \infty$ but these points are interchanged in
the interchange of J and J'. Finally in fact $\tau\tau' = 49$ for, at the
zeros of $\tau^2 + 13\tau + 49$, both J and J' are zero so $\tau'^2 + 13\tau' + 49 = 0$.
So $\tau\tau' = 49$.

When $p = 5$ Klein found (§15) that:
$$J: J - 1: 1 = (\tau^2 - 10\tau + 5): (\tau^2 - 22\tau + 125)(\tau^2 - 4\tau + 1)^2: -1728\tau,$$
with J' being a similar function of τ, and $\tau\tau' = 125$.

The icosahedral equation

These considerations formed the basis of Klein's general
approach to the question of modular equations. In sections III and
IV of the paper he turned to consider the significance of the
icosahedral equation. In his terminology the Galois resolvent of
an algebraic equation is a polynomial in the roots which is not altered
by every element of the group of the equation. It seemed to Klein
to be a remarkable fact that all Galois resolvents containing a
parameter and of genus zero could not only be determined a priori
but indeed had also been determined, and were precisely those
equations which possessed a group of linear self-transformations.
The resolvents were, in short, either of the cyclic, dihedral,
tetrahedral, octahedral, or icosahedral type. If the parameter is J,
then the way the resolvent equation branched and the requirement
that the genus be zero restricts the resolvents which can be
connected to modular functions to be only of the tetrahedral, octahedral
icosahedral types. On the other hand the Galois group of the modular

(transformation) equation at the prime $n(> 2)$ is of order $\frac{1}{2}n(n^2-1)$.

J' is branched over $J = 0$, 1, ∞ in 3's, 2's, and n's respectively,

so the genus is $p = \frac{(n-3)(n-5)(n+2)}{24}$ which is zero only when $n = 3$ or 5.

It is zero for a similar reason when $n = 2$ or 4, but for no other

values.

Klein said (§4): "At $n = 3$, 4, 5 we have the same branching

which the tetrahedral, octahedral, and icosahedral equations display

.... These equations are thus the simplest forms which one can give

the Galois resolvents of the transformations for $n = 3$, 4, 5. In

this way the significance which above all the icosahedral equation,

to which my attention in this work is particularly directed,

possesses for the theory of transformations, is made as sharply

recognisable as one can." (emphasis in original).

In this spirit, Klein observed, the equation for the cross ratio

$\sigma = K^2$ of the roots of $(1 - x^2)(1 - k^2x^2)$ appears as the first of

a series of equations, for $J = \frac{4}{27}\frac{(1-\sigma+\sigma^2)^3}{\sigma^2(1-\sigma)^2}$ is obtained from

the equation $\mu^3 + \mu^{-3} = \frac{2J-4}{J}$ by the linear substitution

$$\mu = \frac{\sigma+\alpha}{\sigma+\alpha^2} \, , \, \alpha = e^{2\pi i/3}.$$

Since the genus of J' over J is zero the Riemann surface for

J' is a sphere branched over the complex J-sphere. The domains

mapping onto a hemisphere are triangles (since there are three

branch points) forming the familiar nets of the regular solids.

So in Klein's interpretation the modular equation at the prime

5 is naturally connected with the quintic, since in this case the

Riemann surface of J' over J is a sphere in which the faces of a naturally inscribed icosahedron are mapped onto the upper and lower half planes.

In the fourth and final section of the paper Klein explained the connection with the solution of the quintic equation by modular functions. He did not regard the occurrence of A_5 in the Galois group of the equations as the complete answer but sought to illuminate the question geometrically in the following way. Consider the quintic equation $x^5 + ax^4 + bx^3 + cx^2 + dy + e = 0$. The substitution $\tilde{x} = x + a/5$ reduces it to an equation of degree 5 in which the coefficient of $\tilde{x}^4$ is zero, so we need consider only equations of the form $x^5 + ax^3 + bx^2 + cx + d = 0$. Tschirnhaus was the first to consider a transformation of the form $\tilde{x} = \alpha + \beta x + \gamma x^2 + \delta x^3$. Eliminating x from these equations yields a quintic equation for x, in which the coefficient of $\tilde{x}^{5-k}$ is a homogeneous function of degree k in the indeterminates α, β, γ, and δ. In particular, the coefficients of $\tilde{x}^4$ and $\tilde{x}^3$ are, respectively, linear and quadratic in these indeterminates, and can both be made to vanish simultaneously. When this is done we obtain a quintic of the form $\tilde{x}^5 + \tilde{a}\tilde{x}^2 + \tilde{b}\tilde{x} + \tilde{c} = 0$. Let us prove these claims geometrically, following Klein.

Notice first that a quintic polynomial is specified completely by its five roots x_0, x_1, x_2, x_3, x_4, whose order does not matter. These five roots x_0, x_1, x_2, x_3, x_4 define in general 120 points in the projective space $\mathbb{C}P^4$, corresponding to the 120 arrangements $[x_0; x_1; x_2; x_3; x_4]$. Consider the lines joining them to $[1, 1, 1, 1, 1]$.

They meet the hyperplane H ($\Sigma x_i = 0$) in 120 points. Moreover, the group S^5 acts on $\mathbb{C}P^4$ by permuting the coordinates, the points [1, 1, 1, 1, 1] is fixed under the action, and the hyperplane H is mapped to itself. In the space $\mathbb{C}P^4/S^5$ the 120 points are all identified to a single point, and the 120 lines become a single line. The line is parameterized by $\lambda \in \mathbb{C}$, a typical point on it being $\lambda[1, 1, 1, 1, 1] + (1 - \lambda)[x_0, x_1, x_2, x_3, x_4]$, and for some suitable value of λ it meets the hyperplane H. The point where the line meets H corresponds to a polynomial for which $\Sigma x_i = 0$, i.e. the coefficient of x^4 vanishes, so we have performed the first reduction.

To eliminate the x^3 term is to obtain a polynomial for which Σx_i^2 is also zero. The locus $\Sigma x_i^2 = 0$ is a quadric, Q, which intersects H in a quadric hypersurface that is non-degenerate. Since H is isomorphic to $\mathbb{C}P^3$, we know that Q is doubly-ruled, by two families of lines, traditionally called the A-lines and B-lines. No A-line intersects any other A-line, no B-line another B-line, but each A-line meets each B-line in a unique point (see the exercises below). To perform this elimination, suppose you can find two points in H, say $[p_1, p_2, p_3, p_4]$ and $[q_1, q_2, q_3, q_4]$ whose coordinates are rational functions of $[x_0, \ldots, x_4]$ that are invariant under the action of A_5. The line joining these points meets Q $\cap$ H in two points, and these points correspond to two quintics of the form $x^5 + ax^2 + bx + c = 0$. (We obtain two quintics, corresponding to the quadratic aspect of the Tschirnhaus transformation.)

Next, one would like to eliminate the x^2 term from the quintic.
This can be done in the same way by introducing the cubic
hypersurface C: $\Sigma x_i^3 = 0$. However, C and Q meet in 3.2 = 6 points
of H, so we are led to a sextic equation for the elimination of the
x^2 term, which is not seemingly a worthwhile step to take when
studying quintics. But it would be if instead the process involved
only solving a cubic, and Bring's discovery is that this is indeed
the case. Once Bring's reduction is performed, the polynomial
equation takes the from $x^5 + ax + b = 0$, which can be further reduced
by setting $\tilde{x} = \rho x$ and suitably choosing ρ, to an equation of the
form

$$x^5 + x + \alpha = 0, \ \alpha \in \mathbb{C}.$$

What does it mean to have the solution to a polynomial equation?
It means that given, say $x^5 + ax^4 + bx^3 + cx^2 + dx + e = 0$ (*) one has
a 'function' ϕ of the coefficients a, b, c, d, e whose 5 values are
the roots of the equation. Such expressions are available as
algebraic functions when the polynomial is of degree 2, 3, or 4,
but for no other degrees. What Hermite provided was a transcendental
function of the coefficients which represented the solution to
quintics. More precisely, he accepted a reduction of the problem to
the study of quintics of the form $x^5 + x + \alpha = 0$ (†) and then
exhibited ϕ as a function of α. The reduction was an algebraic
expression which related the roots of (*) to the roots of (†). The
reduced situation is that the five roots of (†) form a five-sheeted
covering of the complex α-sphere, with S^5 as the corresponding
monodromy group.

The task Klein set himself was to make visible the action of S^5 as monodromy group. To this end he stopped short of Bring's reduction, and analysed the equation $x^5 + ax^2 + bx + c$ in terms of functions ϕ of a, b, and c. These functions were the invariants of the icosahedral group. Now, geometrically, this polynomial is represented by a point V on Q ∩ H. Such a point lies on an A-line and a B-line. Moreover, the action of S^5 maps Q to Q, so also Q ∩ H to Q ∩ H, and such maps send lines to lines since the action is projective. Elements of A_5 send A-lines to A-lines, B-lines to B-lines; elements which are odd switch them round. Now a line in Q ∩ H is a $\mathÇP^1$, so under the A^5 action V is sent to another point of $\mathÇP^1$, thus representing A^5 as maps

$$\frac{\lambda_1}{\lambda_2} \to \frac{a\lambda_1 + b\lambda_2}{c\lambda_1 + d\lambda_2} , \quad \frac{\lambda_1}{\lambda_2} = V \in \mathÇP^1.$$ In this way Klein showed

geometrically how the group A_5 enters the problem.

Klein returned to these questions, and treated them from a different point of view, in his famous book on the icosahedron [1884]. I have decided not to compare the treatment given there with the earlier one, but I am in the happy position of being able to refer the reader to the recent analysis of the book by J.-P. Serre, Extensions Icosaédriques, [1980].

Reduction of the modular equation

Klein devoted a short second paper in the Mathematische Annalen [1879] to the question of the reduction of the modular equation. In the cases p = 5, 7, 11 the genus of the reduced equation is zero so

J is a rational function of degree 5, 7 or 11 which Klein proceeded to calculate. Of interest here is the geometric and group theoretic approach Klein took to these problems.

He began, following Betti [1853], with the group which I shall denote $\bar{\Gamma}(p)$ of all 2 x 2 integer transformations $\omega' = \frac{\alpha\omega+\beta}{\gamma\omega+\delta}$ of determinant +1 reduced mod p (p = 5, 7, or 11), which has order 60, 168, or 660 respectively, and sought subgroups H_p of index p. These he listed explicitly. The (coset) representatives for each equivalence class of elements he took as

$$\omega, \omega + 1, \ldots, \omega + (p - 1).$$

He then let y be a function of ω invariant under H_p but not G, so y takes p different values for each value of J, the absolute invariant: $y(\omega)$, $y(\omega+1), \ldots, y(\omega + p - 1)$, and he regarded as a p-leaved Riemann surface over the J-plane. The branching of y over J was obtained as in the previous paper, but the fundamental polygon is now made up of p elementary quadrilaterals, not p + 1. The explicit form of J as a rational function was only calculated for p = 5 and 7. For example, when p = 5, the branch point J = ∞ is a cycle of all 5 leaves, J = 1 two cycles of 3, J = 0 a cycle of 3, so the genus is 0.

Accordingly J is a rational function of y, $J = \frac{\phi(y)}{\psi(y)}$, ϕ and ψ will be a degree of 5 and from the branching behaviour $\phi(y)$ contains a term $(y - 3)^3$, $\phi(y) - 1$ a square of a quadratic term, and so indeed one is led to an explicit relationship between y and J.

Klein did not show that it is only in these cases that the

modular equation is reducible by showing that it is only then that

subgroups of $\bar{\Gamma}(p)$ exist of index p. Rather, he showed that in

these cases it is quite easy to produce equations of degree 5 and

7 of the kind already studied by Brioschi and Hermite.

In the next chapter it will be seen that when Klein was able

to extend his methods to deal with the case of higher genera, he

was able to show that the modular equation of degree 8 had a certain

Galois group of order 168, but when reduced to an equation of degree

7 it was only a special case of such equations. So he solved

Kronecker's conjecture affirmatively: equations of degree 7 with

this Galois group are solvable by means of elliptic functions; and

he showed that other equations of degree 7 are not.

4.6 *A modern treatment of the modular equation*

Let $M = \begin{pmatrix} \alpha & \beta \\ \gamma & \delta \end{pmatrix}$ have integer entries and determinant $n \geq 1$, and

assume that the greatest common divisor of α, β, γ, and δ is 1.

Let $\mathcal{M}_n$ be the set of all such n. As usual, let $\Gamma = SL(2;\mathbb{Z})$. Γ

acts on the left on $\mathcal{M}_n$, $S \in \Gamma$ sends $M \in \mathcal{M}_n$ to SM. We say that M

and $\tilde{M} = SM$ are equivalent, elements of the space of equivalence

classes (or orbits) $\Gamma \backslash \mathcal{M}_n$ are written ΓM. Every M in $\mathcal{M}_n$ is

equivalent to a matrix of the form $\begin{pmatrix} \alpha & \beta \\ 0 & \delta \end{pmatrix}$, and $\begin{pmatrix} \alpha & \beta \\ 0 & \delta \end{pmatrix}$ and $\begin{pmatrix} \alpha_1 & \beta_1 \\ 0 & \delta_1 \end{pmatrix}$

in $\mathcal{M}_n$ are equivalent if $\begin{pmatrix} \alpha & \beta \\ 0 & \delta \end{pmatrix} = \begin{pmatrix} \pm 1 & 0 \\ 0 & \pm 1 \end{pmatrix}\begin{pmatrix} \alpha_1 & \beta_1 \\ 0 & \delta_1 \end{pmatrix}$. There are,

accordingly $\psi(n) = n \prod_{p|n} (1 + \frac{1}{p})$ equivalence classes in $\mathcal{M}_n$. So, in

particular, when n = 5 there are 6 classes:

$$\begin{pmatrix} 5 & 0 \\ 0 & 1 \end{pmatrix}, \begin{pmatrix} 5 & 1 \\ 0 & 1 \end{pmatrix}, \begin{pmatrix} 5 & 2 \\ 0 & 1 \end{pmatrix}, \begin{pmatrix} 5 & 3 \\ 0 & 1 \end{pmatrix}, \begin{pmatrix} 5 & 4 \\ 0 & 1 \end{pmatrix}, \begin{pmatrix} 1 & 0 \\ 0 & 5 \end{pmatrix}.$$

For each M in $\mathcal{M}_n$ the group Γ_M, which is defined to be $\Gamma \cap M^{-1} \Gamma M$, is now called a transformation group of order n. It is an infinite group, the word 'order' refers to the transformations of functions to be introduced below. For example, when

$$M = \begin{pmatrix} 5 & 0 \\ 0 & 1 \end{pmatrix}, \quad \Gamma_M = \left\langle \begin{pmatrix} \alpha & \beta \\ \gamma & \delta \end{pmatrix} \in \Gamma : \gamma = 0 \pmod 5 \right\rangle .$$

The group Γ_M depends only on the equivalence class of M; $\Gamma_M = \Gamma_{\tilde{M}}$ iff M and $\tilde{M}$ are equivalent under the action of Γ on $\mathcal{M}_n$.

There is a homomorphism of Γ onto a transitive permutation group which permutes the orbits ΓM in $\Gamma \backslash \mathcal{M}_n$:

$$S \to \sigma(S), \text{ where } \sigma(S)\Gamma M: = \Gamma MS.$$

The kernel of this map is $\Gamma^*(n) : = \bigcap_{M \in \mathcal{M}_n} \Gamma_M$.

The matrices $M \in \mathcal{M}_n$ are connected with modular transformations of order n as follows. Let f be a modular function, so

$$f(\gamma\tau) = f(\tau) \quad \forall \gamma \in \Gamma, \forall \tau \in H, \text{ the upper half-plane,}$$

then the transformation $f \to f_M$, where $f_M(\tau) := f(M\tau)$, $M \in \mathcal{M}_n$, is a transformation of order n. It depends only on the equivalence class of M, since

$$f_{SM}(\tau) = f((SM)\tau) = f(S(M\tau)) = f(M\tau) = f_M(\tau).$$

Thus, if K_Γ and K_{Γ_M} denote the modular functions and the functions automorphic for Γ_M respectively, one obtains a map $K_\Gamma \to K_{\Gamma_M}$, $f \to f_M$, and, as M runs through a set of inequivalent elements of $\mathcal{M}_n$, the functions f_M are all pairwise distinct. So one can define the polynomial

$$Q_n(x,f) := \prod_{(M)} (x - f_M),$$

where the product is taken over a complete set of inequivalent M, which is irreducible and of degree $\psi(n)$ in x. When the polynomial f is taken to be j, the Dedekind–Klein absolute invariant, the polynomial

$$P_n(x,j) := \prod_{(M)} (x - j_M)$$

is called the modular equation of degree n. It enjoys some remarkable properties: its coefficients are rational integers, it is of degree $\psi(n)$ in j, and indeed it is symmetric in x and j.

Since j_M is invariant under Γ_M, K_{Γ_M} is a Galois extension of K containing j_M and having Galois group Γ/Γ_M. So $K_{\Gamma*(n)}$ is the splitting field of the modular equation, being the smallest field containing all the j_M, and its Galois group is $\Gamma/\Gamma*(n)$. It is possible to describe its elements precisely, they are all the matrices $\begin{pmatrix} \alpha & \beta \\ \gamma & \delta \end{pmatrix} \equiv \begin{pmatrix} \alpha & 0 \\ 0 & \alpha \end{pmatrix}$ mod n in Γ. So, when n is prime, the index of $\Gamma*(n)$ in Γ is $\frac{\mu(n)}{2}$ where $\mu(n)$ is the index in Γ of the so-called principal congruence subgroup

$$\Gamma(n) := \left\langle \begin{pmatrix} \alpha & \beta \\ \gamma & \delta \end{pmatrix} : \begin{pmatrix} \alpha & \beta \\ \gamma & \delta \end{pmatrix} \equiv \begin{pmatrix} 1 & 0 \\ 0 & 1 \end{pmatrix} \text{ mod } n \right\rangle .$$

Indeed $\Gamma*(n) = \Gamma(n)/\{\pm 1\} = PSL(2;\mathbb{Z}/n\mathbb{Z})$.

Exercises on Chapter IV

1. Check that the group generated by $\begin{pmatrix} 1 & -2i \\ 0 & i \end{pmatrix}$ and $\begin{pmatrix} -1 & 0 \\ 2i & -1 \end{pmatrix}$

   consists of matrices of the form $\begin{pmatrix} a & 2bi \\ 2ci & d \end{pmatrix}$, a, b, c, d

   integers.

   The next few exercises, based on Serre [1973, VII], sketch the theory of the action of the modular group $PSL(2, Z)$ on the upper half plane, H.   Let $S = \begin{pmatrix} 0 & -1 \\ 1 & 0 \end{pmatrix}$ and $T = \begin{pmatrix} 1 & 1 \\ 0 & 1 \end{pmatrix}$ generate a group G'.   Let F be the region defined on p. 164.

2. Show that if $g = \begin{pmatrix} a & b \\ c & d \end{pmatrix} \in G'$ then $Im(gz) = \dfrac{Im(z)}{|cz + d|^2}$.

   Deduce that there is a $g \in G'$ for which $Im(gz)$ is a maximum. Deduce that there is an integer n such that $T^n g(z) \in F$.

3. Show that if $z \in F$ and $gz \in F$, $g \in G'$, then either $Re(z) = \pm\dfrac{1}{2}$ and $gz = z \pm 1$, or $|z| = 1$ and $gz = -1/z$. It is possible to suppose $Im(gz) \geq Im(z)$.

4. Show that points $z \in F$ have trivial stabilizers $S(z) = \{g \in G: gz = z\}$ unless $z = i$, $S(i)$ $= \{I, S\}$,

   $\qquad\qquad\qquad z = \rho$, $S(\rho)$ $= \{I, ST, (ST)^2\}$

   $\qquad\qquad\qquad z = -\bar{\rho}$, $S(-\bar{\rho})$ $= \{I, TS, (TS)^2\}$

5. Show that $PSL(2; Z) \subseteq G'$ by considering z in the interior of F and looking at gz.   Deduce that S and T generate $PSL(2; Z)$.

6. Show that PSL(2; Z) has a presentation as $\langle S, T: S^2, (ST)^3 \rangle$, so it is a free product of a cyclic group of order 2 with a cyclic group of order 3.

7. Show that the group
$$\left\{ \begin{pmatrix} a & b \\ c & d \end{pmatrix} : a, b, c, d \in Z, ad - bc = 1, b \equiv c \equiv 0 \bmod 2 \right\}$$
is a free group on two generators.

8. Schläfli's method (p. 159) is best illustrated by an example.
$$\begin{pmatrix} 9 & 22 \\ 2 & 5 \end{pmatrix} = \begin{pmatrix} 0 & 1 \\ 1 & 0 \end{pmatrix} \begin{pmatrix} 0 & 1 \\ 1 & 4 \end{pmatrix} \begin{pmatrix} 0 & 1 \\ 1 & 2 \end{pmatrix} \begin{pmatrix} 0 & 1 \\ 1 & 2 \end{pmatrix}.$$

The 0, 4, 2, 2 are obtained by this procedure: in $\begin{pmatrix} a & b \\ c & d \end{pmatrix} = \begin{pmatrix} 9 & 22 \\ 2 & 5 \end{pmatrix}$ is $|d| > |c|$? Since in this case the answer is 'yes', write $\frac{d}{b} = \frac{5}{22}$ as continued fraction with purely even entries.

$$\frac{5}{22} = 0 + \frac{1}{22/5} = 0 + \frac{1}{4+2/5} = 0 + \frac{1}{4+1/5/2}$$

$$= 0 + \cfrac{1}{4 + \cfrac{1}{2 + \cfrac{1}{2}}}$$

There are the 0, 4, 2, 2. Had $|d|$ been less than $|c|$, the algorithm says work out $\frac{c}{a}$, and write a 0 at the end.

(i) Consider $\begin{pmatrix} 5 & 6 \\ 4 & 5 \end{pmatrix}$

$$\frac{6}{5} = 0 + \cfrac{1}{2} + \cfrac{1}{-2} + \cfrac{1}{2} + \cfrac{1}{-2} + \cfrac{1}{2}$$

and indeed $\begin{pmatrix} 5 & 6 \\ 4 & 5 \end{pmatrix} = \begin{pmatrix} 0 & 1 \\ 1 & 0 \end{pmatrix} \begin{pmatrix} 0 & 1 \\ 1 & 2 \end{pmatrix} \begin{pmatrix} 0 & 1 \\ 1 & -2 \end{pmatrix} \begin{pmatrix} 0 & 1 \\ 1 & 2 \end{pmatrix} \begin{pmatrix} 0 & 1 \\ 1 & -2 \end{pmatrix} \begin{pmatrix} 0 & 1 \\ 1 & 2 \end{pmatrix}$

(ii) $\begin{pmatrix} 9 & 2 \\ 22 & 5 \end{pmatrix}$ requires us to write $\frac{22}{9} = 2 + \frac{1}{2+} \frac{1}{4}$, so the

numbers are 2, 2, 4, 0 and $\begin{pmatrix} 9 & 2 \\ 22 & 5 \end{pmatrix} = \begin{pmatrix} 0 & 1 \\ 1 & 2 \end{pmatrix}\begin{pmatrix} 0 & 1 \\ 1 & 2 \end{pmatrix}\begin{pmatrix} 0 & 1 \\ 1 & 4 \end{pmatrix}\begin{pmatrix} 0 & 1 \\ 1 & 0 \end{pmatrix}$.

Find e.g. $\begin{pmatrix} 9 & 10 \\ 8 & 9 \end{pmatrix}$ as a Schläfli product.

Since $\begin{pmatrix} 0 & 1 \\ 1 & 0 \end{pmatrix}\begin{pmatrix} 0 & 1 \\ 1 & 2 \end{pmatrix} = \begin{pmatrix} 1 & 2 \\ 0 & 1 \end{pmatrix}$ and $\begin{pmatrix} 0 & 1 \\ 1 & 2 \end{pmatrix}\begin{pmatrix} 0 & 1 \\ 1 & 0 \end{pmatrix} = \begin{pmatrix} 1 & 0 \\ 2 & 1 \end{pmatrix}$ Schläfli's

method permits one to write every matrix $\begin{pmatrix} a & b \\ c & d \end{pmatrix}$ $a \equiv d \equiv 1(4)$,

$b \equiv c \equiv 0(2)$, $ad - bc = 1$, as a product of $\begin{pmatrix} 1 & 2 \\ 0 & 1 \end{pmatrix}$ and

$\begin{pmatrix} 1 & 0 \\ 2 & 1 \end{pmatrix}$. Prove that his method must always work by showing

that these last 2 matrices generate $\bar{\Gamma}(2)$. It is enough

in doing this to show that exactly one product $\begin{pmatrix} 1\pm2 \\ 0 & 1 \end{pmatrix}\begin{pmatrix} a & b \\ c & d \end{pmatrix}$

and $\begin{pmatrix} 1 & 0 \\ \pm2 & 1 \end{pmatrix}\begin{pmatrix} a & b \\ c & d \end{pmatrix}$ lowers $|a|$ if $|a| > |c|$, or $|c|$ if

$|a| < |c|$ (why?). In fact, this observation shows that

$\bar{\Gamma}(2)$ is the free product of $\begin{pmatrix} 1 & 2 \\ 0 & 1 \end{pmatrix}$ and $\begin{pmatrix} 1 & 0 \\ 2 & 1 \end{pmatrix}$. Confirm

your earlier calculations by showing

$$\begin{pmatrix} 9 & 22 \\ 2 & 5 \end{pmatrix} = \begin{pmatrix} 1 & 4 \\ 0 & 1 \end{pmatrix}\begin{pmatrix} 1 & 0 \\ 2 & 1 \end{pmatrix}\begin{pmatrix} 1 & 2 \\ 0 & 1 \end{pmatrix},$$

$$\begin{pmatrix} 5 & 6 \\ 4 & 5 \end{pmatrix} = \begin{pmatrix} 1 & 2 \\ 0 & 1 \end{pmatrix}[\begin{pmatrix} 1 & 0 \\ -2 & 1 \end{pmatrix}\begin{pmatrix} 1 & 2 \\ 0 & 1 \end{pmatrix}]^2,$$

$$\begin{pmatrix} 9 & 2 \\ 22 & 5 \end{pmatrix} = \begin{pmatrix} 1 & 0 \\ 2 & 1 \end{pmatrix}\begin{pmatrix} 1 & 2 \\ 0 & 1 \end{pmatrix}\begin{pmatrix} 1 & 0 \\ 4 & 1 \end{pmatrix},$$

$$\begin{pmatrix} 9 & 10 \\ 8 & 9 \end{pmatrix} = [\begin{pmatrix} 1 & 2 \\ 0 & 1 \end{pmatrix}\begin{pmatrix} 1 & 0 \\ -2 & 1 \end{pmatrix}]^4\begin{pmatrix} 1 & 2 \\ 0 & 1 \end{pmatrix}.$$

9. Every quadric Q in $\mathbb{C}P^3$ can be diagonalised so that it is

described by an equation of the form $y_0 y_3 = y_1 y_2$. It follows

that Q is the image of $\mathbb{C}P^1 \times \mathbb{C}P^1 \to Q \subset \mathbb{C}P^3$ by $[x_0, x_1] \times [x_2, x_3] \to$

$[x_0 x_2, \ x_0 x_3, \ x_1 x_2, \ x_1 x_3]$. For a fixed $[x_2, x_3]$ the image is a projective line in Q, called an A-line, and for a fixed $[x_0, x_1]$ the image is a projective line in Q called a B-line. Show that no A-line meets any other A-line, nor any B-line another B-line, but that each A-line meets each B-line in a unique point.

CHAPTER V   SOME ALGEBRAIC CURVES.

This chapter discusses a topic which was studied from various
points of view throughout the nineteenth century and which presented
itself in such different guises as: the 28 bi-tangents to a quartic
curve, the study of a Riemann surface of genus 3 and its group of
automorphisms, and the reduction of the modular equation of degree 8.
These studies, which began separately, were drawn together by Klein
in 1878 and proved crucial to his discovery of automorphic functions.

It is only possible to sketch the early developments of each
part of this topic in the space available; I hope to return to the
matter more fully elsewhere. This treatment is divided schematically
into two parts: the first on algebraic curves, particularly quartics,
and Riemann surfaces; the second on the modular equation.

5.1 Algebraic curves, particularly quartics.

An algebraic plane curve is, by definition, the locus in the
plane[1] corresponding to a polynomial equation of some degree,
n: $f(x,y) = 0$. For example $f(x,y) := x^3y + y^3 + x = 0$ represents a
quartic. The equation may be written in homogeneous coordinates
$(x;y;z)$ by defining $F(x;y;z) = z^n f(\frac{x}{z}, \frac{y}{z})$, when the curve is considered
to lie in the projective plane. The example above becomes
$F(x;y;z) = x^3y + y^3z + z^3x = 0$ in homogeneous form. In nineteenth
century usage an algebraic curve of degree n was often called a C_n.

The study of higher plane curves, as C_n's were called when $n > 2$,
goes back at least as far as Newton, who in 1667-68 made a thorough
study of cubics Newton([M.P. II, 10-89]), but for present purposes a

start can be made with Plücker, who, in his System der analytischen

Geometrie [1834, 264], showed that every C_n in the projective plane

has in general $3n(n-2)$ inflection points. By 'in general' he meant

that the C_n has no multiple points or cusps. His argument was that

at the inflection points of a curve $f(x,y) = 0$, a line $x = \kappa y + \gamma$ has

three-fold intersection with the curve. So $\dfrac{d^2}{dy^2} f(\kappa y + \gamma, y) = 0$,

ie $\dfrac{\partial^2 f}{\partial y^2} \kappa^2 + 2\kappa \dfrac{\partial^2 f}{\partial x \partial y} + \dfrac{\partial^2 f}{\partial x^2} = 0$, where κ is homogeneous of order n-1,

since it satisfies $\dfrac{\partial f}{\partial y} \kappa + \dfrac{\partial f}{\partial x} = 0$ (any three-fold intersection is

automatically two-fold). Accordingly the true   inflection points

lie at the intersection of $f(x,y) = 0$, of degree n, and $\dfrac{d^2 f}{dy^2} = 0$ of

degree 3n-4. Of these   he showed, however, that 2n are only points

at infinity, so there are $3n(n-2)$ inflection points altogether. He

also showed (p.283) that the 9 inflection points of a cubic curve lie

on four systems of three lines, each line containing three points,

and pointed out that, as a result, only three of the inflection

points can be real[2].

Some of these theorems were subsequently proved again in Hesse

[1844] in a way which enables the geometric significance of the

Hessian to be explained; it was in this context indeed that Hesse

introduced it[3]. Adjacent normals to a curve will meet at the

appropriate centre of curvature, and Hesse argued, as had Plücker,

that at a point of inflection the adjacent normals will be parallel,

and the corresponding radius of curvature infinite. Its reciprocal,

the mean curvature, is therefore zero, but this just is

$\dfrac{\partial^2 f}{\partial x^2} \dfrac{\partial^2 f}{\partial y^2} - (\dfrac{\partial^2 f}{\partial x \partial y})^2$. Hesse introduced homogeneous coordinates x_1, x_2, x_3

to simplify the treatment of the points "at infinity" (p.131), and so

found that the Hessian $|\dfrac{\partial^2 F}{\partial x_i \partial x_j}|$ was of degree 3(n-2), if f is of

degree n, and the Hessian of f meets f in its $3n(n-2)$ inflection points.

Poncelet had suggested in his [1832] that a C_n could have only

finitely many bitangents (lines which touch the curve in two distinct

places). It seems that this theorem was first proved by Jacobi [1850]

and the number found to be $\frac{1}{2}n(n-2)(n^2-9)$ in general. This calculation

solved an intriguing problem raised by Plücker in his Theorie der

algebraischen Curven [1839, Ch 4], as Jacobi explained. If $F(x;y;z) = 0$

is the equation of a curve, C_n, of degree n, and $(x_0;y_0;z_0) = p$ is a

point on the curve, then the tangent to the curve at that point has

the equation $x\frac{\partial F}{\partial x}(p) + y\frac{\partial F}{\partial y}(p) + z\frac{\partial F}{\partial z}(p) = 0$. Conversely, a tangent

through the point $(x_1;y_1;z_1)$ touches the curve at a point p which

satisfies the same equation, $x_1\frac{\partial F}{\partial x}(p) + y_1\frac{\partial F}{\partial y}(p) + z_1\frac{\partial F}{\partial z}(p) = 0$. Given

F and (x_1,y_1,z_1), the locus of points P satisfying this equation is

called the first polar of F with respect to the given point; it is a

curve of degree n-1. As such it meets C_n in n(n-1) points, so one

immediately obtains the result that through any given point there are

in general n(n-1) tangents to a given curve. Following Möbius and

Poncelet, nineteenth century geometers invoked a principle of duality,

so that the tangents were regarded as points in a dual projective

space. Algebraically this can be done by regarding projective space

as made up of lines and interpreting each equation as determining an

envelope. Geometrically this can be done by picking a circle in the

plane and replacing each line by its polar with respect to the circle.

Either way, the n(n-1) tangents become n(n-1) points on the dual

curve, to be defined, and because the tangents all passed through

the point $(x_1;y_1;z_1)$ the n(n-1) points lie on a line (the dual of

$x_1;y_1;z_1$)). So the dual curve will have degree n(n-1). To obtain it

one makes the original point $(x_1;y_1;z_1)$ run along the curve, and

considers the tangents at each point; their polars define the dual

curve. Plücker's paradox is this: plainly the dual curve of the dual curve is the original curve. Yet the degree of the dual to a C_n is n(n-1), so the degree of the dual of the dual is n(n-1)[(n(n-1)-1], which is not n unless n=2.

Plücker's solution rested on two observations. First, a tangent to a curve is simply a line meeting it in two coincident points, so any line through a double point is a tangent, and the first polar therefore passes through the double point. Consequently, the number of lines through a given point which are truly tangent to a curve is diminished by 2 for each double point, for if the curve is regarded as: and there are two spurious tangents: and (Fig 5.2) Second, if the original curve has a cusp then the first polar not only passes through the cusp but is tangent there. So each inflection point reduces the degree of the dual by 3. Plücker argued accordingly that the dual curve should have α double points and β cusps, where $2\alpha + 3\beta = n(n-1)[n(n-1)-1] - n = n^3(n-2)$. Since the dual of a double point is a bitangent and of a cusp is an inflection point, Plücker could also speak of the α bitangent points and β inflection points of the original curve. He then stated that

$$\alpha = \frac{1}{2}n(n-2)(n^2-9),$$

$$\beta = 3n(n-2),$$

[1839, §330]

arguing that β was already known to be 3n(n-2). These formulae connecting the degree of the dual curve with the degree of the old curve and the number of its double points and cusps are nowadays known as Plücker's formulae.

Jacobi regarded the formula for α as more of a conjecture in need of a proof, and proved it directly using the condition that the equation which a line must satisfy in order to be a tangent (to the given C_n) has a repeated root when the line is a bitangent. This condition yields an equation of degree $(n-2)(n^2-9)$ and hence a curve meeting the original C_n in $n(n-2)(n^2-9)$ points, the points of contact of $\frac{1}{2}n(n-2)(n^2-9)$ bitangents[4]. So. for example, there are 28 bitangents to a quartic curve, a result obtained earlier by Hesse [1848, §3].

Plücker had made a study of the bitangents to a quartic in his [1839, Chapter V]. If two of the bitangents are chosen as axes the equation of the curve may be written in the form $pqrs - \mu\Omega_2^2 = 0$, where p,q,r, and s are linear terms, Ω_2 is a quadratic term, and μ is a constant. Plücker claimed that this reduction may be performed in $\frac{28.27.26}{2.3.4} = 819$ ways, but went on to deduce incorrectly the number of conics meeting the quartic at the eight bitangent points associated to p,q,r, and s. He correctly established that all 28 bitangents may be real [1839, §115] by considering deformations of the curve

$$\Omega_4 = (y^2 - x^2)(x - 1)(x - \tfrac{3}{2}) - 2(y^2 - x(x - 2))^2 = 0 \text{ to } \Omega_4 \pm \kappa = 0.$$

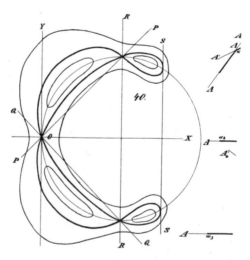

Fig. 5.3 [Plücker, Theorie, 247, Fig. 40.]

The curve made up of the 4 meniscas has 28 real bitangents. He also showed that quartics may be found having 16 or 8 real bitangents but no other values (greater than 8) (§122).

Jacob Steiner took up the study of quartics in 1848. He was an enthusiast for synthetic methods; Kline [1972, 836] records that he threatened to stop submitting articles to Crelle's Journal if his friend Crelle continued to publish Plücker's analytic papers. He also preferred to withhold the proofs of his theorems, leaving them as challenges for his colleagues. On this account, Cremona, who responded particularly to Steiner's work on cubic surfaces, called him "this Celebrated Sphinx".

In his papers [1848] and [1852] Steiner considered the inter-relationships of the 28 bitangents, and stated several conclusions. These include the following two[5]:

The $\frac{1}{2}(28.27) = 378$ pairs of bitangents may be grouped into 63 groupings ("Gruppen") of 6 distinct pairs so that the 6 intersection points of each pair lie on a conic; and the eight contact points of any two pairs of bitangents in the same grouping lie on a conic, so there are $\frac{1}{2}.6.5 = 15$ conics associated to each grouping. There are $\frac{1}{3}.63.15 = 315$ such conics altogether, since each is counted 3 times. (4 bitangents yield 3 sets of two pairs.)

The study of the 28 bitangents brought to light a great deal of rich mathematics which even the complexity of this survey can scarcely suggest. Their intimate connection with the 27 lines on a cubic surface was first made plain by Geiser [1869], who argued that given a cubic surface S and a point P not on it, the tangents through P to S form a cone of degree 6 (any plane through P cuts S in a cubic curve, to which there are n(n - 1) = 6 tangents from P). If P is on S the cone becomes the tangent plane, E, at P to S and a quartic surface, Q. Any plane through Q cuts it in a quartic curve - take the plane E and consider the curve C = E ∩ Q. If g is one of the 27 lines on S then any plane e through g meets S again in a conic which meets g in two points, P_1 and P_2. The lines PP_1 and PP_2 lie in Q, and define a plane which, furthermore, cuts E in a line that is a bitangent to C. In this way the 27 lines on S give rise to 27 bitangents on C, the 28th is obtained from the principal tangents to S at P. A more complicated argument enabled Geiser to show that the 27 lines can be obtained from the 28 bitangents[6].

The group-theoretic aspects of this connection were explored by Jordan in his Traité [1870, 329-333]. These configuration of special points on a plane quartic were of much interest in the 1850's and 1860's, and it was possible to look forward to a rewarding study of higher plane curves if only suitably powerful enough techniques could be developed. The next crucial development was to come from Riemannian function theory, with the introduction of the generalization of elliptic functions (which are appropriate to cubic equations) to Abelian functions. An indication of how this development was connected by Jordan to the work of Hesse is given in Exercises 15 and 16 at the end of the Chapter.

5.2 *Function-theoretic Geometry.*

The studies of Hesse and Steiner showed how intricate the geometry of plane curves could be. At about the same time, Riemann was developing the function theory of such curves along the lines of his Inaugural dissertation. The basic problem in the subject was to understand the integral of a rational function $R(x,y)$,

$$\int_0^z R(x,y)\,dx,$$

on an algebraic curve $F(x,y) = 0$. Jacobi had shown [1834] that if F is a non-singular curve of degree greater than three then there will be more than 2 periods to such an integral, and so it does not define a tractable function of its upper end-point. In the simplest case, where $F(x,y) = y^2 - f(x)$, and f is of degree 5 or 6, he had suggested taking pairs of integrals and pairs of end-points together. This approach was successfully carried out by Göpel [1847] and Rosenhain [1851] independently and was then generalized magnificently by Weierstrass [1853, 1856] to the case $F(x,y) = y^2 - R(x) = 0$, $R(x) = (a_o - x) \ldots (a_{2n} - x)$ being of degree 2n+1, and the a_i's being constant. By analogy with the elliptic integrals $\int \frac{dx}{\sqrt{\text{cubic}}}$, this case was called hyperelliptic; Weierstrass's analysis of it did much to secure his invitation to Berlin. His method was quite novel: he used Abel's theorem and a system of simultaneous differential equations to study n functions each of n variables obtained by inverting sums of n integrals, ie

$$U_i = \int_{a_1}^{x_1} \phi_i(x)\,dx + \int_{a_3}^{x_3} \phi_i(x)\,dx + \ldots + \int_{a_{2n-1}}^{x_{2n-1}} \phi_i(x)\,dx,$$

where $\phi_i(x) = \dfrac{(x-a_1)(x-a_2)\ldots(x-a_n)}{2(x-a_{2i+1})\sqrt{R(x)}}$, $1 \leq i \leq n$.

The integrands only vary from row to row. As Weierstrass showed, the special case of elliptic functions looks like this in his approach. The integral

$$u = \int_0^x \frac{dx}{[(1 - x^2)(1 - k^2 x^2)]^{\frac{1}{2}}}$$

defines the function x = sn(u), which satisfies the differential

equation $\dfrac{d^2 \log x}{du^2} = k^2 x^2 - \dfrac{1}{x^2}$. Conversely, starting from the

differential equation, the substitution $x = P_1/P$ results in the

equation $\dfrac{d^2 \log P_1}{du^2} - \dfrac{d^2 \log P}{du^2} = k^2 \dfrac{P_1^2}{P^2} - \dfrac{P^2}{P_1^2}$, which can be factorized to

give two equivalent equations

$$\frac{d^2 \log}{du^2} P_1 = \frac{-P^2}{P_1^2} , \qquad \frac{d^2 \log}{du^2} P = -\frac{k^2 P_1^2}{P^2} .$$

These equations, Weierstrass showed, can be solved to yield uniformly

convergent power series expansions of P and P_1 on some suitable

domain, and if one fixes the initial conditions at u = 0:

$$P = 1, \frac{dP}{du} = 0, \ P_1 = 0, \ \frac{dP_1}{du} = 1 ,$$

then indeed snu = $\dfrac{P_1}{P}$. [Weierstrass 1856 = Werke, I, 297]

When R is of degree 2n+1 the differential equations gave

Weierstrass uniformly convergent power series in n variables for n

functions which he called $Al_i(x_1, \ldots, x_n)$, $1 \le i \le n$, in honour of

Abel. The mathematical world was greatly excited by Weierstrass's

work, which went a long way to solve a problem that had been out-

standing for a generation. In only the next year, Riemann was able

to solve the problem completely, but his methods were so novel that

they did not command general assent. Until the work of Hilbert and

Weyl other methods were preferred to those of Riemann for dealing

with the new functions. Riemann, his students, and Clebsch, also

sought to use the new functions to understand the geometry of

algebraic curves. Weierstrass himself was so struck by Riemann's

paper that he withdrew a paper of his own, which he had nearly finished, on Abelian functions, and did not feel able to lecture on it satisfactorily until[7] 1869.

Riemann divided his paper [1857c] into two halves. In the first part, as he said, he discussed algebraic functions and their integrals without using theta-series. In §§1-5 he used his version of Dirichlet's principle to define algebraic functions in terms of their branching behaviour and their poles; in §§6-10 he expressed algebraic functions and their integrals as rational functions on a (Riemann) surface; in §§11-13 he looked at equivalent ways of expressing a given function (introducing the ideas of birational equivalence and moduli) and at the simplest expressions for a given function. In §14-16 he discussed Abel's addition theorem as a solution to a system of differential equations. In the second half of the paper he used θ-functions to solve the general Jacobi inversion problem, thus generalizing Weierstrass's treatment of the hyper-elliptic case.

To understand the geometrical implications of this work it is enough to know (1) that Riemann succeeded in defining the genus, p, of a surface defined by an algebraic equation in purely topological terms[8], and showing that such a surface has a linear space of holomorphic differentials of complex dimension p. (This is the paradigm case of an index theorem.) Furthermore (2) he solved the problem of Jacobi inversion, which is:

given the values $e_1, \ldots, e_p$ of p sums of p integrals of p linearly independent holomorphic differentials ω_i, $1 \leq i \leq p$,

$$(\sum_i \int_{a_i}^{x_i} \omega_1, \ldots, \sum_i \int_{a_i}^{x_i} \omega_p) \equiv (e_1, \ldots, e_p)$$

which are defined modulo periods, to find the corresponding values of $x_1, \ldots, x_p$. This inverts the integrals $u_j = \int \omega_j$ and makes $x_1, \ldots, x_p$ functions of $u_1, \ldots, u_p$. His solution introduced the infinitely many-valued function $\theta\,(u_1 - e_1, \ldots, u_p - e_p)$ of p variables on the (Riemann) surface, which was quasi-periodic and had a well-defined set of p zeros on the surface. These zeros, which came to be known classically as the theta-null values, were, he showed, the sought-for values of $x_1, \ldots, x_p$. (I have described some of Riemann's work and the responses it drew in more detail elsewhere [Gray 1984b]. A good account of it has been given in Scholz [1980], and a lucid treatment of Riemann's theta function is given in Chapter VI of the book by Farkas and Kra.)

This formulation, while not providing an explicit expression for $(x_1, \ldots x_p)$ as a function of $(e_1, \ldots, e_o)$, at least connects Jacobi inversion for arbitrary curves with Jacobi inversion for elliptic curves, and reduces the general problem to the vanishing of a scalar function on the Riemann surface. It thus gives a hold on the pre-image of a specified point in $\mathbb{C}^p$ (mod periods) which Riemann was able to exploit. It is comparable to the use of the symmetric functions in solving, say, the problem of finding three numbers x, y, and z, when x + y + z, xy + yz + zx, and xyz are given, and one can write down a cubic equation.

The responses to Riemann's paper were various, but most mathematicians agreed that it was extremely important and difficult. The high degree of generality and the use of a transcendental method for constructing functions militated against its immediate application, and for a while it was only Clebsch and Riemann's own students who advanced his ideas. Foremost amongst these was Gustav Roch, who died in 1866 of tuberculosis, at the age of only 29.

The geometric study of curves based on the theory of algebraic
functions formed part of Riemann's lectures during February 1862,
and some of this material was published in the Werke (second edition,
no. XXXI) based on notes taken by Roch. More was published in the
Nachträge [Werke, 1-66]. Riemann considered functions which have p-1
second order infinities and p-1 double zeros on a curve $F(s,z) = 0$,
which he obtained as follows. A basis for the holomorphic integrands

on the curve is, say $\left\{ \dfrac{\phi_i(s,z)}{\frac{\partial F}{\partial s}} \right\}_{1 \le i \le p}$, and he had shown in [1857c,

§10] that each $\phi_i(s,z)/\phi_j(s,z)$ is a function with 2p-2 simple infinities
and 2p-2 simple zeros.

Now he showed that it was possible to find a finite number of
ϕ_i such that their quotients ϕ_i/ϕ_j have p-1 double zeros and p-1
double poles as required. His construction involved associating
characteristics to the θ-function, that is, expressions of the form

$\begin{pmatrix} \varepsilon_1, & \cdots, & \varepsilon_p \\ \varepsilon_1', & \cdots, & \varepsilon_p' \end{pmatrix}$. The ε's are integers, and could be taken to be

either 0 or 1 without loss of generality. Riemann said the

characteristic $\begin{pmatrix} \varepsilon_1, & \cdots, & \varepsilon_p \\ \varepsilon_1', & \cdots, & \varepsilon_p' \end{pmatrix}$ was even if $\varepsilon_1\varepsilon_1' + \ldots + \varepsilon_p\varepsilon_p' \equiv 0$

(mod 2), and otherwise odd. The zeros of the θ-function are a p-tuple
of values of integrals of the holomorphic integrands, whose upper end
points are the zeros of the functions ϕ_i, and a complicated argument
enabled Riemann to show that the relation between the two sets of
zeros was such that the odd characteristics corresponded to the ϕ_i
with repeated zeros. There are $2^{p-1} \cdot (2^p - 1)$ odd characteristics and
$2^{p-1} \cdot (2^p - 1)$ such functions accordingly, so when p = 3 there are 28
interesting odd characteristics.

ON THE BITANGENTS OF A QUARTIC, BY PROFESSOR CAYLEY.

THE equations of the 28 bitangents of a quartic curve were obtained in a very elegant form by Riemann in the paper "Zur Theorie der Abelschen Functionen für den Fall $p = 3$," Werke, Leipzig, 1876, pp. 456–472; and see also Weber's "Theorie der Abelschen Functionen vom Geschlecht 3," Berlin, 1876. Riemann connects the several bitangents with the characteristics of the 28 odd functions, thus obtaining for them an algorithm which it is worth while to explain, but they will be given also with the algorithm employed p. 231 et seq. of the present work, which is in fact the more simple one. The characteristic of a triple θ-function is a symbol of the form

$$\begin{array}{ccc} \alpha & \beta & \gamma, \\ \alpha' & \beta' & \gamma', \end{array}$$

where each of the letters is $= 0$ or 1; there are thus in all 64 such symbols, but they are considered as odd or even according as the sum $\alpha\alpha' + \beta\beta' + \gamma\gamma'$ is odd or even; and the numbers of the odd and even characteristics are 28 and 36 respectively; and, as already mentioned, the 28 odd characteristics correspond to the 28 bitangents respectively.

We have x, y, z trilinear coordinates, and then $\alpha'', \beta'', \gamma''$ determinate constants, such that the equations

$$x + y + z + \xi + \eta + \zeta = 0,$$
$$\alpha x + \beta y + \gamma z + \frac{\xi}{\alpha} + \frac{\eta}{\beta} + \frac{\zeta}{\gamma} = 0,$$
$$\alpha' x + \beta' y + \gamma' z + \frac{\xi}{\alpha'} + \frac{\eta}{\beta'} + \frac{\zeta}{\gamma'} = 0,$$
$$\alpha'' x + \beta'' y + \gamma'' z + \frac{\xi}{\alpha''} + \frac{\eta}{\beta''} + \frac{\zeta}{\gamma''} = 0,$$

are equivalent to three independent equations; this being so, they determine ξ, η, ζ each of them as a linear function of (x, y, z); and the equations of the bitangents of the curve $\sqrt{(x\xi)} + \sqrt{(y\eta)} + \sqrt{(z\zeta)} = 0$ (see Weber, p. 100) are

No.	Char.	Equation
18	$\frac{111}{111}$	$x = 0,$
28	$\frac{001}{011}$	$y = 0,$
38	$\frac{011}{001}$	$z = 0,$
23	$\frac{010}{010}$	$\xi = 0,$
13	$\frac{100}{110}$	$\eta = 0,$
12	$\frac{110}{100}$	$\zeta = 0,$
48	$\frac{101}{100}$	$x + y + z = 0,$
14	$\frac{010}{011}$	$\xi + y + z = 0,$
58	$\frac{100}{101}$	$\alpha x + \beta y + \gamma z = 0,$
15	$\frac{011}{010}$	$\frac{\xi}{\alpha} + \beta y + \gamma z = 0,$
68	$\frac{110}{010}$	$\alpha x + \beta' y + \gamma' z = 0,$
16	$\frac{001}{101}$	$\frac{\xi}{\alpha'} + \beta' y + \gamma' z = 0,$
78	$\frac{010}{110}$	$\alpha'' x + \beta'' y + \gamma'' z = 0,$
17	$\frac{101}{001}$	$\frac{\xi}{\alpha''} + \beta'' y + \gamma'' z = 0,$
24	$\frac{100}{111}$	$x + \eta + z = 0,$
34	$\frac{110}{101}$	$x + y + \zeta = 0,$
25	$\frac{101}{110}$	$\alpha x + \frac{\eta}{\beta} + \gamma z = 0,$
35	$\frac{111}{100}$	$\alpha x + \beta y + \frac{\zeta}{\gamma} = 0,$
26	$\frac{111}{001}$	$\alpha' x + \frac{\eta}{\beta'} + \gamma' z = 0,$
36	$\frac{101}{011}$	$\alpha' x + \beta' y + \frac{\zeta}{\gamma'} = 0,$
27	$\frac{011}{101}$	$\alpha'' x + \frac{\eta}{\beta''} + \gamma'' z = 0,$
37	$\frac{001}{111}$	$\alpha'' x + \beta'' y + \frac{\zeta}{\gamma''} = 0,$
67	$\frac{100}{100}$	$\frac{x}{1-\beta\gamma} + \frac{y}{1-\gamma\alpha} + \frac{z}{1-\alpha\beta} = 0,$
57	$\frac{110}{011}$	$\frac{x}{1-\beta'\gamma'} + \frac{y}{1-\gamma'\alpha'} + \frac{z}{1-\alpha'\beta'} = 0,$
56	$\frac{010}{111}$	$\frac{x}{1-\beta''\gamma''} + \frac{y}{1-\gamma''\alpha''} + \frac{z}{1-\alpha''\beta''} = 0,$
45	$\frac{001}{001}$	$\frac{\xi}{\alpha(1-\beta\gamma)} + \frac{\eta}{\beta(1-\gamma\alpha)} + \frac{\zeta}{\gamma(1-\alpha\beta)} = 0,$
46	$\frac{011}{110}$	$\frac{\xi}{\alpha'(1-\beta'\gamma')} + \frac{\eta}{\beta'(1-\gamma'\alpha')} + \frac{\zeta}{\gamma'(1-\alpha'\beta')} = 0,$
47	$\frac{111}{010}$	$\frac{\xi}{\alpha''(1-\beta''\gamma'')} + \frac{\eta}{\beta''(1-\gamma''\alpha'')} + \frac{\zeta}{\gamma''(1-\alpha''\beta'')} = 0.$

The whole number of ways in which the equation of the curve can be expressed in a form such as $\sqrt{(x\xi)} + \sqrt{(y\eta)} + \sqrt{(z\zeta)} = 0$ is 1260; viz. the three pairs of bitangents entering into the equation of the curve are of one of the types

			Type	No. is
12.34,	18.24,	14.23	⊠	70
12.34,	13.24,	56.78	◻	" 630
18.28,	14.24,	15.26	◈	" 560
				1260

and it may be remarked that selecting at pleasure any two pairs out of a system of three pairs the type is always ◻ or ||||, viz. (see p. 283) the four bitangents are such that their points of contact are situate on a conic.

From Salmon's Higher Plane Curves 387–389.

In his study of curves $F(s,z) = 0$ of genus $p = 3$, Riemann took
new coordinates $\xi: = \phi_1/\phi_3$ and $\eta: = \phi_2/\phi_3$. Since ξ and η each take
every value $2p - 2 = 4$ times the curve is now expressed as a quartic
$F(\xi, \eta) = 0$. The integrands $\phi/\frac{\partial F}{\partial \xi}$ are everywhere finite, so ϕ is of
degrees $2p - 5 = 1$ in ξ and η and is therefore a linear function
containing 3 constants, i.e. the dimension of the space of holomorphic
integrands is 3. But then F is non-singular for, if it had r singular
points, then a line $\phi = 0$ through a singular point would have a double
zero there, and so the space of lines having double zeros on F would
have dimensions $p + r$. But then $p + r = 3$ and $p = 3$, so $r = 0$. By
construction each ϕ vanishes to the second order on F, so $\phi = 0$
represents a bitangent to the curve. For, the restriction of a
rational function σ in (s,z) space to a curve $F(s,z) = 0$ gives a
function on a curve, and the zero locus of σ is itself a curve which
meets F where $\sigma(s,z) = 0$. Repeated zeros on F of the function σ
correspond to points where σ touches F.

Riemann then introduced homogeneous coordinates x, y, z
$(\xi = \frac{x}{z}, \eta = \frac{y}{z})$, so each ϕ was of the form $cx + c'y + c''z$. A further
linear change of variable allowed him to consider x, y, and z as
bitangents, in which case he showed (see the Exercises) that
$F(x,y,z) = f^2(x,y,z) - xyzt$, where t is linear in x, y and z.

Riemann then argued that the equation can be reduced to the
form $f^2 - xyzt$ in 6 different ways once x and y are given. For
$F = f^2 - xyzt = \psi^2 - xypq$ implies $(f + \psi)(f - \psi) = xy(zt - pq)$, so xy
divides, say, $f - \psi$, i.e. $\psi - f = \alpha xy$, α constant. Then $\alpha(f + \psi) = pq - zt$,
or, eliminating ψ, $a\alpha f + \alpha^2 xy + zt = pq$. The left hand side is
reducible, so its discriminant must vanish, and its discriminant is
of degree three in its coefficients and thus of degree 6 in α.

Accordingly, there are 6 decompositions of F in the form $f^2 - xyzt$; $\alpha = 0$
giving precisely the decomposition in which pq = zt, and $\alpha = \infty$ the
decomposition pq = xy.

Salmon [1879, 227] gave a pleasing geometric interpretation of
this. When the equation of the curve is written in the form
$xyU = (z^2 + ayz + by^2 + cxz + dx^2)^2$, as it may always be by the above
argument, it may thus be written in the form

$\quad xyU = V^2$, or

$\quad xy(\lambda^2 U + 2\lambda V + xy) = (xy + \lambda V)^2$

where U = 0 and V = 0 represent conics. Six conics in the family
$\lambda^2 U + 2\lambda V + xy = 0$ are reducible, i.e. are line pairs. One, $\lambda = 0$,
gives the line pair xy = 0, the other 5 other line pairs, zt = 0 say,
such that the equation of the curve becomes, on setting
$xy + \lambda V = f$, $xyzt = f^2$.

Thus, "through the four points of contact of any two bitangents
we can describe five conics, each of which passes through the four
points of contact of two other bitangents", and, since there are
$\frac{1}{2}28.27 = 378$ pairs of bitangents, but each conic is counted six times,
"there are in all $\frac{5}{6}$ (378) or 315 conics, each passing through the
points of contact of four bitangents of a quartic". This result was
given by Hesse, Steiner and, incorrectly, by Plücker.

Riemann now confronted two tasks: first, given a curve to find
the equation of its 28 bitangents; second, to express the mutual
relationships of the bitangents. In reverse order, the second task
was accomplished by means of the notation for the characteristics,
and then the symmetries between the bitangents enabled Riemann to
find their equations. This is an attractive piece of work, and it
was taken up by Clebsch in an important paper of 1864. Other papers

in this spirit were written by Roch and Weber. It is interesting to note, however, that Klein did not work in this tradition when he turned to the geometry of quartic curves and sought to connect that subject with his earlier work on modular transformations. It is very likely that he was unaware of Riemann's lectures, which were not published until 1892, and he may not have understood the papers by Clebsch and Roch, which inevitably appealed to Riemann's obscure theory of Jacobi inversion. Even the self-appointed heir to Riemann was to admit he found the master very difficult.

5.3 *Klein*.

In 1878 Klein began to publish a series of papers generalizing his earlier work to equations and transformations of higher degree than five. These works mark a considerable development in the theory of Riemann surfaces, and are the start of the systematic study of modular functions, Klein's greatest contribution to mathematics. They also form, as has been suggested, an important stage in the development of Galois Theory, being the origin of that part of the subject which concerns fields of rational functions on an algebraic curve. Klein distinguished in his papers between the function-theoretic and the purely algebraic directions in which his research was proceeding, concentrating chiefly on the former, which will be dwelt on accordingly and the latter discussed only in footnote 9.

The most important paper on the function-theoretic side is Klein's "Über die Transformation siebenter Ordnung der elliptischen Functionen" [1878/79]. Klein described the path he took in this paper as going from a thorough description of a group of 168 linear substitutions $\omega' = \dfrac{\alpha\,\omega + \beta}{\gamma\,\omega + \delta}$ which permute the roots of the modular equation, to a study of a certain function η as a branched function of J, where it turns out that η forms a surface of genus 3. This leads to the study of a curve in the projective plane having 168 symmetries which is, indeed a quartic. Klein devoted quite some time to describing the curve as vividly as possible.

The group of the modular equation of order 7 is, as Galois had known, made up of the 168 linear substitutions $\omega' = \dfrac{\alpha \omega + \beta}{\gamma \omega + \delta}$ with coefficients in the integers reduced mod 7, and determinant + 1. [For, the column $\binom{\alpha}{\gamma}$ in one of the two matrices representing the transformation $\omega^1 = \dfrac{\alpha \omega + \beta}{\gamma \omega + \delta}$ can have any of 8.6 different entries, the number of vectors through the origin in the field $\mathbb{Z}/7\,\mathbb{Z}$, and $\binom{\beta}{\delta}$ can lie in one of 7 other directions. So there are 8.6.7 = 336 matrices, and 168 linear transformations.] The group was usually referred to as the group of order 168, and later denoted G_{168}, as it will be here. It is the only simple group of that order[10], and is isomorphic to PSL(2;$\mathbb{Z}/7\,\mathbb{Z}$). Klein described its elements accordingly to their conjugacy class, which essentially determines their geometric character, referring to conjugacy by the word "gleichberechtigt": S_1 and S_2 are conjugate if there is an element S of G_{168} such that $S_1 = S^{-1} S_2 S$. He found there were:

a)   21 conjugate substitutions of period two, such as $\dfrac{-1}{\omega}$, characterized by the fact that $\alpha + \delta = 0$ [i.e. they have zero trace];

b)   28 conjugate pairs of substitutions of period 3, such as $\dfrac{-2\omega}{3}$ and $\dfrac{-3\omega}{2}$, characterized by $\alpha + \delta = \pm\,1$;

c)   48 substitutions of period 7 coming in 8 conjugate sets of 6, such as $\omega + 1$, $\omega + 2$, ... $\omega + 6$, characterized by $\alpha + \delta = \pm\,2$ and excluding the identity $\omega' = \omega$; and

d)   21 conjugate pairs of substitutions of period 4 corresponding to
each substitution of period 2, such as $\dfrac{2\omega + 2}{-2\omega + 2}$, $\dfrac{2\omega - 2}{2\omega + 2}$
corresponding to $\dfrac{-1}{\omega}$, characterized by $\alpha + \delta = \pm 3$.

There are accordingly the following subgroups of G_{168}:

1)   The identity;

2)   21 G_2's of order 2;

3)   28 G_3's of order 3;

4)   21 G_4's of order 4;

5)   8 G_7's of order 7;

6)   14 G_4''s of order 4 each containing two substitutions of order
two which commute with ("sind vertauschbar ... mit") a given
substitution of order 2, for example ω, $\dfrac{-1}{\omega}$, $\dfrac{2\omega + 3}{3\omega - 2}$, $\dfrac{3\omega - 2}{-2\omega - 3}$
in two families of 7 conjugates;

7)   28 conjugate G_6''s of order 6, containing the three substitutions
of order 2 commuting with a given G_3;

8)   21 conjugate G_8''s of order 8, the centralizers of the G_4's;

9)   8 conjugate G_{21}''s of order 21, the centralizers of the G_7's;

10)   14 G_{24}'''s of order 24 in two families of 7 conjugates, arising
from the families of G_4''s, and which are octahedral groups
(this paper, §14).

Klein claimed that this list was complete, but in footnote 6
in the Werke mentions that each G_{24}'' contains a normal subgroup G_{12}''
of order 12 isomorphic to the even permutations on 4 things. The
G_4''s introduced casually here have come to be known as examples of
Klein's four-group.

Klein next considered the subgroup of PSL(2 ; $\mathbb{Z}$) consisting of
those maps $\omega' = \dfrac{\alpha\, \omega + \beta}{\gamma\, \omega + \delta}$ which are conjugate to the identity mod 7

[in his later work he called this group the principal congruence subgroup of level 7, it is usually denoted $\overline{\Gamma}$ (7)] and introduced a single-valued function η on the upper half plane with the property that

$$\eta(\omega) = \eta\left(\frac{\alpha\,\omega\,+\,\beta}{\gamma\,\omega\,+\,\delta}\right) \text{ if and only if } \omega' = \frac{\alpha\,\omega\,+\,\beta}{\gamma\,\omega\,+\,\delta} \text{ is in } \overline{\Gamma}\ (7).$$

The fundamental region for η is 168 copies of the fundamental region for J, since $\overline{\Gamma}$ (7) has index 168 in PSL(2 ; $\mathbb{Z}$), and PSL(2 ; $\mathbb{Z}$)$/\overline{\Gamma}\ (7) \cong G_{168}$. So for each value of $J(\omega)$ there are 168 different values of $\eta(\omega)$. As for the branching of η over J, it followed from an earlier paper of Klein's that only J = 0, 1, and ∞ can be branch points. It follows from the classification of the substitutions and their fixed points that at J = 0 the branching is of order 3 and the leaves hang in 56 cycles; at J = 1 the branching is of order 2 in 84 cycles; and at J = ∞ the order is 7 (typically $\omega' = \omega + 1$). The genus of the surface is thus

$$p = \frac{1}{2}\ (2\ -\ 2.168\ +\ 2.56\ +\ 1.84\ +\ 6.24)\ =\ 3,$$

and η forms a Riemann surface of genus 3 admitting 168 conformal self-transformations. [$\overline{\Gamma}$ (7) moves the fundamental region around en bloc, and so provides the identifications of the sides to yield a closed surface; G_{168} provides the self-transformations of the surface.]

To understand this surface better Klein introduced certain special points corresponding to the branching. The 24 points $\underline{a}$ correspond to J = ∞; the 56 points $\underline{b}$ to J = 0; the 84 points $\underline{c}$ to J = 1. Since each $\underline{a}$ is fixed by a G_7, of which there are 8, each G_7 fixes 3 $\underline{a}$'s. Likewise each G_3 fixes 2 $\underline{b}$'s, and each G_2 four $\underline{c}$'s. No G_4 fixes any point.

To obtain an equation for the curve Klein (quoting Clebsch-Gordan
[1866] p.65 and Clebsch-Lindemann [1876] pp.687, 712, but only in the
Werke) argued that the equation would be either a plane quintic with
triple points, if the curve was hyperelliptic, or else a non-singular
quartic. The curve in question could not be hyperelliptic [because
G_{168} is simple] so it can be represented by a non-singular quartic,
C_4, and G_{168} becomes a group of 168 plane collineations. Klein
remarked here that G_{168} had seemingly been omitted by Jordan in his
list of finite subgroups of PGL(3,$\mathbb{C}$) [1877/78], and only included in
the Correction [1878]. But now the a's, b's, and c's can be
immediately connected with distinguished point sets on a quartic which
correspond to certain projective properties. The a's are the 24
inflection points; the 28 pairs of b's the points of contact of the
28 bitangents; the 21 quadruples of c's the sextactic points, at which
a cubic has threefold contact with the quartic (a sextactic point is
one where a conic has 6-fold contact with a curve, the term is due to
Cayley [1859b]).

Furthermore, the triples of a's (inflection points) preserved by
a G_3 can be regarded as follows. Each inflection tangent to a C_4 meets
it again in 1 point, so there are 24 thus distinguished points, which
must be the inflection points themselves (since they form the only
distinguished set of 24 points). No inflection tangent is a
bitangent, so each inflection tangent meets the curve in a different
inflection point. A collineation fixing a given a must also fix its
inflection tangent and so the new inflection point, and by iterating
this argument one sees that the triple a's preserved by the G_3 are
the vertices of a triangle whose sides are inflection tangents, and
so there are 8 such inflection triangles.

The 21 quadruples of $\underline{c}$'s are invariant under collineations of order 2, i.e. perspectivities, which thus occur having 21 centres and 21 axes, each quadruple lying on an axis. There are 4 axes through each centre, 4 centres on each axis (since each transformation of order two fixes 4 $\underline{c}$'s and 4 other points). Each bitangent carries 3 centres, through each of which pass 4 bitangents. Each G_{24}'' permutes a distinguished set of 4 bitangents.

If an inflection triangle is taken as triangle of reference, so the sides are $\lambda = 0$, $\mu = 0$, $\nu = 0$, it is soon clear (§ 4) that the equation of the curve is $f(\lambda, \mu, \nu) = \lambda^3 \mu + \mu^3 \nu + \nu^3 \lambda = 0$. Klein gave a series of explicit calculations for the equations of bitangents with respect to this triangle of reference and others[11] adapted to display the symmetry and so invariance with respect to particular subgroups of G_{168}. He also gave a representation of the group G_{168} as projective collineations of the following forms

$$\lambda' = A\lambda + B\mu + C\nu$$
(1) $$\mu' = B\lambda + C\mu + A\nu$$
$$\nu' = C\lambda + A\mu + B\nu$$

where $A = \dfrac{\gamma^5 - \gamma^2}{\sqrt{-7}}$, $B = \dfrac{\gamma^3 - \gamma^4}{\sqrt{-7}}$, $C = \dfrac{\gamma^6 - \gamma}{\sqrt{-7}}$

and $\gamma = e^{2\pi i/7}$ so $\gamma + \gamma^4 + \gamma^2 - \gamma^6 - \gamma^3 - \gamma^5 = \sqrt{-7}$, or

(2) $\lambda' = \mu$, $\mu' = \nu$, $\nu' = \lambda$, or

(3) $\lambda' = \gamma\lambda$, $\mu' = \gamma^4 \mu$, $\nu' = \gamma^2 \nu$,

and compounds of (1), (2), and (3).

The associated invariants and covariants of the curve are fairly easy to study geometrically. Its Hessian $\nabla = 5\lambda^2 \mu^2 \nu^2 - (\lambda^5 \nu + \nu^5 \mu + \mu^5 \lambda)$ of course meets it in the 24 inflection points, and since this is the smallest set of distinguished points there is no invariant polynomial of degree less than 6, and, up to constant multiples, only the Hessian of degree 6. Similarly the next invariant polynomial will be of degree 14 and its intersection with the curve will be the bitangent points. There are many such, but all are sums of $f^2 \nabla$ and any one which is not a constant multiple of $f^2\nabla$. Klein chose

$$
C = \frac{1}{9}
\begin{vmatrix}
\dfrac{\partial^2 f}{\partial \lambda^2} & \dfrac{\partial^2 f}{\partial \lambda \partial \mu} & \dfrac{\partial^2 f}{\partial \lambda \partial \nu} & \dfrac{\partial \nabla}{\partial \lambda} \\[2ex]
\dfrac{\partial^2 f}{\partial \mu \partial \lambda} & \dfrac{\partial^2 f}{\partial \mu^2} & \dfrac{\partial^2 f}{\partial \mu \partial \nu} & \dfrac{\partial \nabla}{\partial \mu} \\[2ex]
\dfrac{\partial^2 f}{\partial \nu \partial \lambda} & \dfrac{\partial^2 f}{\partial \nu \partial \mu} & \dfrac{\partial^2 f}{\partial \nu^2} & \dfrac{\partial \nabla}{\partial \nu} \\[2ex]
\dfrac{\partial \nabla}{\partial \lambda} & \dfrac{\partial \nabla}{\partial \mu} & \dfrac{\partial \nabla}{\partial \nu} & 0
\end{vmatrix}
$$

For an invariant of degree 21 he chose the functional determinant, K, of f, ∇, and C:

$$
K = \frac{1}{14}
\begin{vmatrix}
\dfrac{\partial f}{\partial \lambda} & \dfrac{\partial \nabla}{\partial \lambda} & \dfrac{\partial C}{\partial \lambda} \\[2ex]
\dfrac{\partial f}{\partial \mu} & \dfrac{\partial \nabla}{\partial \mu} & \dfrac{\partial C}{\partial \mu} \\[2ex]
\dfrac{\partial f}{\partial \nu} & \dfrac{\partial \nabla}{\partial \nu} & \dfrac{\partial C}{\partial \nu}
\end{vmatrix}
$$

Since K is the only invariant polynomial of degree 21 it must represent the 21 axes of the perspectivities. Finally to represent the 168 corresponding points on f = 0, he considered the family of curves $\nabla^7 = k \, C^3$, where k is a constant. There is a linear relation between ∇^7, C^3, and K^2 precisely as in the icosahedral case, in this case a comparison of coefficients in the explicit representations of ∇, C, and K in terms of λ, μ, ν shows

$$(-\nabla)^7 = (\frac{C}{12})^3 - 27(\frac{K}{216})^2 ,$$

i.e. $J : J - 1 : 1 = (\frac{C}{12})^3 : 27(\frac{K}{216})^2 : - \nabla^7,$

or, in terms of the invariants of elliptic integrals

$$g_2 = \frac{C}{12} ; \; g_3 = \frac{K}{216} ; \; \sqrt[7]{\Delta} = -\nabla,$$

and f, ∇, C, K represent a complete system of covariants[12] of f.

The equation of degree 168 is the resolvent of the modular equation. Since G_{168} has subgroups of indices 7 and 8 the equation has resolvents of degrees 7 and 8, which Klein also found. The resolvent of degree 8 he connected to the 36 systems of contact cubics of even characteristic via the 8 inflection triangles. This led him to an explicit solution to the equation of degree 168 in terms of elliptic functions, and he remarked in footnote 21 "[The equation] must also be solvable by means of a linear differential equation of the third order; how can one construct it?" This problem was solved by Hurwitz and Halphen[13].

Klein then turned, in the concluding sections of the paper, to the task of describing these resolvents "as graphically as possible with the aid of Analysis Situs". The function J maps one half of its fundamental region onto the upper half-plane, the other onto the lower half-plane. Let the triangle mapping onto the upper half-plane be shaded, then the fundamental region for μ contains 168 shaded and 168 unshaded triangles. The vertices of these triangles, in accordance with the branching of μ, are the a, b, and c points; at a point a 14 triangles meet, at a b 6 triangles and at a c 4 triangles. As originally presented the points a are at infinity, and the vertical angle is 0, whereas the vertical angles at the b's are $\pi/3$, and at the

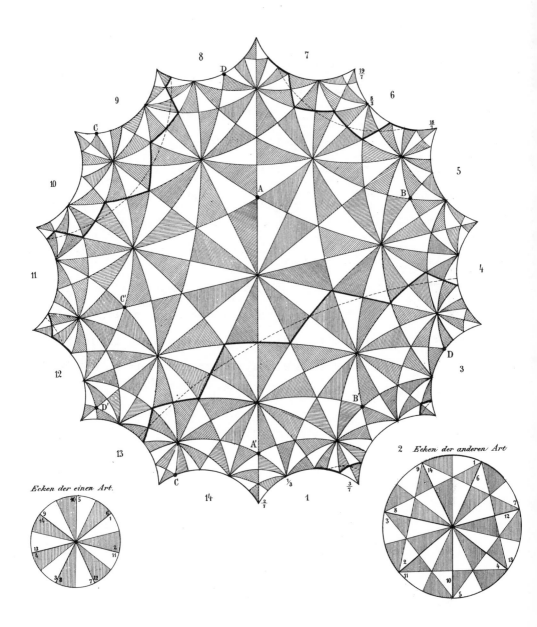

Fig. 5.4.

c's $^{\pi}/2$, and Klein found it more attractive to work with triangles

having angles $^{\pi}/7$, $^{\pi}/2$, $^{\pi}/3$. He did not explain how this transition

could be made; his student M. W. Haskell wrote a thesis on this curve

in 1890 and pointed out that the Schwarzian s-function $s(\frac{1}{3}, \frac{1}{2}, \frac{1}{7}, J)$

can be used [Haskell, 1890, p. 9n]. An examination of elements in

PSL(2;$\mathbb{Z}$) congruent to the identity mod 7 provides the identification

of the edges on the boundary needed to make the curvi-linear 14-gon

into a closed surface : if the edges are numbered anticlockwise then 1

is identified with 6, 3 with 8, 5 with 10, and so on, until 13 is

identified with 4.

The beautiful figure of 168 shaded and 168 unshaded triangles not

only displays the inflection points, bitangent points, and sextactic

points of the curve as the vertices of the triangles. It also

contains several interesting lines, which run straight through the

a's, b's, and c's in this order:

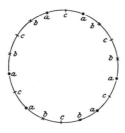

Fig. 5.5

They are the curves on the surface on which J is real-valued. On the

closed surfaces the lines are closed curves, i.e. projective lines.

There are 28 symmetry lines, and they can be traced on the figure,

some care is needed at the vertices of the 14-gon, which are of two

kinds. For example, the vertical line $0_1 0_8$ continues as the edge 5

(which is, of course, identified now with the edge 10). Klein remarked

(§13) "These symmetry lines are for many purposes the simplest means of orientation on our surface; I will use them here to define the mutually corresponding points a, b, c on the surface. It is then easy to picture the corresponding self transformations of our surface." The seven symmetry lines emanating from an a meet again in the two corresponding a's. The three symmetry lines through a point b meet in the corresponding b; and the two through a c meet again in a c. Furthermore, if the surface is cut open along the seven lines through a given a, then, because the genus is 3, it appears as a simply-connected region with boundary curve: the 14-gon is recovered. Two corresponding b's give rise to three symmetry lines which meet another symmetry line through two c's, so there is a one to one correspondence between the b's and the symmetry lines. Finally the two lines through a pair of corresponding c's meet all the other symmetry lines but two, which themselves meet in two c's, giving rise to the quadruples of c-points.

Klein wanted to display the figure in as regular a way as possible, but he know that there is no solid in three dimensional space whose symmetry group is G_{168}. However, he pointed out, G_{168} contains several octahedral subgroups G_{24}''. Each permutes a set of 4 pairs of b-points. It is possible to surround a b-point with 3 14-gons (strictly 7-gons, because of the adjacent vertical angles of $\pi/2$) made up of the 14 triangles meeting at an a-point, thus dividing the figure so that it is seen as containing $4.2.(3.14) = 336$ triangles. The G_{24}'' acts then on a true octahedron, with a b point at the mid-point of each face, the a points as vertices, and some c-points as mid-edge points, provided diametrically opposite points are identified.

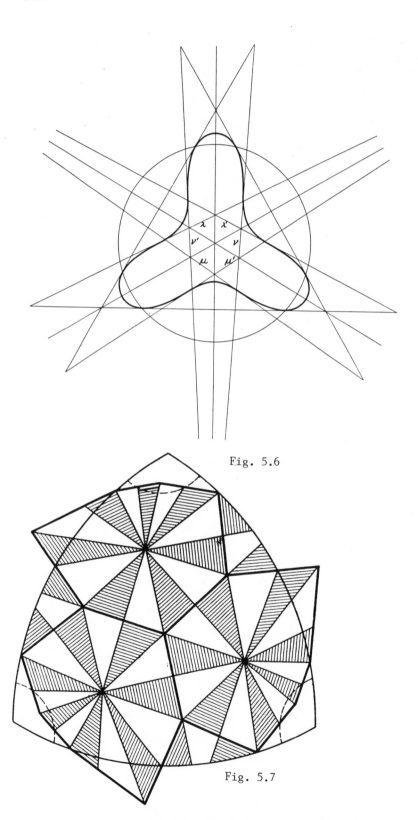

Fig. 5.6

Fig. 5.7

The aspects of the figure which cannot be realized in three-space have this interpretation: one imagines the octahedron as composed of three hyperboloids of one sheet with axes crossing at right angles, and with opposite edges identified at infinity, thus representing a surface of genus 3. The axes of the hyperboloids may be said to pass through the vertices of the octahedron.

Klein concluded the paper with a discussion of the real part of the curve, in keeping with an earlier interest of his, and supplied the following attractive figure shown opposite.

This paper must have convinced Klein, if he needed convincing, of the power of his new methods. A year later he reported on this progress in a paper suggestively entitled "Towards a theory of elliptic modular functions". In this work he set himself the task of showing how the different forms in which one encountered the modular equation were very special cases of a simple general principle. The theory should begin, he said, with an analysis of all the subgroups of PSL(2; $\mathbb{Z}$), he proposed in this paper only to deal with subgroups of finite index, which he admitted was a great restriction. He pointed out that this use of groups as a classifying principle went beyond Galois in that it invoked infinite as well as finite groups. to be precise, infinite normal ("ausgezeichneten") subgroups of PSL(2; $\mathbb{Z}$). His second classifying principle, introduced on empirical grounds, was that the subgroups should be defined arithmetically by means of congruences, but, he pointed out sternly, "it must be clearly understood that not all subgroups are congruence groups". Thirdly, he argued that a fundamental polygon should be introduced for each group, and the function theory of the corresponding Riemann surface brought into consideration. In particular, the genus of the surface should be evaluated, above all to see whether or not it is zero.

For a given subgroup, he considered the functions invariant under the substitutions of the subgroup; such functions he called Moduln, here translated as modules (moduli would also be confusing). When p = 0 a module can be found taking each value precisely once on the fundamental polygon. For p > 0 two modules are needed to specify a point, and they will be related by an algebraic equation of degree p. The case p = 0 includes, he pointed out, the vth roots of Legendre's modules κ^2 and of $\kappa^2 \kappa'^2$ for integers v. Up until now only the following had been studied in the usual theory: κ^2, κ, $\sqrt{\kappa}$, $\sqrt[4]{\kappa}$, $\kappa^2 \kappa'^2$, $\kappa \kappa'$, $\sqrt{\kappa \kappa'}$, $\sqrt[4]{\kappa \kappa'}$, $\sqrt[3]{\kappa \kappa'}$, $\sqrt[6]{\kappa \kappa'}$, $\sqrt[12]{\kappa \kappa'}$, and this was, he said, because only these are congruence modules.

For a modular transformation $w' = {}^w/n$ or $w' = {}^{-n}/w$ Klein proposed to consider the equation between $J(w)$ and $J(w')$ as the prototype of all modular equations, and to investigate it by finding its degree, its Galois group, and subgroups thereof.

One notices very clearly here the role of invariant functions and the sense in which Galois groups appear as indices of the relation between two fields of functions. It is in this sense that Klein, and Gordan also, have been described in Chapter III as extending or going beyond the Galois theory of their day[14]. Also, Klein was clearly aware how many groups are excluded from his approach by the arithmetic restriction. It was to be his chief concern in the next two years to find some way of eliminating it from the theory of Riemann surfaces and modular functions, as will be discussed in the final chapter.

Exercises on Chapter V.

The first 13 exercises concern a fascinating paper by Hesse,
which I briefly describe. The geometry of plane curves is very rich.
They may be touched not only by lines but also by curves of various
degrees. Hesse took up the crucial notion of systems of curves
touching a given C_n in his paper [1855a]. He defined two C_{n-1}'s which
touch a given C_n to be in the same system if their contact points lay
on a common C_{n-1}. He claimed (§8) there were 36 systems of cubics
touching a quartic in six different points, but deferred the proof
until his [1855b]. For these cubics the six points of contact do not
lie on a conic, but he also found 28 systems of cubics whose contact
points do lie in 6's on conics (§9), and showed that any conic through
two bitangent points meets the C_4 again in 6 points which can be taken
as the contact points of the C_4 and a C_3. He also found there are 63
systems of conics which touch a given C_4 (§10).

To discuss the bitangents themselves Hesse, in his [1855b], passed
to the consideration of figures in space, observing that eight points
in general position in space give rise to $\frac{1}{2}.8.7 = 28$ lines. He gave a
thorough treatment of the geometry of a plane quartic Δ in terms of the
geometry of a related sextic curve in space, κ, in particular he
showed that a line joining any two of the points cuts κ in two points
which correspond to bitangent points on Δ. So the 28 bitangents
correspond to 28 lines in space. These 28 lines through 8 points have
various pleasing inter-relationships which formed the subject of
sections 9-12 and 15-16 of Hesse's paper, and which it will be
convenient to list.

Three bitangent pairs give rise to 6 lines in space. If these
three pairs of lines are the opposite edges of a tetrahedron then the

four contact points of one pair and the four intersection points of
the remaining two pairs lie on a conic (§9).

Four bitangents give rise to 8 points on Δ, which lie on a conic
if the four corresponding lines in space either form a quadrilateral
(§9) or join the eight points a_1, ..., a_8 in pairs (§12).

If four points on Δ are the points at which a conic, C_2, touches
Δ then any other conic through those points meets Δ in four more points
which may also be taken as the points in which Δ touches a conic, C_2'
say. C_2 and C_2' belong to the same system of conics touching Δ (§10).

Three bitangents to Δ whose corresponding lines in space form a
triangle or meet in a common point give rise to 6 bitangent points
(§10, 12).

The 12 contact points of 6 bitangents lie on a cubic if the
corresponding lines in space form either two disjoint triangles or
two triangles with at most a point in common (§10).

In sections 15 and 16 Hesse counted the configurations of each
kind which can arise, and found:

i) There are 2016 triples of bitangents whose 6 points do not lie
on a conic, and 1260 triples whose 6 points do.

ii) There are 315 quadruples of bitangents whose 8 points lie on a
conic.

iii) There are 1008 + 5040 sextuples of bitangents whose 12 points lie
on a cubic, of which the 1008 arrangements contain no tuples
whose 6 points lie on a conic. These occur when the corresponding
lines in space form one of the following configurations:

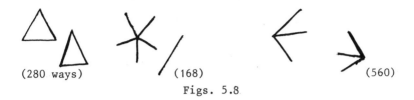

(280 ways)   (168)   (560)

Figs. 5.8

The 5040 (= 7!) arrangements each yield 12 points lying on 4
conics, each pair of conics cutting Δ in 2 more bitangent pairs, so
producing 4 more bitangents and 8 points lying on a conic. The
corresponding arrangements of lines in space are:

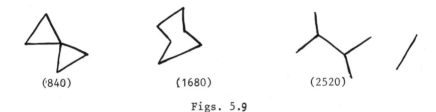

('840)   (1680)   (2520)

Figs. 5.9

Hesse showed that there are 36 different ways (i.e. unrelated
by linear transformations) of representing the same quartic Δ. These
are obtained by dividing the eight points a_1, ..., a_8 up into two sets
of 4, which can be done in $\frac{8.7.6.5}{4!2}$ = 35 ways; each division will produce
a new representation, so there are 35 + 1 = 36 in all. He listed the
corresponding arrangement of bitangents in a table at the end of §14.
There are therefore 36 systems of cubics associated to a given quartic,
whose contact points do not lie on a conic, as Hesse had claimed in
the earlier paper.

1.   Suppose $F(x, y, z) = ax^2 + by^2 + cz^2 + 2fyz + 2gzx + 2hxy = 0$
     represents a conic in the projective plane. Let $p_1 = (x_1, y_1, z_1)$
     and $p_2 = (x_2, y_2, z_2)$ be two points, so any point on the line
     $\lambda x + \mu y + \nu z = 0$ joining them has coordinates

$(\ell_1 x_1 + \ell_2 x_2, \; \ell_1 y_1 + \ell_2 y_2, \; \ell_1 z_1 + \ell_2 z_2)$. Show that the line

meets the conic where $\ell_1^2 F_1 + 2\ell_1 \ell_2 F_{12} + \ell_2^2 F_2 = 0$, where

$F_1 = F(p_1)$, $F_2 = F(p_2)$ and

$F_{12} = ax_1 x_2 + by_1 y_2 + cz_1 z_2 + f(y_1 z_2 + y_2 z_1) +$

$g(z_1 x_2 + z_2 x_1) + h(x_1 y_2 + x_2 y_1)$.

2. Show that, if p_1 lies on the conic then the line meets the conic where

$$F_2 x_1 - 2F_{12} x_2 = 0, \; F_2 y_1 - 2F_{12} y_2 = 0,$$

and $F_2 z_1 - 2F_{12} z_2 = 0$. Deduce that the equation of the tangent to the conic at z_1 is $F_{12} = 0$, interpreted as a locus in x_2, y_2, z_2.

3. Recall that if the two tangents from a point q to a conic meet it in points lying on a line $\underline{1}$ then $\underline{1}$ is called the polar of q and q the pole of $\underline{1}$. Deduce that the polar of the point (x_2, y_2, z_2) is the line $f_{12} = 0$ interpreted as an equation in (x_1, y_1, z_1).

4. Show that the pole of $\lambda x + \mu y + \nu z = 0$ is the point (x_1, y_1, z_1) which satisfies $F_{x1} = \lambda$, $F_{y1} = \mu$, $F_{z1} = \nu$, where

$$F_{x1} = \frac{\partial F}{\partial x}_{(x_1, \, y_1, \, z_1)} \quad \text{and } F_{y1} \text{ and } F_{z1} \text{ are defined similarly.}$$

5. Show that if $F = 0$ and $G = 0$ are the equations of two conics through four points, then any other conic through those points has an equation of the form $\alpha F + \beta G = 0$.

6. Show, given a line $\underline{1}$ and the family on conics through four points which form a quadrilateral in a plane, that the locus of the poles of $\underline{1}$ with respect to the conics is itself a conic which passes through the diagonal points of the quadrilateral.

7. Generalize this to obtain Chasles' theorem: given a plane and the (one-parameter) family of quadrics through 8 points in general position, the locus of the pole of the plane is a space curve of the third order. This curve will be denoted σ in the sequel.

8. Obtain Hesse's theorem, that the locus of the pole of a plane with respect to the two parameter family of quadrics through 7 points is a cubic surface, Σ.

9. Consider the one-parameter family of cones through 7 points, and show that the locus of their vertices is a space sextic, K.

10. The surface Σ depends on the choice of plane in (8), but K does not. Deduce that any two Σ meet in K and variable cubic curve in space.

11. Show that K corresponds to a plane quartic as follows. Let f_1, f_2, and f_3 be any three quadrics through 7 points, then any other quadric through those points has equation $\lambda_1 f_1 + \lambda_2 f_2 + \lambda_3 f_3 = 0$. Let $2F = \Sigma u_{ij} x_i x_j = 0$ be a cone in this family, with vertex (x_1, x_2, x_3, x_4). The u's are linear in the λ's. Eliminate x_1, x_2, x_3, x_4 from $2F = 0$ and obtain $\Delta = \det(u_{ij}) = 0$, which is to be thought of as a quartic in $\lambda_1, \lambda_2, \lambda_3$.

12. Show that a line meets Δ in four points corresponding to four points on K which are the vertices of 4 cones meeting in a common curve, and establish the converse.

13. Deduce from (12) that two lines meeting Δ yield 8 points on K which can be taken as the 8 intersection points of 3 quadrics.

14. Verify Klein's calculation that the following octahedral group

G''_{24} is contained in $PSL(2; Z/7Z)$. Take a Klein four-group

$G'_4 = \{w, -\dfrac{1}{w}, \dfrac{2w + 3}{3w - 2}, \dfrac{3w - 2}{-2w - 3}\}$, find 6 elements of order 4 whose

squares are in G'_4.

Find 6 elements of order 2 which do not lie in G'_4 but commute

with one of the elements of it, and lastly find 4 pairs of

commuting elements of order 3.

Klein gave this solution:

$\dfrac{2w + 2}{-2w + 2}$ and $\dfrac{2w - 2}{2w + 2}$ whose squares are $-1/w$,

$\dfrac{w + 1}{w + 2}$ and $\dfrac{-2w + 1}{w - 1}$ whose squares are $\dfrac{2w + 3}{3w - 2}$

$\dfrac{3w - 3}{-3w + 1}$ and $\dfrac{w + 3}{3w + 3}$ whose squares are $\dfrac{3w - 2}{-2w - 3}$;

$\dfrac{2w - 3}{-3w - 2}$ and $\dfrac{3w + 2}{2w - 3}$ which commute with $-1/w$

$\dfrac{-w + 1}{-2w + 1}$ and $\dfrac{w + 2}{-w - 1}$ which commute with $\dfrac{2w + 3}{3w - 3}$

$\dfrac{3w - 1}{3w - 3}$ and $\dfrac{-3w - 3}{w + 3}$ which commute with $\dfrac{3w - 2}{-2w - 3}$;

and the following commuting pairs of elements of order 3:

$\dfrac{-3w - 1}{2}$, $\dfrac{-2w - 1}{3}$; $\dfrac{2w}{w - 3}$, $\dfrac{3w}{w - 2}$; $\dfrac{2}{3w + 1}$, $\dfrac{-w + 2}{3w}$; $\dfrac{-w + 3}{2w}$, $\dfrac{-3}{2w + 1}$

Show that the G''_{24} exhibited above permutes the following pairs of

projective points in the projective line over the field of 7

elements:

$\{0, -2\}$, $\{1, 2\}$, $\{3, -1\}$, $\{\infty, -3\}$.

Show that $\dfrac{2w - 3}{-3w - 2}$ switches $\{1,2\}$ and $\{-1,3\}$ but fixes $\{0,-2\}$, $\{\infty,-3\}$;

that $\dfrac{-w + 1}{-2w + 1}$ switches $\{1,2\}$ and $\{0,-2\}$ but fixes $\{-1,3\}$ and $\{\infty,-3\}$;

and that $\dfrac{3w - 1}{3w - 3}$ switches $\{1,2\}$ and $\{\infty,-3\}$ but fixes $\{-1,3\}$ and $\{0,-2\}$.

This means that the group generated by those three elements certainly

contains S_4. Show that they generate precisely S_4 by verifying that

they satisfy the appropriate relations.

15. Interpret the 64 characteristics introduced by Riemann as vectors

in the 6 dimensional vector space over the field of 2 elements,

and show how the symplectic form corresponding to the 6 x 6 matrix

$$\begin{pmatrix} 0 & I \\ -I & 0 \end{pmatrix} \equiv \begin{pmatrix} 0 & I \\ I & 0 \end{pmatrix} \pmod 2 \quad \text{picks out the odd characteristics. Hence}$$

relate the group of the bitangents to $Sp(6, \mathbb{Z}/2\mathbb{Z})$. Hesse's

analysis of the bitangents as re-derived by Clebsch enabled

Jordan in his Traité (§332) to show that the group of the

bitangents is actually isomorphic to $Sp(6;\mathbb{Z}/2\mathbb{Z})$.

16. By considering generators and relations, show that G_{168} is also

isomorphic to $SL(3,Z/2Z)$, a result Weber (Lehrbuch der Algebra,

1896, II p.539) tells us Kronecker discovered while working on

the modular equation at the prime 7. This isomorphism well

exhibits the action of G_{168} on the bitangents.

17. By using techniques similar to those in Exercise 14, prove

Hermite's theorem that $PSL(2;\mathbb{Z}/5\mathbb{Z})$ is isomorphic to A_5, the

alternating group on 5 symbols.

CHAPTER VI AUTOMORPHIC FUNCTIONS

This final chapter is, naturally, concerned with the triumphant accomplishments of Poincaré: the creation of the theory of Fuchsian groups and automorphic functions. These developments brought together the theory of linear differential equations and the group-theoretic approach to the study of Riemann surfaces, so this account draws on all of the preceding material. It begins with a significant stage intermediate between the embryonic general theory and the developed Fuchsian theory: Lamé's equation.

6.1 *Lamé's Equation*

Lamé's equation[1] may be written variously as

$$\frac{d^2 y}{dx^2} + \frac{1}{2} \left[\frac{1}{x-e_1} + \frac{1}{x-e_2} + \frac{1}{x-e_3} \right] \frac{dy}{dx} - \frac{(n(n+1)x + B)y}{4(x-e_1)(x-e_2)(x-e_3)} = 0, \quad (6.1.1)$$

which exhibits it algebraically as an equation of the Fuchsian type, as

$$\frac{d^2 y}{du^2} - (n(n+1)\wp(u) + B)y = 0 \tag{6.1.2}$$

which may be called the Weierstrassian form, or as

$$\frac{d^2 y}{dv^2} - (n(n+1)k^2 \, sn^2 v + A)y = 0 \tag{6.1.3}$$

which may be called the Jacobian form. Form (6.1.1) is essentially how it was introduced by Lamé [1845] in connection with a study of triply orthogonal coordinates in space. The substitution $x = \wp(u)$, where $\wp$ is Weierstrass's elliptic function with half-periods e_1, e_2, and e_3 transforms (1) into (2); in this form the equation was studied by Halphen [1884]. The substitution $r = u(e_1 - e_3)^{1/2}$ transforms (2) into (3), where $A = \dfrac{B + e_3 n(n+1)}{(e_1 - e_3)}$ and $k = \left(\dfrac{e_2 - e_3}{e_1 - e_3}\right)^{1/2}$; in this form it was studied by Hermite [1877] and Fuchs [1878].

As an equation of the Fuchsian type it has four regular singular points: e_1, e_2, and e_3, at which the exponents are 0 and $\frac{1}{2}$ in each case, and infinity, at which the exponents are $-\frac{1}{2} n$, $\frac{1}{2} (n + 1)$. Its properties, however, come out more clearly when it is written in either of the elliptic forms. Hermite [1877 = Oeuvres III, 266] summarized work of his predecessors as follows. Lamé had shown that for suitable values of the constant, A, in (6.1.3) one solution can be written as a polynomial of degree n in snx. Liouville [1845] and Heine [1845] independently had studied the second solution and Heine established a connection, via what he called higher order Lamé functions, with spherical harmonics. In a series of papers Hermite [1877-1882 = Oeuvres III, 266-418] considered the case when the constant A may be arbitrary, and showed that the solutions were always elliptic functions of the second kind. These are functions F such that

$$F(x + 2K) = \mu F(x)$$
$$F(x + 2iK') = \mu' F(x)$$

$$(6.1.4)$$

for some constants μ and μ'. Hermite showed that they may always be written in terms of Jacobi's functions Θ and H. The designation of the second kind was introduced by Picard [1879], it conforms with Legendre's classification of elliptic integrals. Hermite gave many examples of how elliptic functions, of both the old and the new kinds, could solve problems in applied mathematics, and initiated a considerable amount of research in that area[2].

Fuchs, in his [1877c] observed that the explicit forms of the solution follow from the elementary fact that, if y_1 is one solution of an equation $\frac{d^2 y}{dx^2} + p_1 \frac{dy}{dx} + p_0 y = 0$, then there is another, linearly independent, solution of the form $y_2 = y_1 \int \frac{e^{-\int p_1 \, dx}}{y_1^2} \, dx$. Hermite's work connected with cases studied by Fuchs [1876a, §13], when working on

the all-solutions-algebraic problem, of equations integrable by elliptic

or Abelian functions: the equation $\dfrac{d^2y}{dx^2} = Py$ has a solution of the form

$y = \phi(z)^{1/2} \; e^{\left(\sqrt{-\frac{\lambda}{4}} \int \frac{dz}{\phi(z)}\right)}$, where $\phi^2(z)$ is rational in z, λ is a constant,

and $P = \dfrac{1}{4}\left(\dfrac{d}{dz}\log\phi\right) + \dfrac{1}{2}\dfrac{d^2}{dz^2}\log\phi - \dfrac{\lambda}{4\phi^2}.$ Fuchs observed [1878a] that

$\lambda = 0$ led to Heine's Lamé functions of higher order, and that the

integrals are elliptic if $\phi^2(z) = (1 - z^2)(1 - k^2z^2)$, when the

differential equation reduces to Lamé's.

In a third paper [1878c] Fuchs investigated what conditions must be

imposed on the coefficient P so that $\dfrac{d^2y}{dx^2} = Py$ has a basis of solutions

consisting of two elliptic functions of the second kind whose poles are

the poles of P. Since $P = \dfrac{d^2y_i}{y_i dx^2}$ for any solution y_i of the equation, P

must be a single-valued doubly periodic function of x, which, Fuchs,

went on to show, can be written explicitly as a sum of Jacobian

functions H (equation 10).

The converse to this theorem was established by Picard, who showed

[1879b, 1880a, b] that every differential equation $\dfrac{d^2y}{dx^2} + p\dfrac{dy}{dx} + qy = 0$

of the Fuchsian class with doubly periodic coefficients and single-valued

solutions, has as its general solution a sum of two doubly periodic

functions of the second kind, with an analogous result for equations of

higher order. He also showed how the solutions can in general be written

explicitly; the exceptional cases were treated by Mittag-Leffler [1880].

Picard's [1881] presented his results to a German audience with

extensions to systems of first-order linear equations. The result

about the basis of solutions followed easily from Picard's observation

that, if it is false, then there are equations of the form

$$f(x + 4K) = Af(x) + Bf(x + 2K)$$

$$f(x + 4iK') = A'f(x) + B'f(x + 2iK')$$

where f is a solution of the differential equation and 2K and 2iK' are the periods of p and q. But then a suitable linear combination of f(x) and f(x + 2K) can be found, say $\alpha f(x) + \beta f(x + 2K)$, such that

$$\alpha f(x + 2K) + \beta f(x + 4K) = \lambda(\alpha f(x) + \beta f(x + 2K))$$

for some λ, and so $\alpha f(x) + \beta f(x + 2K)$ is the sought-for solution. Once it has been found, and because $e^{-\int p dx}$ is doubly periodic since the general solution is single-valued, the function $\dfrac{e^{-\int p dx}}{f^2(x)}$ is also doubly-periodic of the second kind. There is then an independent doubly periodic solution (of the second kind): $f(x) \displaystyle\int \dfrac{e^{-\int p dx}}{f^2(x)}$. The quasi-periods of the solutions are the periods of the coefficients. It may, of course, happen that the solutions are themselves doubly-periodic functions of the first kind.

A thorough study of equations with doubly-periodic coefficients was made by Halphen in 1880 in his prize-winning essay (published as [1884] and again in [1921]). He showed (p55) that any linear differential equation with single-valued[3] doubly-periodic coefficients and a pair of independent solutions whose ratio is only undefined at infinity can be transformed into one of the same form for which the solutions are single-valued. The periods of the coefficients will in general be changed by this transformation. Halphen drew on the earlier results of Hermite and Picard, and on Weierstrass's theory of elliptic functions as presented in [Kiepert, 1874] and [Mittag-Leffler, 1876][4]. Halphen also showed how to solve the hypergeometric equation when its solutions are elliptic functions, and how to solve Lamé's equation: e.g. (93, equation 44)

$$\frac{d^2y}{du^2} = \frac{3}{4} \quad (u)y$$

has the solutions

$$y = (' \, (\tfrac{u}{2}))^{-1/2} \text{ and } y = (' \, (\tfrac{u}{2}))^{-1/2} \quad (u/2).$$

Halphen's presentation also showed how one might proceed from equations with rational coefficients to those with elliptic coefficients and then to those with more general algebraic coefficients via the theory of invariants (Chapter IV above). Nevertheless, for the most part he confined himself to cases integrable by elliptic functions. The successful treatment of the more general cases was inspired by the work of Fuchs which will now be discussed, and presented to the Academy in an essay which took second place to Halphen's.

In a series of paper [1880 a, b, c, d, e, 1881 a, b, c] (this summary follows [1880 a, b]) Fuchs took the equation

$$\frac{d^2y}{dz^2} + P(z) \frac{dy}{dz} + Q(z)y = 0, \tag{6.1.5}$$

where P and Q are rational functions of z, took functions $f(z)$ and $\phi(z)$ as a basis of solutions for it, and considered, by analogy with Jacobi inversion, the equations

$$\int_{\zeta_1}^{z_1} f(z)dz + \int_{\zeta_2}^{z_2} f(z)dz = u_1$$

$$\int_{\zeta_1}^{z_1} \phi(z)dz + \int_{\zeta_2}^{z_2} \phi(z)dz = u_2$$

$$\tag{6.1.1}$$

as defining functions of u_1 and u_2: $z_1 = F_1(u_1, u_2)$, $z_2 = F_2(u_1, u_2)$. Plainly by varying the paths of integration one obtains

$$F_i(\alpha_{11}u_1 + \alpha_{12}u_2 + \alpha_1 c, \ \alpha_{21}u_1 + \alpha_{22}u_2 + \alpha_{22}c) = F_i(u_1, u_2), \ i = 1, 2,$$ for integrers α_y which describe the monodromy of u_1 and u_2 as the paths cross the cuts joining the singularities of (6.1.5) to ∞ and α_1 and α_2 are analogous to the periods of an elliptic integral. What follows is obscure in Fuchs.

He let a_i, $i = 1, \ldots, n$, and ∞ be the singular points of the differential equation, and took the roots of the associated indicial equations to be $r_1^{(i)}$, $r_2^{(i)}$, $i = 1, \ldots, n$, and s_1 and s_2. He let a and b be two distinct points, possibly infinite, and let u_1, u_2 be such that $z_1(u_1, u_2) = a$ and $z_2(u_1, u_2) = b$ and let either a, b, or both be singular points, but insisted that $f(z_2)\phi(z_1) - f(z_1)\phi(z_2) \neq 0$. (Call these conditions A.) Then the four derivatives $\dfrac{\partial z_i}{\partial u_j}$ are holomorphic functions of z_1, z_2 near $z_1 = a$, $z_2 = b$. Furthermore every value $(z_1, z_2) \in \phi^2$ is to be attainable with finite $(u_1, u_2) \in \phi^2$.

For this it is necessary and sufficient that at each finite singular point a_i, $r_1^{(i)}$ and $r_2^{(i)}$ satisfy $r_1^{(i)} = -1 + \dfrac{1}{n_i}$, $r_2^{(i)} = -1 + \dfrac{k_i}{n_i}$, $1 < k_i < n_i$, n_i, k_i positive integers and at ∞ $s_1 = 1 + \dfrac{1}{n}$, $s_2 = 1 + \dfrac{k}{n}$, $1 < k$; n, k, positive integers.

Extra conditions on the roots of the indicial equation ensure that $\dfrac{f(z)}{\phi(z)} = \zeta$ defines z as a single valued functions of ζ and that the equation $f(z_1)\phi(z_2) - f(z_2)\phi(z_1) = 0$ has only the trivial solution $z_1 = z_2$. They are that either $r_2^{(i)} - r_1^{(i)} = 1$ or $\dfrac{1}{n_i}$ and either $s_2 - s_2 = 1$ or $\dfrac{1}{n}$, together with the extra condition that the solutions to the differential equation do not involve logarithmic terms (conditions B). If furthermore $r_2^{(i)} - r_1^{(i)} = \dfrac{1}{n_i}$ or $r_2^{(i)} = -r_1^{(i)} = \dfrac{1}{2}$, and $s_2 - s_1 = \dfrac{1}{n}$ or $s_2 = \dfrac{5}{2}$, $s_1 = \dfrac{3}{2}$ (conditions C) and if ∞ is not a regular point, but Fuchs is obscure here, then z_1, z_2 are roots of a quadratic equation, whose coefficients are single-valued holomorphic functions of u_1, u_2. In this case the number of finite singular points cannot exceed, but can equal, 6. In the example where 6 finite singular points occur the functions

$F_1(u_1, u_2)$, $F_2(u_1, u_2)$ are necessarily hyperelliptic, but generally they will not be even Abelian functions, since the differential equation will not be algebraically integrable.

Fuchs's proofs of these assertions proceeded by a case by case analysis of each kind of singularity that could occur in terms of the local power series expansions of the functions. Poincaré pointed out (see below) that the analysis rapidly becomes confusing and was, in any case, incomplete.

The condition that no logarithmic terms appear in the solutions to the differential equation even though $r_2 - r_1 = 1$, an integer, is a strong restriction on the kind of branching that can occur. Fuchs also seems to be assuming, or perhaps is only interested in, the case when ζ takes very value in C, not merely in some disc.

The curious condition that there are at most six finite singular points was obtained simply, as follows (§7). From the general theory of differential equations of the Fuchsian type, if the number of finite singular points is A then

$$\sum_i (r_1^{(i)} + r_2^{(i)}) + s_1 + s_2 = A - 1.$$

If there are A' where $r_1 = \frac{-1}{2}$ and $r_2 = \frac{1}{2}$, and A" others, where necessarily $r_1^{(i)} + r_2^{(i)} = -2 + \frac{k_i}{n_i}$, $1 < k_i < n_i$, then

$$\sum_{i=1}^{A''} \frac{3}{n_i} + s_1 + s_2 = A' + 3A'' - 1 \quad \text{(summing only over the second kind)}.$$

But $s_1 + s_2 \leq 5$, and $\frac{1}{n_i} \leq \frac{1}{2}$ implies

$$3 \sum_1^{A''} \frac{1}{n_i} \leq \frac{3A''}{2},$$

so $A' + 3A'' - 1 \leq \frac{3A''}{2} + 5$ or

$$A + \frac{1}{2} A'' \leq 6$$

so $A \leq 6$, as was to be shown.

This case can arise when, say, ∞ is an ordinary point, $s_1 = 2$ and $s_2 = 3$. These were the conclusions drawn from conditions (C).

As an example of the case when there are 6 singular points Fuchs (§8) adduced the hyperelliptic integrals

$$y_1 = \int \frac{g(z)}{\sqrt{\phi(z)}}, \quad y_2 = \int \frac{h(z)}{\sqrt{\phi(z)}}, \quad \text{where } \phi(z) = (z - a_1) \ldots (z - a_6)$$

and ∞ is not a singular point, so $s_1 = 2$, $s_2 = 3$, and $g(z)$ and $h(z)$ and linearly independent polynomials of degree 0 or 1 (say $g(z) := 1$, $h(z) := z$). In this case $z_1 = F_1(u_1, u_2)$ and $z_2 = F_2(u_1, u_2)$ are hyperelliptic functions of the first kind, but (§9) non-Abelian functions may arise. Fuchs gave as an example an equation with finite singular points

a_1, at which $r_1^{(1)} = -\frac{2}{3}$, $r_2^{(1)} = -\frac{1}{3}$

a_0, at which $r_2^{(1)} = -\frac{5}{6}$, $r_2^{(2)} = -\frac{4}{6}$, and

∞, at which $s_1 = \frac{3}{2}$, $s_2 = 2$ (These satisfy conditions (C)

with $n_1 = 3$, $n_2 = 6$, $n = 2$.)

The original differential equation $\frac{d^2 y}{dz^2} + P(z) \frac{dy}{dz} + Q(z)y = 0$ is transformed by the substitution $y = (z - a_1)^{-1}(z - a_2)^{-5/4} \omega$ into one with exponents $\frac{1}{3}, \frac{2}{3}$ at a_1, $\frac{5}{12}, \frac{7}{12}$ at a_2, and $\frac{-3}{4}, \frac{-1}{4}$ at ∞ and so is of the form $\frac{d^2 \omega}{dz^2} = P\omega$. But, by Fuchs's test (Chapter III, p.119) the denominators of the exponents at a_2 are $12 > 10$, so the equation is not algebraically integrable unless its solutions (or a second degree homogeneous polynomial in the solutions) is the root of a rational function. But this is not possible either, since the denominators at a_1 and a_2 are neither 1, 2, nor 4.

Fuchs was chiefly concerned to study the inversion of equations
(6.1.6) and only slightly interested in the function $\zeta = \dfrac{f(z)}{\phi(z)}$. His
obscure papers rather confused the two problems but they were soon to
be disentangled.

On the 29th May 1880, a young Frenchman at the University of Caen
wrote a letter to Fuchs expressing interest in the subject of Fuchs's
paper in the Journal für Mathematik, but seeking clarification of some
points. The author was very interested in the global theory of
differential equations, whether first-order real or linear and complex.
Within two years his work was to transform both subjects completely,
opening up whole new aspects of research in the one, and in the other
leaving little, it has been said, for his successors to do. His name
was Jules Henri Poincaré.

6.2 Poincaré

Poincaré was born at Nancy on 29 April, 1854. His father was
professor of medicine at the university there, his mother, a very active
and intelligent women, consistently encouraged him intellectually, and
his childhood seems to have been very happy.[5] He did not at first show
an exceptional aptitude for mathematics, but towards the end of his
school career his brilliance became apparent, and he entered the École
Polytechnique at the top of his class. Even then he displayed what
were to be life-long characteristics: a capacity to immerse himself
completely in abstract thought, seldom bothering to resort to pen and
paper, a great clarity of ideas, a dislike for taking notes so that he
gave the impression of taking ideas in directly, and a perfect memory
for details of all kinds. When asked to solved a problem he could
reply, it was said, with the swiftness of an arrow. He had a slight

stoop, he could not draw at all, which was a problem more for his examiners than for him, and he was totally incompetent in physical exercises. He graduated only second from the École Polytechnique because of his inability to draw, and proceeded to the École des Mines in 1875. In 1878 he presented his doctoral thesis to the faculty of Paris on the subject of partial differential equations. Darboux said of it that it contained enough ideas for several good theses, although some points in it still needed to be corrected or made precise. This fecundity and inaccuracy is typical of Poincaré; ideas spilled forth so fast that, like Gauss, he seems not to have had the time to go back over his discoveries and polish them. On the first of December 1879 he was in charge of the analysis course at the Faculty of Sciences at Caen, there, perhaps, as a result of a rare attempt by the ministry of education not to allow Paris to recruit talent at the expense of the provinces.

1880 was a busy time for him. On the 22nd of March, he deposited his essay "Mémoire sur les courbes définies par une équation différentielle" with the Académie des Sciences as his entry in their prize competition. That essay considered first-order non-linear differential equations $\frac{dx}{X} = \frac{dy}{Y}$, where X and Y are real polynomial functions of real variables x and y, and investigated the global properties of their solutions. He later withdrew the essay, on 14 June 1880, without the examiners reporting on it, perhaps wanting to concentrate on the theory of complex differential equations.

Poincaré's question to Fuchs concerned the nature of inverse function $z = z(\zeta)$. Fuchs had claimed that z is meromorphic function of $\zeta = \frac{f(z)}{\phi(z)}$, whether z is an ordinary or a singular point of the differential equation. Indeed, z is finite at ordinary points and infinite at singular points. Poincaré observed[6] that z is meromorphic at $\zeta = \infty$, so $z = z(\zeta)$ seems to be meromorphic on the whole ζ-sphere, and so is a rational function of ζ. But then the original differential equation must have all its solutions algebraic, which Fuchs had expressly denied. Poincaré suggested that there were three kinds of ζ value: those reached by $\dfrac{f(z)}{\phi(z)}$ as z traced out a finite contour on the z-sphere; those reached on an infinite contour, and those which are not attained at all. *A priori*, he said, all three situations could occur, and indeed the last two would if the differential equation did not have all its solution algebraic. Fuchs's proof would only work for ζ-values of the first kind; however, Poincaré went on, he could show that $z(\zeta)$ was meromorphic even if the other kinds occurred, and he was led to hypothesize that (1) if indeed all ζ-values were of the first kind then z would be a rational function; (2) if there are values of only the first and second kinds but z is monodromic at the values of the second kind then Fuchs's theorem is still true; (3) if z is not monodromic or (4) if the values of the third kind occur and so the domain of z is only a domain D on the ζ-sphere, then z is single-valued on D. In this case the ζ-values of

Fig. 6.1

the first kind occur inside D, as shown. Those of the second kind lie on the boundary of D, and the unattainable values lie outside D. There is, finally, a fifth case, when all three kinds of ζ occur, but D has this form

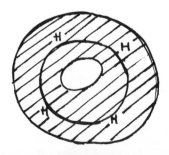

Fig. 6.2

values of the first kind filling out the annulus. Now, said Poincaré, z will not return to its original value on tracing out a closed curve HHHH in D.

Fuchs replied on the 5th June: "Your letter shows you read German with deep understanding, so I shall reply in it". He admitted that his Theorem I was imprecisely worded, and suggested that the hypothesis of his earlier Göttingen Nachrichten articles were to be preferred, namely that the exponents at the ith singular point satisfy either $r_2^{(i)} = r_1^{(i)} + 1$ or $r_1 = -1 + \frac{1}{n_i}$, $r_2 = -1 + \frac{2}{n_i}$, and the exponents at infinity satisfy $\rho_2 = \rho_1 + 1$ or $\rho_1 = 1 + \frac{1}{\nu}$, $\rho_2 = 1 + \frac{2}{\nu}$ for integers n_i, ν. He added a few words on the meaning: he excluded paths in which $f(z)$ and $\phi(z)$ both become infinite, and then the remaining ζ-values filled out a simply connected region of the ζ-plane with the excluded values on the boundary.

Poincaré replied on the 12th, apologising for the delay in doing so, but he had been away. He found some points were still obscure, and suggested the following argument. Suppose the singular points of the differential equation are joined to ∞ by cuts, and z moves without crossing the cuts. Then ζ traces out a connected region F_0. Let z cross the cuts, but no more than m times, then the values of ζ fill out a connected region F_m. As m tends to infinity F_m tends to the region Fuchs called F, and F will be simply connected if F_m is simply connected for all m. Now, asked Poincaré," is that a consequence of your proof? One needs to add some explanation." He agreed that F_m could not cover itself as it grew in this fashion:

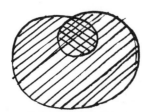

Fig. 6.3

but the proof left open this possibility:

Fig. 6.4

He said that when there were only two finite singular points it was true that z was a single-valued function, "that I can prove differently" and he went on "but it is not obvious in general. In the case that there are only two finite singular points I have found some remarkable properties of the functions you define, and which I intend to publish. I ask your permission to give them the name of Fuchsian functions". In conclusion, he asked if he might show Fuchs's letter to Hermite.

Fuchs replied on the 16th, promising to send him an extract of his forthcoming complete list of the second order differential equations of the kind he was considering. This work, he said, makes any further discussion superfluous. He was very interested in the letters, and very pleased about the name. Of course his replies could be shown to Hermite.

The reply points to an interesting difference of emphasis between the two men. For Fuchs the main problem was to study functions obtained by inverting the integrals of solutions to a differential equation, thus generalizing Jacobi inversion. As a special case one might also ask that the inverse of the quotient of the solutions is single valued, which imposed extra conditions. For Poincaré, interested in the global nature of the solutions to differential equations, it is only the special case which was of interest, and he gradually sought to emancipate it from its Jacobian origins. There is also some humour in the situation of the young man gently explaining about analytic

continuation and the difference between single-valued and unbranched
functions, to one who had consistently studied and applied the technique
for fifteen years.

Poincaré's reply of the 19th June (one is struck by the efficiency
of the postal service almost as much as by the rapidity of his thought)
pointed up this difference of emphasis. Taking the condition on the
exponents to be

$$r_1^{(i)} = -1 + \frac{1}{n_i}, \ r_2^{(i)} = -1 + \frac{2}{n_i} \text{ or } r_1^{(i)} = -\frac{1}{2} = -r_2^{(i)}, \text{ and}$$

$$s_1 = 1 + \frac{1}{n}, \ s_2 = 1 + \frac{2}{n} \text{ or } s_1 = \frac{3}{2}, \ s_2 = \frac{5}{2} \text{ (at infinity)},$$

he wrote that he had found that when the differential equation was put
in the form $y'' + qy = 0$, the finite singular points with exponent
difference 1 vanished. Thus he found that at all the finite singular
points the exponent difference was an aliquot part of 1 and not equal
to 1, and that there were no more than 3 singular points. If there was
only one, then z was a rational function of ζ. If there were two, and
the exponent differences were ρ_1, ρ_2, and ρ_3 at infinity, then either
$\rho_1 + \rho_2 + \rho_3 > 1$, in which case z is rational in ζ, or $\rho_1 + \rho_2 + \rho_3 = 1$,
in which case z was doubly periodic. Even in this case there were
difficulties, as he showed with this example. Let $z = \Lambda(u)$ be a doubly
periodic function, and set $\eta = (\frac{dz}{du})^{1/2} e^{\alpha u}$. Then η and $(\frac{dz}{du})^{1/2} e^{-\alpha u}$
satisfy a second-order linear differential equation and $\zeta = e^{2\alpha u}$. So
for z to be single valued in ζ, it must have period $\frac{i\pi}{\alpha}$ as a function of u,
which, he pointed out, was not the case in general. Finally, if there
were three finite singular points then the exponents would have to be
$-\frac{1}{2}$ and 0, and at infinity they would be $\frac{3}{2}$, 2. But although these
satisfied Fuchs's criteria z was not a single-valued function of ζ, so
the theorem is wrong. Poincaré proposed to drop the requirement that

Fuchs's functions $z_1 + z_2$, $z_1 \cdot z_2$ be single-valued in u_1 and u_2. He
went on to say that this give a "... much greater class of equations than
you have studied, but to which your conclusions apply. Unhappily my
objection requires a more profound study, in that I can only treat two
singular points". Dropping the conditions on $z_1 + z_2$, $z_1 \cdot z_2$ admits the
possibility that the exponent differences ρ_1, ρ_2, and ρ_3 satisfy
$\rho_1 + \rho_2 + \rho_3 < 1$. Now z is neither rational nor doubly periodic, but is
still single valued. "These functions I call Fuchsian, they solve
differential equations with two singular points whenever ρ_1, ρ_2, and
ρ_3 are commensurable with each other. Fuchsian functions are very
like elliptic functions, they are defined in a certain circle and are
meromorphic inside it". On the other hand, he concluded, he knew
nothing about what happened when there were more than two singular
points.

We do not have Fuchs's reply, but Poincaré wrote to him again on
the 30th of July to thank him for the table of solutions "which lifts
my doubts completely", although he went on to point out a condition
on some of the coefficients of the differential equations which Fuchs
had not stated explicity in the formulation of his theorems. As to
his own researches on the new functions, he remarked that they
"... present the greatest analogy with elliptic functions, and can be
represented as the quotient of two infinite series in infinitely many
ways. Amongst those series are those which are entire series playing
the role of Theta functions. These converge in a certain circle and
do not exist outside it, as thus does the Fuchsian function itself.
Besides these functions there are others which play the same role as
the zeta functions in the theory of elliptic functions, and by means
of which I solve linear differential equations of arbitrary orders with

rational coefficients whenever there are only two finite singular points and the roots of the three determinantal equations are commensurable. I have also thought of functions which are to Fuchian functions as abelian functions are to elliptic functions and by means of which I hope to solve all linear equations when the roots of the determinantal equations are commensurable. Finally functions precisely analogous to Fuchsian functions will give me, I think, the solutions to a great number of differential equations with irrational coefficients."

Poincaré's last letter (20th March, 1881) merely announces that he will soon publish his research on the Fuchian functions, which partly resemble elliptic functions and partly modular functions, and on the use of zeta Fuchsian functions to solve differential equations with algebraic coefficients. In fact, his first two articles on these matters already appeared in the Comptes Rendus, and these will be discussed below.

Poincaré had not had long to study Fuchs's work by the time he wrote him his first letter on 29th May 1880, for he had only received the journal at the start of the month. Yet, incredibly, he had already written up and presented his findings in the form of an essay for the prize of the Paris Académie des Sciences. Indeed, he had submitted this entry on the 28th May 1880, the day before he wrote to Fuchs. His essay was awarded second prize, behind Halphen's, but it was not published until Nörlund edited it for Acta Mathematica (vol 39, 1923, 58-93 = Oeuvres, I, 578-613). In awarding it second prize, Hermite said of the essay, which also discussed irregular solutions, that:

"... the author successively treated two entirely different questions
of which he made a profound study with a talent by which the commission
was greatly struck. The second ... concerns the beautiful and important
researches of M. Fuchs, ... The results ... presented some lacunas in
certain cases that the author has recognized and drawn attention to in
thus completing an extremely interesting analytic theory. This theory
has suggested to him the origin of transcendents, including in particular
elliptic functions, and which has permitted him to obtain the solutions
to linear equations of the second order in some very general cases.
A fertile path is there that the author has not entirely gone down,
but which manifests an inventive and profound spirit. The commission
can only urge him to follow up his researches in drawing to the
attention of the Academy the excellent talent of which they give proof."
(Quoted in Poincaré, Oeuvres, II, 73.)

In the essay Poincaré considered when the quotient $z = \dfrac{f(x)}{\phi(x)}$ of
two independent solution of the differential equation $\dfrac{d^2 y}{dx^2} = Qy$ defines,
by inversion, a meromorphic function x of z. He found Fuchs's
conditions were not necessary and sufficient. It was necessary and
sufficient for x to be meromorphic on some domain that the roots of
the indicial equation at each singular point, including infinity, differ
by an aliquot part of unity (i.e. $\rho_1 - \rho_2 = 1/n$, for some positive
integer n). If the domain is to be the whole complex sphere then this
condition is still necessary, but it is not longer sufficient. He
found that Fuchs's methods did not enable him to analyse the question
very well, as special cases began to proliferate, and sought to give
it a more profound study, beginning with Fuchs's example ([Fuchs 1880b,
168 = 1906, p.210], above p.281) of a differential equation in which
there are two finite singular points a_1 and a_2, where the exponent

differences are $\frac{1}{3}$ and $\frac{1}{6}$, and the exponent difference at ∞ is $\frac{1}{2}$. In this case he found the change in z was of the form

$$z \to z'', \quad \frac{z''-\alpha}{z''-\beta} = e^{2\pi i/3} \left(\frac{z-\alpha}{z-\beta}\right) \text{ upon analytic continuation around } a_1,$$

$$z \to z'', \quad \frac{z''-\gamma}{z''-\delta} = e^{\pi i/3} \left(\frac{z-\gamma}{z-\delta}\right), \text{ upon analytic continuation around } a_2,$$

and $\frac{1}{z} \to -\frac{1}{z}$ (around ∞).

Accordingly x is a meromorphic single-valued function of z mapping a parallelogram composed of eight equilateral triangles onto the complex sphere, and $z = \infty$ is its only singular point, so z is an elliptic function. The differential equation, Poincaré showed, has in fact an algebraic solution $y_1 = (x - a_1)^{1/3} (x - a_2)^{5/12}$ and a non-algebraic solution y_2 such that

$$\frac{y_2}{y_1} = \int (x - a_1)^{-2/3} (x - a_2)^{-5/6} dx.$$

This result agrees with Fuchs's theory.

Poincaré next investigated when a doubly-periodic function can give rise to a second order linear differential equation, and found that one could always exhibit such an equation having rational coefficients for which the solution was a doubly periodic function having 2 poles. If furthermore the periods, h and K, were such that

$$2i\pi = 0 \ (\text{mod } h, \ K)$$

then x would be a monodromic function of z with period $2i\pi$.

His reasoning is too lengthy to reproduce, but the condition on h and K derives essentially from the behaviour of z under analytic continuation: if z is monodromic it must reproduce as $z \to z' = \frac{az+b}{a'z+b'}$ (for some a, b, a', b'), and thus can be written as

$$\frac{z'-\alpha}{z'-\beta} = \lambda \frac{z-\alpha}{z-\beta},$$

and $\lambda^n = 1$. After a further argument Poincaré concluded (p.79) that (i) there were cases when one solution of the original differential equation was algebraic, and then Fuchs's theory was correct, but (ii) there were cases when the differential equation had four singular points, elliptic functions were involved, and then extra conditions were needed.

However, it might be that the domain of x could not be the whole z-sphere. Poincaré showed that this could happen even when the differential equation had only two finite singular points. For example if the exponent differences were $\frac{1}{4}$, $\frac{1}{2}$ and $\frac{1}{6}$ at ∞, then as long as x crosses no cuts z stays within a quadrilateral (Figure 2, p.86)α $0\alpha'$ γ.

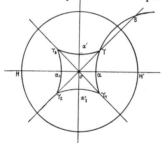

Fig. 6.5

Furthermore, however x is conducted about in its plane, z cannot escape the circle HH'. Poincaré described the quadrilateral as 'mixtiligne', the circular-arc sides meet the circle HH at right angles. This geometric picture is quite general, curvilinear polygons are obtained with non-re-entrant angles and circular arc sides orthogonal to the boundary circle. Thus the domain of x is $|z| < OH$, and Poincaré then investigated whether x is meromorphic. This reduces to showing that, as x is continued analytically, the polygons do not overlap. This does not occur if the angles satisfy conditions derived from Fuchs's theory, unless the overlap is in the form of an annular region:[7]

Figs. 6.6a, b

However, if the angles are not re-entrant, this cannot happen, and so x is meromorphic. Poincaré's proof of this is of incidental interest. He projected the circle HH' stereographically onto the southern hemisphere of a sphere, and then projected the image orthogonally back onto its original plane. The circular arcs orthogonal to HH' become straight lines, which renders the theorem trivial.

This result virtually concluded Poincaré's essay. As he said in his letter to Fuchs, his understanding was limited essentially to the case of two finite singular points. But one notices in this last argument that the final image of the disc is the Beltrami picture of non-Euclidean geometry. Poincaré did not recognize it at this stage, but he soon did.

Poincaré himself has left us one of the most justly celebrated accounts of the process of mathematical discovery, which concerns exactly his route to the theory of Fuchsian functions. Although it is very well known, it is not apparent from the usual sources (Darboux [Poincaré, Oeuvres II, 1vii, 1viii], Hadamard [1954,12-15]) precisely how and when it connects with the correspondence with Fuchs; even the year, 1880, is left unstated. Poincaré gave this account in a lecture he gave to the Société de Psychologie in Paris 1908, and it was later published as the third essay in his volume Science et Méthode [1909]. Recently, I discovered that there is still more evidence about Poincaré's work in 1880, which does not appear to have been considered before. The Comptes Rendus record three anonymous supplements to the essay bearing the motto 'Non inultus premor', received by the Academy on 28th June, 6th September, and 20th December. Prize essays were submitted anonymously and only identified by a motto, this motto identifies the supplements as Poincaré's (it is in fact the motto of his home town of Nancy). The supplements are to be found in the

Poincaré dossier in the Académie des Sciences, but for some reason Nörlund did not publish them when he published the essay in Acta Mathematica, nor have they been included in Poincaré's Oeuvres. The supplements confirm and greatly amplify what Poincaré said in the lecture 28 years later.[8]

He began by doubting that Fuchsian functions could exist, but shortly came to the opposite view. He tells us in the lecture that:

"For two weeks I tried to prove that no function could exist analogous to those I have since called the Fuchsian functions: I was then totally ignorant. Every day I sat down at my desk and spent an hour or two there: I tried a great number of combinations and never arrived at any result. One evening I took a cup of coffee, contrary to my habit; I could not get to sleep, the ideas surged up in a crowd, I felt them bump against one another, until two of them hooked onto one another, as one might say, to form a stable combination. In the morning I had established the existence of a class of Fuchsian functions, those which are derived from the hypergeometric series. I had only to write up the results, which just took me a few hours." ([1909], 50.)

This account is consistent with his knowledge of Fuchs's problem as presented in the prize essay, and plainly marks his realization of the fundamental invariance properties of the new functions. It is most likely therefore that it refers to the period between 29 May and 12 June 1880 (the date of the second letter). He may also have accomplished the second stage of his discoveries by that time:

"I then wanted to represent the functions as a quotient of two series; this idea was perfectly conscious and deliberate; the analogy with elliptic functions guided me. I asked myself what must be the property of these series, if they exist, and came without difficulty to construct the series that I called theta-fuchsian." ([1909], 51.)

His probable ignorance of Riemann's theory of θ-functions may well have been a blessing to him here, for the analogy is with the Jacobian theory of the θ-function of a single variable.

Next comes his marvellous, almost Proustian, donné:

"At that moment I left Caen when I then lived, to take part in a geological expedition organized by the École des Mines. The circumstances of the journey made me forget my mathematical work; arrived at Coutances we boarded an omnibus for I don't know what journey. At the moment when I put my foot on the step the idea came to me, without anything in my previous thoughts having prepared me for it; that the transformations I had made use of to define the Fuchsian functions were identical with those of non-Euclidian geometry. I did not verify this, I did not have the time for it, since scarcely had I sat down in the bus than I resumed the conversation already begun, but I was entirely certain at once. On returning to Caen I verified the result at leisure to salve my conscience." ([1909], 51, 52.)

The letter of 12th June records that he had been away from Caen, and speaks of the remarkable properties of these new functions. One may surely conclude that this alludes to the connection with non-Euclidean geometry. It would be possible that the reference to being away is misleading, and that the crucial bus journey took place between the 12th and the 19th. This would explain the more independent tone of the third letter, and its explicit reference to the domain of the functions being a circle. Nonetheless it is the second letter which speaks of the dramatic nature of the discoveries, whereas the third is more of a methodical reworking of the material guided by the new insights. It is hard to account for the excitement of the second letter if the progress is only arriving 'without difficulty' at the θ-fuchsian series. That no-one had had such an idea before does not seem to have a particular source of joy to Poincaré, who happily lacked the personally competive streak of some mathematicians.

The first supplement sheds considerable light on Poincaré's grasp of Fuchsian functions and non-Euclidean geometry at this time. It is 80 pages long, and was received by the Académie on 28th June 1880. Poincaré began by reviewing the tessellation of the disc by 'mixtiligne' quadrilaterals obtained by successively operating on one, which he called Q, by transformations M and N. He observed (p.9) that these transformation form a group, and remarked:

"There is a direct connection between the preceding considerations and the non-Euclidean geometry of Lobachevskii. What indeed is a geometry? It is the study of a group of operations formed by the displacements one can apply to a figure without deforming it. In Euclidean geometry this group reduces to rotations and translations. In the pseudo-geometry of Lobachevskii it is more complicated.

"Well, the group of operations which are combinations of M and N is isomorphic ('isomorphe') to a group contained in the pseudo-geometric group. To study the group of combinations of M and N is therefore to have to do with the geometry of Lobachevskii. The

pseudogeometry, as a consequence, is going to furnish us with a
convenient language for expressing what we will have to say about this
group".

(Poincaré's emphasis.)

Poincaré's realization on boarding the bus at Coutances can be
described very simply. He realized that the straighted version of the
'mixtiligne' figures described at the end of his Prize essay were
identical with the figures in Beltrami's description of non-Euclidean
geometry; that therefore the original figures were conformally accurate
representations of non-Euclidean figures; and finally that this meant
the transformations formed from M and N were non-Euclidean isometries.
Beltrami's detailed discussion of the non-Euclidean differential
geometry of the disc (described in Appendix 4) enabled Poincaré to give
a new meaning to his previously analytical transformations. Consequently,
on p.20, he remarked that:

"The Fuchsian functions are to the geometry of Lobachevskii what the
doubly periodic functions are to that of Euclid."

Fuchsian functions only illuminate the study of a differential
equation if they can be defined independently of the equation. This
Poincaré did by introducing the Fuchsian and theta-Fuchsian series.
He attempted to prove their convergence by an ingenious argument which
again appealed to non-Euclidean geometry. The theta-Fuchsians are
perhaps easier to understand. Poincaré took an arbitrary rational
function, H, and considered $\Sigma_K H(zK) \frac{d}{dt} (zK)^m$, where the summation is
taken over all elements K in the group under consideration. The
convergence argument involved considering the partial sum over n
elements K_1, ..., K_n, and estimating the non-Euclidean area and

Euclidean perimeter of the region $QK_1 \cup \ldots \cup QK_n$. Poincaré showed that the sum converged if $m > 1$, but was unable to show that his series defining the Fuchsian functions converged. However, he remarked (p.66):

"The quotient of two theta-Fuchsian series (corresponding to the same value of m) is a rational function of a Fuchsian function."

He also introduced zeta-Fuchsian functions (p.49) to solve differential equations where the exponents are arbitrary rationals, and not simply reciprocals of integers. All this will be discussed in more detail below.

Poincaré remained stuck on the case of the hypergeometric equation at least until his fourth letter, 30th July. Liberation came from an unexpected source, arithmetic, just as arithmetical considerations had earlier enriched Klein's work. But here the response was to be entirely different:

"I then undertook to study some arithmetical questions without any great result appearing and without expecting that this could have the least connection with my previous researches. Disgusted with my lack of success, I went to spend some days at the sea-side and thought of quite different things. One day, walking along the cliff, the idea came to me, always with the same characteristics of brevity, suddenness, and immediate certainly, that the arithmetical transformations of ternary indefinite quadratic forms were identical with those of non-Euclidean geometry.

"Once back at Caen I reflected on this result and drew consequences from it; the example of quadratic forms showed me that there were Fuchsian groups other than those which correspond to the hypergeometric series; I saw that I could apply them to the theory of theta-fuchsian series, and that, as a consequence, there were Fuchsian functions other than those which derived from the hypergeometric series, the only ones I knew at that time. I naturally proposed to construct all these functions; I laid siege systematically and carried off one after another all the works begun; there was one however, which still held out and as the chase became involved it took pride of place. But all my efforts only served to make me know the difficulty better, which was already something. All this work was quite conscious." ([1909], 52, 53.)

The second supplement, of 23 pages, was received by the Académie on 6th September 1880. Most of it is given over to a more rigorous description of non-Euclidean geometry, and to tessellations of the disc by polygons with angles π/m for integers m. When the polygon is a triangle, he also discussed more carefully the ways of constructing Fuchsian functions in this case and was led to conjecture a result which he said he was not yet in any state to prove - it is the Riemann mapping theorem! Then, on p.17, he abruptly stated the connection with the theory of quadratic forms, a subject he had studied under Hermite.

He let T be a matrix ('substitution') with integer coefficients which preserved an indefinite ternary quadratic form Φ, and S be a substitution sending $\xi^2 + \eta^2 - \zeta^2$ to Φ. Then STS^{-1} preserves $\xi^2 + \eta^2 - \zeta^2$ and sends (ξ, η, ζ) to (ξ', η', ζ'), say. The quantities

$$z = \frac{\xi}{\zeta} + \sqrt{-1}\,\frac{\eta}{\zeta}, \quad z' = \frac{\xi'}{\zeta'} + \sqrt{-1}\,\frac{\eta'}{\zeta'}$$

are related by transformation $z' = zK$ of the non-Euclidean plane provided $\xi^2 + \eta^2 - \zeta^2 < 0$. He did not prove that a sheet of the

hyperboloid of two sheets provides a model of non-Euclidean geometry[9] –
which is easy enough to prove (see p. 395) – and remarked only that (p.19):

"All the points zK are the vertices of a polygonal net obtained
by decomposing the pseudogeometric plane into polygons pseudogeometrically
equal to each other."

Next he observed that the corresponding Fuchsian functions (obtained
as quotients of theta-Fuchsian series as before) can be used to solve
differential equations with algebraic coefficients (see below). It
becomes clear that Poincaré has not heard of the Schwarzian derivative.

The third supplement, of only 12 pages, was received on 20 December.
Its main result is the extension of the method of polygonal decomposition
to include cases where the angles are zero, and the roots of the indicial
equation differ by integers. The notable example is Legendre's equation.
Poincaré's method is to push the polygons outwards until one or more
vertices are 'at infinity', i.e. are on the boundary of the disc, and
the corresponding angles vanish. Since the polygons hitherto studied
had angles which were only rational multiples of π, Poincaré's argument
relies heavily on its geometrical plausibility.

The unpublished work makes abundantly evident the astounding
clarity of Poincaré's mind, coupled to an almost equally dramatic
ignorance of contemporary mathematics. There is no mention of the work
of Schwarz on the hypergeometric equation, so naturally and even
pictorially connected with the crucial case $\rho_1 + \rho_2 + \rho_3 < 1$. Nor
is there any mention of the work of Dedekind or Klein, and even Hermite's
work on modular functions, which he must have known, seems to have
been forgotten. We shall see that these omissions are not mere

oversights; Poincaré genuinely did not then know the German work. The contrast with the deliberately well-read Felix Klein could not be more marked.

6.3 *Klein*

Although Klein was far from idle in 1880, he did not make the dramatic progress that Poincaré did. He reworked his 1879 paper on eleventh order transformations of elliptic functions, connecting it, and his work on the transformations of orders 5 and 7, with division values of θ-functions, in a paper he finished on January 3, 1881, [1881]. He reported to the Akademie der Wissenschaften in Munich on the significance of his work for the study of the normal forms of elliptic integrals of the first kind, and concerned himself with the work of his students Gierster, Dyck, and Hurwitz. Bianchi stayed in Munich during the year autumn 1879 to autumn 1880, and Klein [1923, 6] described the work he did with him as "more transient ..., but as very essential for me" for it showed him how to connect his work with the developments of the Weierstrassian school, including that of his friend Kiepert. In the autumn of 1880 he moved to become professor at the University of Leipzig, and gave his inaugural address on the 25th of October on the relationship of the new mathematics to applications, a theme dear to his heart. Carl Neumann stimulated his interest in applied mathematics, and he wrote a short paper in mid-January 1881 on the Lamé functions. Even so he was, he said later, so deeply immersed in geometric function theory that he began to lecture on it as soon as he arrived at Leipzig. He gave two series of lectures, the second has become very well-known in the slightly re-worked form of his book "Über Riemanns Theorie der algebraischen Functionen und ihre Integrale".

Klein wanted to conceive of an algebraic curve F(s, z) not as spread out over the z-plane, but as a closed surface in its own right, and a complex function as a pair of flows, with singularities, on the surface. In so doing, he argued that he was only following Riemann's own approach, and he sought out Riemann's students and colleagues to ask them how true was it that Riemann had thought this way. At first Prym agreed with Klein, but later correspondence between Prym and Betti makes it clear that Riemann had not[10], so it seems that Klein's description must be regarded not as historically accurate, but as an inspired response to reading Riemann. Although his presentation of Riemann's topological ideas is highly attractive, it can perhaps be doubted if it helped Klein with his "old problem ...: *to construct all discontinuous groups in one variable, esp. the corresponding automorphic functions*" ([1923a], 581, italics Klein's) for it avoids the problem of how the corresponding Riemann surfaces are best constructed. One sees here very clearly how Klein wanted to advance a particular view of mathematics, on the one hand visual and intuitive, on the other naturally algebraic, group theoretic or invariant theoretic, and linked to projective geometry. The unity of these domains and the connection with number theory delighted him, and also I shall suggest, bound him too closely to one view of the problem.

Klein had been working on Riemann's ideas since 1874, when he began a series of four papers which showed how to display Riemann surfaces in a more intuitively intelligible way. He only began to study Riemann's work more intensively in his later years in Munich. This brought him up against a style of mathematics that was new to him, for Riemann's approach to geometry was much more metrical, even topological, and markedly less algebraic then Klein's. Klein later wrote [1923a, 477] that although Clebsch had introduced him to Riemann's results: "the geometrical-physical considerations by which Riemann had come to his conclusions were properly not seen to be adequate or extendable. In this respect the development which led me to Riemann was in contradiction with the tradition in which I was brought up". Klein took the chief importance of Riemann's ideas to be their implications for complex function theory, and although Klein had been called to Leipzig to lecture specifically on geometry he took the opportunity to deepen his understanding of function theory. He wrote in his autobiography [1923b, 20f]: "But I did not conceive of the word geometry onesidedly as the subject of objects in space, but rather as a way of thinking that can be applied with profit in all domains of mathematics. Correspondingly, despite many contradictions, I began my Leipzig professorship with a lecture on geometric function theory in which I went further with the ideas that had stirred me in Munich... At the time I began a cycle of lectures on geometry that comprised analytic geometry, projective geometry, and differential geometry". It is only on arrival at Leipzig that Klein began to don the Riemannian mantle. At that time, as he was willing to admit, he found understanding Riemann's work extremely difficult, and we should be careful not to see him in any simple way as Riemann's true successor.

6.4   1881

Klein's attention was first drawn to Poincaré when he read Poincaré's three notes "Sur les fonctions Fuchsiennes" in the Comptes Rendus which will now be described. The papers, and the ones which followed them, are a jumble of promises, allusions, examples, and, sometimes, mistakes, within which various themes can be detected. The main object of study is what Poincaré called Fuchsian functions,

functions defined on a disc or half plane and invariant under certain subgroups of SL(2, ℝ). The main method of study is the use of fundamental regions or polygons, in the sense of the non-Euclidean geometry intrinsic to the disc. The main results concern the relations between Fuchsian functions associated to a given group, and the use of these related functions to solve linear differential equations with rational or algebraic coefficients.

In the first paper (14 February 1881) Poincaré announced the discovery of a large class of functions generalising elliptic functions and permitting the solution of a differential equations with algebraic coefficients, which he proposed to call Fuchsian "in honour of M. Fuchs, whose works have been very useful to me in these researches". These functions were invariant under discontinuous subgroups of the group $z \rightarrow \frac{az + b}{cz + d}$ which leaves a "fundamental" circle fixed; discontinuous meaning no z is infinitesimally near any transform of itself. As a result, the group, called by Poincaré a Fuchsian group, can be studied by looking at how it transforms a region R bounded by arcs of circles perpendicular to the fundamental circle. He said non-Euclidean geometry would be helpful here, but he did not explain how. Instead he gave an example of a triangular polygon R_0 = ABCD vertical angles $\hat{BAC} = \hat{BDC} = \frac{\pi}{\alpha}$, $\hat{CBA} = \hat{CBD} = \frac{\pi}{\beta}$, $\hat{BCA} = \hat{BCD} = \frac{\pi}{\gamma}$ where α, β, γ are positive integers or ∞ and $\frac{1}{\alpha} + \frac{1}{\beta} + \frac{1}{\gamma} < 1$. The transforms of such a region would provide an example of a Fuchsian group, and Poincaré asked if any polygon would do. As other examples he gave the special case $\alpha = 2$, $\beta = 3$, $\gamma = \infty$, which gives SL(2; Z), and any linear substitution group preserving an indefinite ternary form $(px^2 - qy^2 - rz^2)$. Finally he introduced functions he called theta-fuchsians, by analogy with the theta-functions occurring in theory of elliptic functions. Suppose z is a point inside the fundamental circle, K_i an element of the Fuchsian group and zK_i the transform of z by K_i. The single-valued function θ is a theta-function if, for some integer m, it satisfies $\theta(zK_i) = \theta(z)\left(\dfrac{dzK_i}{dz}\right)^{-m}$, or equivalently, if $\theta\left(\dfrac{az + b}{cz + d}\right) = \theta(z)(cz + d)^{2m}$, assuming $ad - bc = 1$.

The existence of theta-fuchsian functions follows, he said, from the convergence of the series

$$\sum_{i=1}^{\infty} H(zK_i) \left(\frac{dzK_i}{dz}\right)^{m},$$ m an integer greater than 1,

where H is a rational function of z. Two cases arise, one in which every point of the fundamental circle is an essential singular point of the theta-fuchsian, the other in which the essential singular points, although infinite in number, are isolated. Only in this case can the function be extended to the whole plane.

Next week's installment, so to speak, [1881b], began by pointing out that the quotient of two theta-fuchsian functions corresponding to the same Fuchsian group was a Fuchsian function. Poincaré then observed that between any two Fuchsian functions corresponding to the same group there existed an algebraic relation, but he gave no proof, and that every Fuchsian function F permitted one to solve a linear differential equation with algebraic coefficients.[11] For, if $x = F(z)$ is a Fuchsian function, then $y_1 = \left(\dfrac{dF}{dz}\right)^{\frac{1}{2}}$ and $y_2 = z\left(\dfrac{dF}{dz}\right)^{\frac{1}{2}}$ are the solutions of the equation $\dfrac{d^2y}{dx^2} = y\phi(x)$, where ϕ is algebraic in y. He gave the hypergeometric equation as an example, and pointed out that one variant of that,

$$\frac{d^2y}{dx^2} = y\left[\frac{x(x - 1) - 1}{4x^2(x^2 - 1)}\right],$$

gives the periods of sinam as functions of the square of the modulus. Zeta-functions (to be defined below) were also introduced, and Poincaré claimed that such functions provided a basis of solutions to any linear differential equation having rational coefficients and two finite singular points.

The third paper [1881c] suggested a connection between Fuchsian functions x(z) and y(z) corresponding to the same group and abelian integrals u(x, y). Regarding u as a function of z, and operating on z by transformations in the given group which send (x, y) round a cycle, Poincaré was led to relate the number of generators, 2p + 2, of the group to the periods of the abelian integral, and consequently obtained an upper bound, p, for the genus of the algebraic equation connecting x and y. Only a truly great mathematician can achieve this degree of visionary imprecision.

Klein read these notes on 11th June 1881 and wrote to Poincaré the next day. He had, he said[12], considered these topics deeply in recent years, and had written about elliptic modular functions in several articles, which "naturally are only a special case of the relations of dependence considered by you, but a closer comparison would show you that I had a very general point of view". After listing these works, and referring to related work of Halphen and Schwarz, Klein went on to say that the task of modern analysis was to find all functions invariant under linear transformations, of which those invariant under finite groups and certain infinite groups, as the elliptic modular functions, were examples. He had talked to other mathematicians about these questions, without coming to any definite result, and now he wondered if he should not have got in touch earlier with Poincaré or Picard; he hoped his letter would start a correspondence. He admitted that at the moment other duties kept him from working on the problem, but he would return to it soon for he was to lecture on differential equations in the winter.

Klein was rather exaggerating his achievements, and seems a little concerned to impress Poincaré with what he had already done, perhaps even to over-awe him. We have seen that he was, in fact, at an impasse with his research in this direction, and had moved off onto other topics. Now that a rival had entered the field, he would return to it by the winter.

Poincaré's reply (15 June) was characteristically more modest, and he was more willing to admit to ignorance, even quite astounding ignorance. He immediately conceded priority over certain results to Klein, but said he was not at all surprised "... for I know how well you are versed in the study of non-Euclidean geometry which is the veritable key to the problem which occupies us". He would do justice in that matter when he next published his results, meanwhile he would try to find the relevant Mathematische Annalen, which were not in the library at Caen. But, since that would take time, could Klein please explain some things straight away. Why speak of modular functions in the plural, when there was only one, the square of the modulus as a function of the periods? And had Klein found all the circular-arc polygons which give rise to discontinuous groups and found the corresponding functions?

Klein got the letter on the 18th and replied the next day, enclosing reprints of his own articles and promising to send those of his students Dyck, Gierster, and Hurwitz. Then he warmed to the theme of using Fuchs's name. All such research, he said, was based on Riemann. His own was closely related to that of Schwarz, which he urged Poincaré to read "if you do not already know it". The work of Dedekind had shown how modular functions could be represented geometrically, which had already become clear to him (Klein), whereas Fuchs's work was ungeometric. He did not criticize the rest of Fuchs's work on

differential equations, but here he had made a fundamental mistake which
Dedekind had had to correct. On the subject of polygons Klein pointed
out that any polygon was equivalent to a half plane, so repeated
reflection or inversion in the sides generated a group and invariant
functions in the manner of Schwarz and Weierstrass, without the need
to return to general Riemannian principles. But some polygons gave
rise to discontinuous groups which did not preserve a fixed circle,
so "the analogy with non-Euclidean geometry (which is in fact very
familiar to me) does not always hold". He gave this polygon as an
example

Fig. 6,7

Poincaré replied on the 22nd, before the reprints had arrived,
seeking permission to quote the passage about the group in his next
publication and defending the name 'Fuchsian function' on the grounds
that, even if "... the viewpoint of the geometric savant of Heidelberg
is completely different from yours and mine, it is also certain that
his work served as a point of departure ..." and so it was only just
that his name should stay attached to those functions.

Klein's brief reply (25 June) demurred, directing Poincaré back
to Fuchs's original publication in the Journal für Mathematik for
the purposes of comparison.

Poincaré wrote on the 27th to say that finally the reprints had
arrived, having been sent via the Sorbonne and the Collége de France
even though they were correctly addressed (so the postal service was
not always so efficient). He now admitted that "I would have chosen
a different name [for the functions] had I known of Schwarz's work,

but I only knew of it from your letter after the publication of my
results, ..." and he could not change the name now without insulting
Fuchs. As to the mathematics, had Klein determined the fundamental
polygons for all the principal congruence subgroups? And what, in
that connection, was the _Geschlecht_ in the sense of Analysis situs?
Was it the same as the _genre_ that he, Poincaré, had defined, for he
only knew that they both vanished simultaneously. Poincaré asked
for a definition of the topological genus, or, if it was too long to
give in a letter, a reference to where it could be read. As for the
polygon, presumably its sides should not meet when extended, and
finally, he asked what Klein understood by general Riemannian principles?

On July 2nd Klein answered these questions as well as he could.
Dyck had established the polygons for congruence subgroups for prime n,
composite n had not been considered. Genus in the sense of the analysis
situs was the maximal number of closed curves which can be drawn
without disconnecting the surface, and was materially the same number
(Klein's emphasis) as the genus of the algebraic equation representing
the surface. He went on "I have only conjecturally a freer representation of a Riemann surface and the definition of p based on it."

The furthest he had gone on the question of connection of
polygons and curves was this. Map a half-plane onto a polygon such as

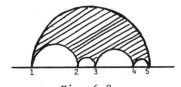

Fig. 6.8

this one, and let the points corresponding to 1, 2, 3, 4, 5 be I, II,
III, IV, V respectively, which can be arbitrary. Then I, II etc.

are the branch points of an algebraic function, w(z), and w(z) can have no other branch points. So w and z are single-valued functions of the right kind, and if all the branchpoints lie on a circle in the z-plane there is nothing more to be said. But in the other cases, and here the unsymmetric polygon of the last letter is relevant, reflections generated a fundamental space. But did they together cover only one part of the plane? "I find myself already brought to a halt on this difficulty for a long time". In this conclusion Schottky's study of inversion in families of non-intersection circles was interesting.

Riemann's principles he went on, do not tell you how to construct a function, so for that reason these consequences are somewhat uncertain, and Weierstrass and Schwarz have done a lot of work on circular arc polygons. Riemann's principles map a many-leaved surface to a given polygon and enable one to prove the existence of a function having prescribed infinities and real parts to their periods. "This theorem, which by the way I have only completely understood recently, includes, so far as I can see, all the existence proofs of which you speak in your notes as special cases or easy consequences."

Finally, referring to another of Poincaré's papers ("Sur les fonctions abéliennes", CR 92, 18 April 1881) Klein asked why the moduli were presumed to be 4p + 2 in number when they are really only 3p − 3. "Haven't you read the relevant passage of Riemann? And is the entire discussion of Brill and Noether...unknown to you?"

Poincaré's reply of 5 July apologised profusely for asking about the topological genus when it was defined "on the next page of your memoir". Not having had a refusal of his request to quote Klein he

had done so, taking silence for consent. As for the branchpoints of
algebraic functions, that had been proved by him too, he said, and
published on May 23rd, but where was it in the work of his predecessors?
Finally, the number 4p + 2 had only been needed as an upper bound and
was easy to obtain. He made no mention of his deficient reading.

This part of the correspondence closed with Klein's letter of
9 July, in which he pointed out that Riemann's use of Dirichlet's
principle was not conclusive, but that one could find stronger proofs
in e.g. Schwarz's work. By now Poincaré had published several more
short papers, and these should be described.

The notes of 18 April, 23 and 30 May [1881e,g,h], published
before the correspondence with Klein had begun, dealt in more detail
with the role of circular-arc polygons in generating groups. Poincaré
supposed the fundamental circle to be bisected by the real axis, O_x,
and considered the polygon with sides C_1, ..., C_n as follows:

i) All the C_i meet the fundamental circle at right angles;

ii) C_1 meets O_x at α_1 and β_1, making an angle of $\frac{\lambda_1}{2}$;

iii) C_i meets C_{i-1} at α_i and β_i, making an angle of λ_i:

iv) C_n meets O_x at α_{n+1} and β_{n+1}, making an angle of $\frac{\lambda_{n+1}}{2}$;

where he supposed each λ_i is an aliquot part of 2π and
$$\lambda_1 + 2\lambda_2 + \ldots + 2\lambda_n + \lambda_{n+1} < 2\pi(n-1)$$ (so that the construction is
possible).

The transformation sending z to z_j by

$$\frac{z_j - \alpha_j}{z_j - \beta_j} = e^{i\lambda_j} (\frac{z - \alpha_j}{z - \beta_j}) \qquad j = 1, \ldots, n + 1$$

leaves α_j and β_j fixed, and rotates the family of coaxal circles surrounding α_j and β_j by λ_j. Since one may suppose α_j is inside the fundamental circle and β_j outside, this has the effect of rotating the polygon about the vertex α_j so that its new position lies alongside its old one. Even so it is not true, as Poincaré claimed, that one obtains a discontinuous group in this way unless the sides of the polygons have the same length (i.e. vertex α_{j-1} is rotated onto α_{j+1}). The idea is clear nonetheless, successive transformations move the polygon around crabwise and a function defined on the original domain extends to (meromorphic) Fuchsian function accordingly, defined on the interior of the unit disc, and satisfying $F(z) = F(z_1) = \ldots$ $= F(z_{n+1})$, so F is invariant under all composites of the above transformations.

Since two such functions x and y will be related algebraically one can consider the genus of the equation connecting them, $f(x, y) = 0$. Should it be zero, so f is a rational function, then taking $x = F(z)$, $y = (\frac{dF}{dz})^{\frac{1}{2}}$, one obtains

$$\frac{d^2y}{dx^2} = y\phi(x).$$

in which ϕ is also a rational function, and the singular points of the equation, being the infinities of ϕ, are at $F(\alpha_1)$, $\ldots$, $F(\alpha_n)$. Suppose these to be all arbitrary but real, and $F(z)$ to be real along the sides C_1, $\ldots$, C_n of the polygon, then the case considered is that of an algebraic function with arbitrary real branch points, as discussed in the correspondence. Suppose furthermore that $\lambda_1 = \ldots$, $\lambda_{n+1} = 0$, so α_1, $\ldots$, α_{n+1} lies on the fundamental circle itself. Then $F(z)$ fails to take the values $F(\alpha_1)$, $\ldots$, $F(\alpha_{n+1})$. So if one has a linear differential equation with coefficients rational in x and real singular points at $x = F(\alpha_1)$, $\ldots$, $F(\alpha_{n+1})$, then one

sets $x = F(z)$ and the solutions of the equation are zeta fuchsian
functions of z. In this way a large class of differential equations
are solved.

On 30th May Poincaré clarified this a little: when $\lambda_{n+1} = 0$ the
appropriate $F(z)$ is the limit of the original $F(z)$ as the λ's tend
to 0, and it is only in this case that the values of
$F(\alpha_1), \ldots, F(\alpha_{n+1})$ can be arbitrary. In this paper he broached a
continuity argument to show that if a given differential equation
has $2n$ singular points, these can perhaps be allowed to tend to $2n$
real points and the solutions to the real case allowed to deform
into those in the general case:

"If I succeed in showing that these equations always have a real
solution I will have shown that all linear equations with algebraic
coefficients can be solved by Fuchsian and zeta-fuchsian
transcendents".

On the 27th June, after he had heard from Klein, Poincaré [1881i]
returned to the description of non-Euclidean polygons, this time
defined in the upper half plane (which can be obtained from the
disc by an inversion). He took a and b in the upper half plane,
with conjugates $\bar{a}$ and $\bar{b}$, and defined

$$(a, b): = \frac{(a - \bar{a})(b - \bar{b})}{(a - \bar{b})(b - \bar{a})}$$

Since this is invariant under the projective group $PSL(2:\mathbb{R})$ which
preserves the upper half plane, it can play the role of a metric
invariant in the sense of non-Euclidean geometry. In particular
there will be a transformation sending a to c and b to d if and only if

(a, b) = (c, d). If this condition is stipulated for each of n pairs
of sides of a 2n - gon whose angles are, as before, aliquot parts of 2π,
and whose vertices are made to correspond only if they are either both
above the real axis or both on it (or both segments of it, Poincaré
admitted the infinite case

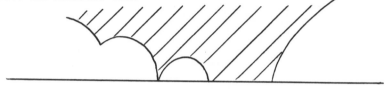

Fig. 6.9

then indeed a discontinuous group is obtained. Poincaré added, without
a hint of a proof, that every Fuchsian group can be obtained in this
way. Turning to Klein's example of an unsymmetric polygon, which does
not preserve a fundamental circle, and to his generalization of that
to a region bounded by 2n circles exterior to one another and possibly
touching externally, he said that successive reflections again generated
a discontinuous group and so also invariant functions. Making amends
for his earlier choice of names, and surely with a twinkle in his eye,
Poincaré added that "...I propose to call [these functions] Kleinian
functions, because it is to M. Klein that one owes the discovery. There
will also be theta-Kleinian and zeta-Kleinian functions analogous to the
theta-fuchsian functions".

In the paper of 11th July entitled "Sur les groupes Kleinéens"
[1881j] Poincaré delightfully restored the analogy between discontinuous
groups and non-Euclidean polygons which Klein had said his example of
unsymmetric polygons broke. His insight was to take the (x, y) plane
as the boundary of the space (x, y, z), z $\geq$ 0, and to regard the polygon
as bounded by arcs which were the intersections of hemi-spheres centre
(x, y, 0) with the (x, y) plane. Thus the groups generated by

reflections on the sides is regarded as acting on 3 dimensional
non-Euclidean space, realised as the space above the (x, y) plane in this
way:

A pseudogeometric plane (Poincaré referred to the pseudogeometry of
Lobachevskii) is a hemisphere;
a pseudogeometric line is the intersection of two such planes; the
distance along a line from p to s is half the logarithm of the cross-
ratio pscd, where c and d are the points where the line ps meets the
xy plane; and
the angle between two intersecting lines is their usual geometric angle.

This is the first appearance in print of Poincaré's conformal
model of non-Euclidean geometry. Oddly enough, it is three-dimensional,
just as the original versions of Lobachevskii and Bolyai were.

Papers continued to stream out of Caen. In another (8 August = 1881k)
he claimed that he had been able to show by a simple polynomial change
of variable that his earlier work (30 May) on differential equations
led to the conclusions that

i)   Every linear differential equation with algebraic coefficients can
     be solved by zeta fuchsians;
ii) The co-ordinates of points on any algebraic curve can be expressed
     as Fuchsian functions of an auxiliary variable.

The second point is the uniformization theorem for algebraic curves:
if $f(u, v) = 0$ is the equation of an algebraic curve then it is claimed
that there are Fuchsian functions $\phi(x)$ and $\psi(x)$ such that $u = \phi(x)$,
$v = \psi(x)$, $f(\phi(x), \psi(x)) = 0$. (For a modern account of this important
theorem, see Abikoff [1981].)

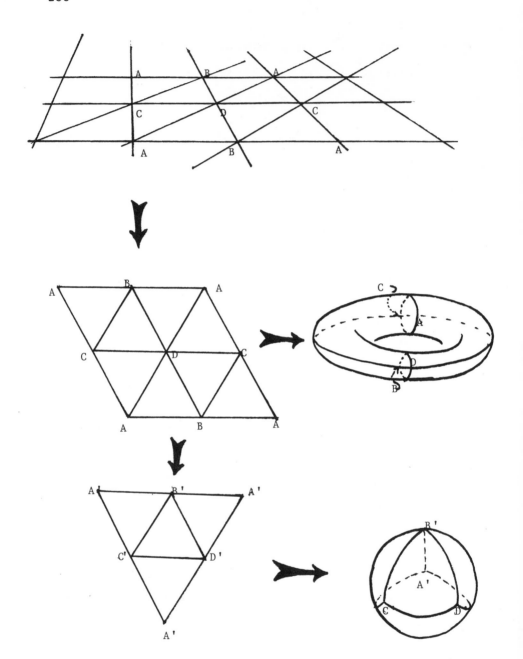

In Poincaré's view the upper half plane (top), acted upon by a discrete subgroup (a lattice) has a quotient space which is a torus. In the view of Riemann and Klein the algebraic curve $y^2 =$ quartic in x is a double cover of the x-sphere branched over four points.

Klein meanwhile had concluded his summer lectures on geometric function theory, and on 7th October he sent the manuscript of Über Riemanns Theorie der algebraischen Funktionen und ihrer Integrale to Teubner in Leipzig. Now he could return to his old problem. He wrote to Poincaré on the 4th of December, congratulating him on solving the general differential equation with algebraic coefficients and the uniformization problem, and asking for an article on this work for Mathematische Annalen, of which he was the editor. He, Klein, would add a letter to it, connecting it with his work on modular functions. Publication would serve two purposes, acquainting the readership of the Annalen with his work, and explaining its relationship to Klein's [one wonders who would gain most by this]. Poincaré, who had moved to Paris by now, agreed, and sent a manuscript of Klein on December 17th, which made the deadline for the last part of volume 19. In content it much resembles Poincaré's report presented to the Mémoirs de l'Académie nationale des Sciences, Arts et Belles-Lettres de Caen, written and published shortly before, but it goes into more details on the topics Klein had requested. Before discussing these two memoirs, which represent Poincaré's first attempts to survey what he had done in that memorable year, one must turn to the sourer discussion of priorities which Klein's letters provoked.

Klein again objected to the name Fuchsian functions, on the grounds that Fuchs had published nothing on the domain of fuchsian functions, whereas Schwarz had. Nor had he, Klein, done anything on Kleinian functions except to bring one special case to Poincaré's attention; in this case Schottky deserved more of a mention, as did Dyck for the group-theoretic emphasis in his work. Klein had warned Poincaré by letter (13 January 1882) that he would protest against the names as much as was in him, and Poincaré indicated that he would like space in the

Annalen to defend his choice. So, in the next issue he pointed out that it was not ignorance of Schwarz that made him select Fuchs's name, but the impossibility of forgetting the remarkable discoveries of Fuchs, based on the theory of linear differential equations. Furthermore, Schwarz was only a little interested in differential equations, and Klein indeed dwelled more on elliptic functions, whereas Fuchs had brought forward a new point of view on differential equations "which had become the point of departure for my researches". As for Kleinian functions, the name belonged rightly to the man who had "stressed their principal importance", said Poincaré, turning Klein's observation on Schottky in his own work [1882a] against him.

Now that Klein had criticized Fuchs publicly, Fuchs came to hear of the dispute and could reply himself, which he did in the Journal für Mathematik [1882] and also privately in a letter to Poincaré.[13] Stung by the allegation that he had published nothing on invariant functions, he cited not only those works of his which had inspired Poincaré, but also his earlier study [1875] of differential equations with algebraic solutions. He observed that Klein himself, in the Annalen, vol II, had said that these works had led to his own developments. Fuchs was not quite correct. Klein had said rather that he should, at the conclusion of a work devoted to differential equation having algebraic solutions, refer to Fuchs's [1875], "which likewise concerns the subject matter here presented, the study of which I was recently induced to do, even to derive the simple method I substitute here" [Klein, 1876a, = 1922, 305]. In that paper he invoked Schwarz's work at the start, and his own work on binary forms. It is entirely plausible that this was his route to the problem, and that his acquaintance with Fuchs's studies was slight and scarcely influential.

It is a story he was to repeat in his book Vorlesungen über das Ikosaeder [1884], although towards the end of his life he conceded that Fuchs had been a particular impulse [1922, 257]. But he was eager to see Poincaré harboured no doubts on the matter. His next letter to Paris (3 April 1882) contains a lengthly restatement of the historical priorities as viewed from Düsseldorf. In Klein's view his study of linear transformations of a variable went back to 1874. When in 1876 he solved the problem thrown up by Fuchs the facts were contrary to what Fuchs had said: "I did not take the ideas from his work, rather I showed that his theme must be handled with my ideas" (Klein's emphasis). He would welcome the new class of functions of several variables being called Fuchsian, but by the way, were they single valued or only unbranched? Then Klein returned to mathematical comments. This remark is the most interesting one: uniformization permits one to show, what is only claimed as probable in the lectures on Riemann's Theory of Algebraic Functions and their Integrals: that a Riemann surface of genus $p > 0$ cannot have infinitely many discrete self transformations. [Klein meant $p > 1$, see his section XIX].

Klein hoped his letter would close the debate, and Poincaré at once agreed (4 April). He wrote that he had not started the debate and he would not prolong it. As for the question of 'Kleinian' he had known of no right to the property before Klein's letter. As for Fuchs's functions of two variables these divided into three classes, two effectively uniform and one only unbranched, and he directed Klein's attention to the smaller works of Fuchs which had followed the main paper. He concluded by wishing the debate about names to be over: "Name is Schall and Rauch" he wrote in German ("Name is sound and fury"). Happily, no more was said, at least on this occasion, but Fuchs remained

acutely sensitive to plagiarism, once accusing Hurwitz of it, in 1887,
once P. Vernier, and once, incorrectly, Loewy (in 1896). In private he
had indeed already registered a similar wound in 1881. He wrote (30
March 1881) to Casorati complaining that Brioschi had referred only to
Klein and not to him on the algebraic solutions question, although his
work preceded Klein's [1876a] (letter quoted by E. Neuenschwander [1978]).

Poincaré's two memoirs on Fuchsian functions have been mentioned
already, and shall now be described, dwelling on the <u>Annalen</u> paper,
which subsumes the Caen one. Poincaré hardly ever supplied proofs
in these article; statements were given bald, supported only by
reference to long calculations or even obscurely qualified by vague
remarks about simple cases. Some of these evasions are unimportant,
others are more interesting.

A Fuchsian group, he said, is a group of substitutions of the
form $S_i = (\zeta, \dfrac{\alpha_i \zeta + \beta_i}{\gamma_i \zeta + \delta_i})$, where α_i, β_i, γ_i, δ_i are all real and
$\alpha_i \delta_i - \beta_i \gamma_i = 1$, which move a region R_0 of the upper half plane around
so that the various $S_i R_0$ form a sort of web. The $S_i R_0$ are to be
non-Euclidean polygons, not necessarily finite in extent, fitting
together along their common edges, and whose angles are aliquot
multiplies of 2π.

Theta Fuchsian function $\theta(\zeta)$ are defined as follows: let $H(\zeta)$ be
any rational function, then the series

$$\theta(\zeta) = \sum_{i=1}^{\infty} H(\frac{\alpha_i \zeta + \beta_i}{\gamma_i \zeta + \delta_i}) \frac{1}{(\gamma_i \zeta + \delta_i)} 2m$$

converges for each integer m greater than 1 and defines a theta-fuchsian
function. Consideration of $\int \dfrac{\theta'(\zeta) d\zeta}{\theta(\zeta)}$ over the boundary of R_0 shows that
the number of its zeros and its infinities in R_0 is finite. A quotient
of two theta-fuchsians is a Fuchsian function $F(\zeta)$, it satisfies

$$F\left(\frac{\alpha_i \zeta + \beta_i}{\gamma_i \zeta + \delta_i}\right) = F(\zeta).$$

If R_0 has a segment of its boundary in common with the real axis F can be extended analytically into the lower half plane, but otherwise, he remarked, it is only defined on the upper half plane.

If the substitutions of the group have all these properties except that of preserving a half-plane or circle, than Poincaré called the group Kleinian, and the functions theta-Kleinian and Kleinian respectively, remarking that their study required processes derived from three dimensional non-Euclidean geometry. He gave no reason for the name in the Annalen, in the Caen essay he said that Klein had been the first to give an example of such groups.

Poincaré stated four main results for the theory of Fuchsian functions and groups. The first concerned the genus of the algebraic equation connecting two Fuchsian functions belonging to the same groups. He called two edges of R_0 conjugate if they were identified by some S_i, and said that vertices which were identified belonged to the same cycle. Then in the case of a finite polygon, if there were 2n sides to R_0 and p cycles, then genus of the equation connecting any two functions was $\frac{n + 1 - p}{2}$. For a polygon with 'sides' along the real axis the formula is $n - p$.

The second theorem concerned the connection between Fuchsian functions and second order differential equations. If x and y are two Fuchsian or Kleinian functions corresponding to the same group then

$$v_1 = \sqrt{\frac{dx}{d\zeta}} \text{ and } v_2 = \zeta\sqrt{\frac{dx}{d\zeta}}$$

are certainly solutions of

$$\frac{d^2 v}{dx^2} = v\,\phi(x, y),$$

where ϕ is an algebraic function. Poincaré claimed that he could state very complicated conditions on the coefficients of ϕ under which the converse was true, and the solutions of $\dfrac{d^2 v}{dx^2} = v \, \phi(x, y)$ were Fuchsian functions, but he did not do so. He gave instead one example of an equation not of the Fuchsian type, in which if the coefficients satisfied certain (unspecified) inequalities the solutions were Kleinian functions, and if they satisfied certain (unspecified) equalities the solutions were Fuchsian. [This brings in the 'notorious' accessory parameters. He had claimed in [1881m] that an equation with specified singularities is Fuchsian for exactly one choice of parameters, a claim he proved in his [1884a].]

The third theorem was uniformization, derived from the existence of a Fuchsian function $F(\zeta)$ omitting n + 1 values a_1, a_2, ..., a_n, ∞ from its range, which rested in turn on a long numerical calculation, which was not given. If $f(x, y) = 0$ is an algebraic equation with singularities at a_1, a_2, ..., a_n, and $x = F(\zeta)$ then y is also a Fuchsian function and the curve $f(x, y) = 0$ has been uniformized.

The fourth theorem concerned zeta-fuchsian functions, i.e. functions

$$\phi_1(\zeta), \ \phi_2(\zeta), \ \ldots, \ \phi_n(\zeta)$$

which satisfy

$$\phi_\lambda \left(\frac{\alpha_i \zeta + \beta_i}{\gamma_i \zeta + \delta_i} \right) = A^i_{1\lambda} \, \phi_1(\zeta) + A^i_{2,\lambda} \, \phi_2(\zeta) + \ldots + A^i_{n,\lambda} \, \phi_n(\zeta)$$

for constants A such that $\det(A^i_{j,\lambda}) = 1$ for all i. Poincaré claimed that his earlier results could be extended to show that every n^{th} order linear differential equation with algebraic coefficients can be solved by Fuchsian and zeta-fuchsian functions, or by Kleinian and zeta-Kleinian ones.

The same volume of the _Annalen_ also contained Klein's first response to Poincaré's ideas. Excited, but also challenged by the sudden emergence of the younger man, he felt driven to use methods which even he regarded as "to some extent irregular" and to publish his results before he had managed to put them in order. This was a departure from his usual practice, which might indeed gloss over special cases and prefer geometric reasoning to the arithmetizing of, say, Weierstrass, but is nonetheless quite precise. The contrast between Klein and the Berlin school shows itself in Klein's preference for generic situations to exceptional cases, however, the generic situation is always treated carefully. The contrast between Klein and Poincaré is between the clear-sighted and the visionary. Poincaré's genius was paradoxically helped by his strange lack of mathematical education; he was making it up as he went along. Klein was reformulating his considerable store of existing knowledge, bringing to the new problems not only techniques he knew had worked in the past but a unifying view of mathematics to which, he felt, the new subject must conform. The result of each comparison is oddly similar. The thoroughness of the Berlin school revealed hidden riches in known fields whereas Klein's study of modular functions was almost a new field of study; Klein's thoroughness in turn lacks the vivid re-creation of the theory of Riemann surfaces and differential equations one finds in Poincaré.

In his first brief paper [1882b] Klein asserted that for each Riemann surface of genus greater than 1, there is always a unique function η which is single-valued on the simply-connected version of the surface, on crossing a cut changes to a function of the form $\frac{\alpha\eta + \beta}{\gamma\eta + \delta}$ and so by analytic continuation maps the cut surface without overlaps onto a 2p-connected region of the η-sphere. The (suppressed) argument to the existence was a continuity one; the uniqueness derives from the manner of the analytic continuation. It has the uniformization of the Riemann surface as a corollary, and Klein give a more general description of the generation of the discontinuous group in this context. Poincaré always at this time used infinite polygons in his discussion of uniformization, as Freudenthal [1954] has pointed out. Klein regarded the analytic continuation of η as $\frac{\alpha\eta + \beta}{\gamma\eta + \delta}$ as moving the image of the cut surface around on the η-sphere. Furthermore, these generating substitutions, although they must satisfy certain inequalities determined by the shape of the fundamental region, contain $3p - 3$ complex variable parameters. He explained this count as follows. The image of the cut surface may be taken to be a polygon whose $2p$ sides, the images of the cut, are identified in pairs. Each identification, being a fractional linear transformation, contains three parameters, so $3p$ parameters specify the group, but then 3 may be chosen arbitrarily since the group does not depend on the initial position of the polygon. Klein's description of the decomposition of a Riemann surface by means of cuts is confused: $2p$ cuts are needed to render a surface of genus p simply connected, when it becomes a polygon of $4p$ sides. In the "Neue Beiträge" he counted more carefully: there are $2p$ lengths and $4p$ angles in such a polygon, so $6p - 3$ independent real co-ordinates, and then, using the upper half plane model, the coefficients of η are real and η may be replaced by $\frac{\alpha\eta + \beta}{\gamma\eta + \delta}$, so these parameters are inessential, and the function η depends on $6p - 6$ real or $3p - 3$ complex parameters. Now, the space of moduli for Riemann surfaces has complex dimension $3p - 3$ and Klein made the audacious claim that every Riemann surface corresponds accordingly to a unique group. As Freudenthal [1954] remarks, the direct route to uniformization is the only point at which Klein surpassed Poincaré. One might even make this point a little more strongly: it is only Klein who

invoked the birational classification of surfaces and so made at all
precise how the correspondence between discontinuous groups and curves
might be expected to work.

Klein went on to describe symmetric surfaces and surfaces whose
equations have real coefficients. In this way he encountered as a
special case the figure bounded by p + 1 non-intersecting circles,
which, he said, Schottky had already described "without emphasizing
its principal importance".

The next developments were vividly described by Klein himself in
a lecture of 1916, reprinted in his Werke (III, 584) and his
Entwicklung (p. 379). He wrote: "In Easter 1882 I went to recover my
health to the North Sea, and indeed to Nordeney. I wanted to write
a second part of my notes on Riemann in peace, in fact to work out the
existence proof for algebraic functions on a given Riemann surface in
a new form. But I only stayed there for eight days because the life
was too miserable, since violent storms made any excursions impossible
and I had severe asthma. I decided to go back as soon as possible to
my home in Düsseldorf. On the last night, 22nd to 23rd March, when I
needed to sit on the sofa because of Asthma, suddenly the 'Grenzkreis
theorem' appeared before me in $2\frac{1}{2}$ hours as it was already quite
properly represented in the figure of the 14-gon in volume 14 of the
Math. Annalen. On the following morning, in the post wagon that used
to go from the North to Emden in those days, I carefully thought through
what I had found once in every detail. I knew now that I had a great
theorem. Arrived in Düsseldorf I wrote all it at once, dated it the
27th of March, sent it to Teubner and allowing for corrections to
Poincaré and Schwarz, and by way of examples to Hurwitz." So Klein
too had a sleepless night of inspiration.

Hurwitz immediately recognized the importance of the theorem, but when Poincaré replied on 4th April he made no mention of it, as we have seen. In his next note for the Comptes Rendus, however, Poincaré modestly remarked that the results "have been obtained by Klein on other grounds". Schwarz's response was more complicated. At first he denied the validity of the Grenzkreis theorem. Then he accepted it, and proposed that it be seen as a conformal representation of an infinite sheeted covering of the Riemann surface on a disc. Later still he derived it directly by showing how concept of distance can be defined on the surface corresponding to the non-Euclidean metric on the disc (see Klein [1923a, 584]).

The 'Grenzkreis theorem' (or boundary circle theorem) is the assertion that every Riemann surface of genus greater than one can be represented in an essentially unique way by an invariant function without branch points, defined on a disc-like region (Klein called it a spherical skull-cap bounded by a plane). What is novel in Klein's theorem is the idea that the function η in the Ruckkehrschnitt theorem can be so chosen that the corresponding image of the Riemann surface on the sphere is a circular arc polygon of 4p sides all perpendicular to a common circle. When this is done, the images of the surface under analytic continuation of η fill out the region bounded by the circle, which is a limit that they approach indefinitely without overlapping. This gives a magnificent correspondence: the space of all groups and the space of all curves correspond one-to-one via the function η. What Klein saw in the figure of the 14-gon was a way of depicting any Riemann surface within a disc-like region upon which a group acts discontinuously. This dispenses entirely with Riemann's approach, which started from the

equation, considered to define a surface spread out over the complex
sphere. This surface was rendered simply connected by means of cuts,
and then functions on the surface were studied in terms of their
behaviour at singularities and on crossing the cuts; in short, the
surface is given intact and sits above the z-sphere. In the new view
it is given opened out and sits underneath the η-domain. Analytic
continuation of functions is given by the group action simultaneously
with the identification of the sides - in modern jargon the surface is
in Riemann's view a total space and in the new view a quotient or base
space[14].

Seen in this light one may say that this was Poincaré's attitude
all along. Totally ignorant of Riemann's work (he did not know even
of Dirichlet's principle, as his question (27 June 1881) about
"general Riemannian principles" makes clear) he had come to construct
Riemann surfaces naturally from discontinuous groups. In his mind
this connected with linear differential equations, a topic much less
interesting to Klein. This insight of Poincaré, so painfully gained
by Klein, testifies to the strong hold the idea of mathematical
unity had upon Klein. The paradox is that Klein, who had done so much
to further non-Euclidean geometry in the 1870's, did not appreciate it
here. This derives from his view of geometry as essentially projective.
He had been able in 1871 to ground non-Euclidean geometry in projective
geometry; the Klein model is a projective model. Thereafter his
attitude to geometry had been connected with invariant theory or with
line geometry, again a projective idea, or else with topological or
visual properties of figures, questions of reality, etc. He does try
to discuss intrinsic metrical ideas on a Riemann surface in his
lectures on Riemann's algebraic functions, but the discussion is

clumsy. Differential geometry was not his forte, and did not fit
centrally into his view of mathematics. On the other hand, Poincaré
went directly to a conformal model of non-Euclidean geometry, and
viewed the transformations as isometries; Klein missed the metrical
force of the idea of reflection and kept it confined to a complex
analytic circle of ideas. This is one way in which Poincaré's ideas
are more flexible and better adapted to the problem at hand. Klein
for once was too committed to his picture of the overall nature of
mathematics. It is a cruel irony to penetrate further into Riemann's
way of doing things than anyone else at just the moment when epoch-
making progress is being made on Riemannian questions in a non-
Riemannian way. One can only admire the heroic and largely successful
effort Klein made to grasp the new ideas against the current of his
recent work.

The fruits of his labour were published in his long paper
"Neue Beiträge zur Riemannschen Functionentheorie", finished in
Leipzig 2 October 1882 and published in vol. 21 of the Math. Annalen
1882/83. It will shortly be compared with Poincaré's two long papers
in volume I of Mittag-Leffler's Acta Mathematica, also published in
1882, but first some more of Poincaré's short notes and the
correspondence of the two men must be pursued.

Poincaré had published two notes on discontinuous groups and
Fuchsian functions while Klein wrestled with the Grenzkreis Theorem.
The one on discontinuous groups (27 March 1882) made explicit the
connection he had been so pleased to discover between number theory
and non-Euclidean geometry. In it he pointed out that substitutions
$(x, y, z; ax + by + cz, a'x + b'y + c'z, a''x + b''y + c''z)$ with real
coefficients, preserving $z^2 - xy$, correspond to substitutions

$$(x,y \frac{ax+by+c}{a''x+b''y+c} , \frac{a'x+b'y+c'}{a''x+b''y+c''})$$

and thence to substitutions

$$(t, \frac{\alpha t+\beta}{\gamma t+\delta})$$

where α, β, γ, δ are real and t is a complex quantity such that $xt^2 + 2zt + y = 0$. To make t lie in a half-plane he assumed $z^2 - xy$ was negative. Accordingly, discontinuous subgroups G of the initial group correspond to discontinuous subgroups G' of the final group, and the same is true for other forms in x, y, z. In particular one might start by assuming a, b, ..., c'' were integers. Furthermore, the polygonal region R' appropriate to G' (when G' is thought of as transforming the interior of a conic, say a circle) corresponds very simply to the polygon R moved around by G. Place a sphere so that stereographic projection from the plane to the sphere lifts the circular region on which G' acts up to a hemisphere. Now project the image vertically down onto the plane. The image of R' so obtained is R, and is rectilinear. This is the now standard process (discussed above) for converting the conformal description of non-Euclidean geometry of the projective one.

In the paper on Fuchsian groups (10 April) he looked again at nth order linear differential equations whose coefficients $p_i(x,y)$ are rational in x and y, and may become infinite at some points, where x and y satisfy an algebraic equation $f(x,y) = 0$ of genus p. He claimed that if there exists two Fuchsian functions $F(z)$ and $F_1(z)$ defined inside some fundamental circle, such that $x = F(z)$ and $y = F_1(z)$ satisfy $f(x,y) = 0$, and the only singular points of the equation are such that the roots of the determinant equation are multiples of $\frac{1}{n}$, and F, F_1 and their first n − 1 derivatives vanish at the singular points, then the equation is solvable by zeta-fuchsian

functions of z. Furthermore the fundamental region has 4p sides and

either opposite sides are identified, or the sides are identified

4q + 1 with 4q + 3 and 4q + 2 with 4q + 4.

Klein was struck by the close resemblance of this paper of

Poincaré's to his own, and wrote to him on the 7th of May to say he

found the methods interchangeable. He went on: "I prove my theorems

by continuity in that I assume the two lemmata: 1. that to any "groupe

discontinu" there corresponds a Riemann surface, and 2. that to a

single necessarily dissected Riemann surface (under the restrictions

of the present theorem) there can correspond only one such group (in

so far as a group does generally correspond to it." (Klein's emphasis.)

He offered Poincaré a new version of his general theorem to accompany

the two notes in the Math. Annalen, for which he said he had very

little time. Let $p = \mu_1 + \mu_2 + ... + \mu_m$, where the μ's are integers

greater than 1. Take m points on the Riemann surface $0_1, ..., 0_m$, and

draw $2\mu_i$ boundary cuts from 0_i. On the η-sphere draw m circles lying

outside one another and draw a circular-arc polygon in the regions

they bound having $4\mu_1$ sides each perpendicular to the first circle,

then another of $4\mu_2$ sides standing on the second circle, and so on

until an m-fold connected polygon is obtained. Order the sides of the

boundary A_1, B_1, A_1^{-1}, B_1^{-1}, A_2, B_2, ..., then the corresponding

linear substitutions satisfy $A_1 B_1 A_1^{-1} B_1^{-1} ... A_{\mu_i}^{-1} B_{\mu_i}^{-1} = 1$.

Then, he said "there is always one and only one analytic function

which represents the dissected Riemann surface on the circular-arc

polygon described in this way."

In his paper [1884a, 331-2] Poincaré gave this criticism of

Klein's argument. Suppose S and S' are two manifolds of the same
dimension, that to each s in S there corresponds a unique s' in S',
and that to each s' in S' there corresponds at most one point in S.
If this correspondence is analytic, then the condition that S is
closed (i.e. without boundary) forces the map from S to S' to be an
onto map. That is, to each s' in S' there always corresponds a unique
s in S. But if S has a boundary, the conclusion no longer follows.
For Klein's argument to be valid it is necessary to show that the
space of groups has no boundary, which said Poincaré, is "not at all
evident a priori". The question is made more difficult by the fact
that the same group may have many very different fundamental polygons,
and the onto nature of the map must be established by a special
discussion which Klein had failed to provide. "This is a difficulty
one cannot overcome in a few lines" [1884a, 332]. But Poincaré only
replied at the time (12 May) that their methods differed less in the
general principle than in the details. As for the lemmas, he had
established the first by considerations of power series, and the
second "presents no difficulty and it is probable we will establish
it in the same way". Once this was done, "and it is in effect there
that I begin, like you I employ continuity, ...".

Klein replied by return of post (14 May). To prove the second
lemma he considered the sets of all groups and of all Riemann surfaces
as forming analytic manifolds with analytic boundaries in Weierstrass's
sense of the term. The correspondence between these manifolds is
analytic and 1 - x in different places, where x is 0 or 1, and the lemma
follows from the result that the mapping from groups to surfaces has a
nowhere vanishing functional determinant. But he admitted this argument was
only advanced in principle, he expected that to carry it out would be a

lot of trouble and it might have to be modified. He had also been to
see Schwarz in Göttingen, and without Schwarz's permission to do so
he offered Poincaré Schwarz's ideas on the problem. Schwarz imagined
the dissected Riemann surface infinitely covered by leaves joined
along the boundary cuts, so as to form a surface. This surface, at
least for η-functions of the 'right' kind, would be simply-connected
and simply bounded (in as much as infinite surfaces can be described
in such ways at all) and so can be mapped onto the interior of a circle.
As he said, "this Schwarzian line of thought is in any case very
beautiful."

Poincaré agreed in that estimation (18 May) but admitted that he
had tried and failed to extend Schwarz's ideas to more complicated
problems, and hoped that Schwarz would have better luck.

As spring turned into summer both men were free to write up their
researches in extended form. Poincaré had been commissioned by
Mittag-Leffler to publish his work in the first issue of the new Acta
Mathematica[15], and Klein had his own Mathematische Annalen. It is
clear from their correspondence that they had each arrived at a
convenient summit, offering a view over a considerable amount of new
mathematical terrain, and with a hint of new peaks in the distance.
In the event both men chose to survey their discoveries, rather than
to continue their assault, and indeed the uniformization theorem and
Kleinian groups were to prove markedly less tractable problems[16]. As
a result, it is possible to describe their long papers quite briefly.

Klein began his "Neue Beiträge zur Riemannschen Funktionentheorie"
with a discussion of a Riemann surface as a closed surface in space,
carrying a metric

$$ds^2 = \hat{E}dp^2 + 2\ Fdpdq + Gdq^2$$

for which

$$\frac{\partial}{\partial p}\left(\frac{F\frac{\partial u}{\partial q}-G\frac{\partial u}{\partial p}}{\sqrt{EG-F^2}}\right) + \frac{\partial}{\partial q}\left(\frac{G\frac{\partial u}{\partial p}-E\frac{\partial u}{\partial q}}{\sqrt{EG-F^2}}\right) = 0$$

This is the approach taken in the earlier "Riemann's Theorie der
algebraischen Funktionen und ihrer Integrale". It leads naturally to
a local theory of Riemann surfaces and then to the global study of the
patches of surfaces obtained from the above equation. Dirichlet's
principle as proved by Schwarz is the crucial tool for representing
surfaces. Klein quoted an extensive passage from a letter Schwarz
wrote to him on 1st February 1882 in which the non-simply connected
case is discussed. Various patches are described including circular
arc-triangles and polygons, and the way in which they might fit
together to form surfaces is discussed, following the lines of the
correspondence. This line of argument is extended in the second
section of the paper to a general discussion of the analytic continuation of a function by means of tiling or polygon nets (Bildungsnetz)
which leads to the central idea of the paper: single valued functions
with linear transformations to themselves.

These functions and their groups are discussed carefully in
section III. Klein pointed out in §5 that even amongst discontinuous
groups there are these which have no fundamental region, the fundamental points are everwhere dense in the complex plane. Then there
are groups which arise when the complex plane is divided up by several
natural boundaries, and those which have no connected fundamental
domain. So, said Klein, the idea of a discontinuous group must be
supplemented by that of the fundamental domain before it becomes
workable. He then devoted several paragraphs to examples where the

fundamental domains are well understood: the regular solids, the
parallelogram lattice for elliptic functions, and the circular arc
triangles of Schwarz. In the last case he introduced considerations
of non-Euclidean geometry as Poincaré had done. He went on to look
at more general examples where the group preserves a fundamental
circle, and showed how Riemann surfaces may be produced by drawing a
certain fundamental polygon and identifying the sides. In §14 he
showed that a bounded Riemann surface of genus p with n branch points
may be produced by identifying the sides of a polygon of 4p + 2n sides.
To each boundary cut correspond 2 sides A_i and A_i^{-1}, B_i and B_i^{-1}.
There are 8 real choices to be made to specify A_1 and A_1^{-1}, B_1 and B_1^{-1};
6 for each subsequent quadruple. The branch points each require two
sides (which are identified by the group, the curve so obtained joins
each branch point to the initial point 0 of A_1) so there are 2n choices
to be made, so there are seemingly 6p + 2n + 2 choices in all. But the
polygon must be closed, so three choices are lost, the position of 0 is
arbitrary, losing two more, and η may be replaced by $\frac{\alpha\eta+\beta}{\gamma\eta+\delta}$ with real
α, β, γ, δ, so the decomposition depends on 6p+2n-6 real parameters.

Fairly general remarks followed, about the situation when there
is no fundamental circle, and the process of continuously varying the
parameters. This gave way to the fourth and penultimate section of
the paper, on the so-called fundamental theorem. This asserted that
to every Riemann surface of genus p, with n branch points joined by
lines ℓ_k to a common point, there corresponded one and only one normal
function η which mapped the polygon onto the dissected surface. The
theorem, said Klein, had the Grenzkreis theorem and the Ruckkehrschnitt
theorem (to give them their later names) as special cases. Its proof
rested on a 1-1 correspondence between two manifolds, M_1 consisting of

all Riemann surfaces of type (p, n, ℓ_k) and M_2 of all the normalized functions η. M_1 and M_2 have the same dimension, $6p + 2n - 6$. Klein argued that each element of M_2 can be associated to exactly one element of M_1. For, given η, one obtained a Riemann surface of a certain type, but any two which could be obtained could be deformed continuously into one another since the dissected surfaces can be mapped onto discs, and their images in the discs deformed continuously into one another. This was, he said, intuitively evident. Conversely, to each element of M_1 there corresponds a unique element of M_2, i.e. given a Riemann surface one can find a unique function η. For, Klein argued, suppose these were two: η and η^1. Then analytic continuation of the same funda-mental region would yield two equal regions, which, together with any isolated singularities and their boundary points, may be mapped onto one another by a harmonic function, which in turn establishes a linear relationship between η and η^1. Finally Klein invoked a continuity argument to establish that the correspondence between M_1 and M_2 was analytic. He admitted that the arguments in this section constituted only grounds for a proof and not a rigorous proof, which is indeed a fair understatement of what still needed to be done.

Klein concluded his paper by remarking on how the theory of elliptic functions fitted into this framework. The group in this case is commutative, and the transformation theory concerns the passage from η-functions for one modulus to η-functions invariant under a sub-group.

As Klein himself said, ironically the price he had to pay for the work was extra-ordinarily high. In autumn 1882, while working on his "Neue Beiträge" his health broke down completely. A sad passage in his Werke [II, 258] records that he had to rest for a long time, and was never able to work again at the same high level; elsewhere he spoke of the centre of his productive thought being destroyed. The competitive element in his personality might be supposed to have driven him too hard, and much has been made of the 'race' between the two men, but the true story is clearly more complicated.

It was not simply, or even chiefly, a race. The generosity of the two men towards each other made it more of a cooperative effort. As has been made clear, doubts and insights are shared in the letters, even to the point of divulging private discussions with other mathematicians. Klein never withheld anything, and cannot be accused of delay in reporting his ideas. But, although Klein had a profound grasp of mathematics, he was not the innovator that Poincaré was, and did not bring to the quest for automorphic functions the same depth of vision that Poincaré did. As a contest it was unequal, but Klein was too fiercely ambitious not to feel it as a challenge.

In 1880 he was the dominant figure of his generation. Only Schwarz could rival him in Germany, there was no one to compare with him in France or England, and there can be no doubts about his ambition.

The magisterial tone of at least his early letters to Poincaré, the ceaseless carping about Fuchs, the displays of erudition betray a man eager to be seen as the leader of his profession. The older generation in Berlin were out of reach, but their likely successors (Fuchs, Schwarz, Frobenius) he saw as rivals as much as colleagues. The familiar picture of Klein in his later years as an autocrat who could not be contradicted merely confirms the portrait of the young man who wanted to know all mathematics, and who felt he must shout at

the Berlin seminars to put across his point of view. And yet it is a
fine achievement to accompany Poincaré for a year as he invented the
theory of automorphic functions, so often reformulating old ideas in
ways that ran counter to the tradition Klein had just struggled for
some years to master. Klein felt called upon to try his utmost to
match Poincaré's achievements, and failed cruelly in the attempt. The
mathematical community today shows fewer signs than ever of resisting
comparing men by comparing their work. Klein's ambition was
sustained by a hierarchical mathematical community which a hundred
years later generates similar pressures; his grievous misfortune was
not his fault.

Poincaré's two papers straddled Klein's. The first, on Fuchsian
groups, was finished in July and printed in September, the second,
on Fuchsian functions, was finished in late October and published at
the end of November 1882. The papers gave the first detailed account
of the theory since the resume published in Annalen, but, despite their
greater length (62 and 102 pages), they are more reticent.

Poincaré's papers of 1882

The first paper describes how a discontinuous group of
transformations of the upper half-plane H, $z' = \dfrac{az + b}{cz + d}$ (ad - bc = 1,
a,b,c,d $\in$ $\mathbb{R}$) may be considered geometrically. If $\alpha, \beta \in$ H, and the
circle through them perpendicular to the real axis meets the real
axis, $\mathbb{R}$, at h,k say, then the cross-ratio $\dfrac{(\alpha - h)}{(\alpha - k)} \dfrac{(\beta - k)}{(\beta - h)} = [\alpha,\beta]$ is
preserved by all elements of the group and is determined by α and β.
It is multiplicative in that, if γ lies on the same circle, then
$[\alpha,\beta][\beta,\gamma] = [\alpha,\gamma]$, and, for infinitesimal points z and z + dz in H,
Poincaré showed that

$$[z, z + dz] = 1 + \frac{|dz|}{y},$$

neglecting higher terms. So $\log([\alpha,\beta])$ may be taken as defining the

distance between α and β, whence the element of arc length is

$ds = \frac{|dz|}{y}$ and the element of area is $dS = \frac{dxdy}{y^2}$, setting $z = x + iy$,

$dz = dx + idy$. Poincare observed that these definitions implied that

the geometry thus introduced into the upper half plane was

non-Euclidean, but decided not to employ that terminology in order to

avoid any confusion. He called the group 'discontinuous' if no

substitution in it could be found, say f_i, for which $\log[z, f_i(z)]$ was

infinitesimally small, and a discontinuous group of real

substitutions he called a Fuchsian group (§3).

To each Fuchsian group he assumed he could associate a region R_0

of H such that the transforms of R_0 by the elements of the group

covered it exactly once with overlaps only on the boundaries of regions.

Two regions with a piece of boundary in common he called limitrophic

(limitrophes). An edge of a region was a piece of boundary in the form

of an arc of a circle perpendicular to the real axis, i.e. a

non-Euclidean line segment; two edges of one region were said to be

conjugate if one edge could be mapped onto the other by an element of

the group. The interior of R_0 was mapped by f_i, an element of the

group, onto the interior of $f_i(R_0)$. R_0 was, by definition, a region

containing only one point in the set $\{\frac{a_i z + b_i}{c_i z + d_i} = f_i(z)\}$ as f_i ran

through the group, for each point $z \in H$. Poincare noted (§4) that

this constraint fell far short of defining R_0 uniquely, and argued that

one could always find an R_0 which was in one piece and without a hole.

For, if R_0 has a hole, it also has a second piece exterior to it, S_0, and a transformation, f_i, exists which maps S_0 into the hole, so R_0 can be replaced by $R_0 + f_i(S_0) - S_0$. Furthermore, the region R_0 may be taken to be bounded by edges and to form a convex region, upon suitably adding and subtracting pieces to R_0 along conjugate sides. Poincaré admitted polygonal regions for which segments of the real axis $\mathbb{R}$ formed part of the boundary (such segments he called vertices). A convex region R_0 bounded by edges Poincaré called a normal polygon, and he claimed that given such a polygon and the pairing of conjugate sides the group was determined completely. As Nörlund noted [1916, 126], this is true unless an edge AB has both vertices on the real axis, when the polygon determines an infinity of groups.

Poincaré was thus led to 7 families of groups, depending on the nature of the vertices of R_0. A vertex was of the first kind if it lay strictly in H, of the second kind if it was a point of $\mathbb{R}$, of the third kind if it was a segment of $\mathbb{R}$. Accordingly the seven kinds of region were obtained by insisting that either (1) all vertices were of the first kind, or (2) all were of the second kind, or (3) all of the third kind, or (4) all were of the second and third kinds, or (5) all of the first and third kinds, or (6) all of the first and second kinds, or (7) were of all kinds.

He said vertices z and z' corresponded if an element of the group took z to z', and the set of all corresponding vertices to a given one formed a cycle. The cycle was of the first category if it was closed under the process of leaving a vertex, tracing an edge and then its conjugate edge until a new vertex is reached (the edges being directed in a standard way in advance), and so on, and all the vertices were of the first kind. Cycles of the second category are closed and only contain vertices of the second kind. Cycles of the third category contain vertices of the second or third kind, and are open. For R_0 to generate a Fuchsian group he observed that the angle at any vertex of the first kind must be an aliquot part of 2π, and that corresponding sides must have the same length (as they evidently will). Conversely, he showed (§6) that these conditions are also sufficient for R_0 to generate a Fuchsian group. At this stage in his argument Poincaré assumed that certain regions R_0 could be found for which the set of all $f_i(R_0)$ did not cover all of H but was a proper subset, which is false.

He then gave several examples before turning to the computation of the genus of the surface obtained by identifying corresponding sides of R_0. He found (by Euler's formula) that if R_0 had $2n$ edges all of the first kind and q closed cycles of vertices then the genus, p, was given by

$$p = \frac{n + 1 - q}{2} .$$

Similarly, if R_0 has n edges of the second kind then Euler's formula implied that the genus was

$$p = n - q.$$

Finally, if R_0 was of the third kind the genus was necessarily 2.

Poincaré discussed how the generating polygon might be simplified, gave more examples and observed that a Fuchsian group has as many fundamental relations between its fundamental substitutions (those which transform R_0 to a limitrophic region) as it has cycles of the first kind. He observed that a Fuchsian group is obtained by letting a discontinuous group of complex substitutions preserve a fixed circle but if a discontinuous group has no fixed circle it is not Fuchsian but Kleinian. Finally he gave some brief historical notes which indicate how much he had learned from Klein while still paying generous tribute to Fuchs.

In his second paper, on Fuchsian functions, Poincaré began by showing that, if $f_i(z) = \dfrac{\alpha_i z + \beta_i}{\gamma_i z + \delta_i}$, then $\dfrac{df_i(z)}{dz} = \dfrac{1}{(\gamma_i z + \delta_i)^2}$,

and $\displaystyle\sum_{i=0}^{\infty} \left|\dfrac{df_i}{dz}\right|^m$ converges for any integer m greater than 1. His domain for z is now the unit circle, and he showed that the series converges inside and outside the circle, and at those points which are not limit points of a sequence $(-\dfrac{\delta_i}{\gamma_i})$, i.e. z belongs to an edge of the second kind (is interior to a segment of the boundary of the unit circle which is part of the boundary of R_0). He gave two proofs of this result, the second rests essentially on the observation that there are not many points $f_1(z)$ within any circle centre $z = 0$ and radius $R < 1$. Furthermore, he said, if the parameters determining R_0 are varied without the angle sum changing then R_0 in it changed form generate isomorphic groups and the series $\displaystyle\sum_i \left(\dfrac{df_i}{dz}\right)^m$ depends continuously on these parameters.

A theta-Fuchsian function he defined by the formula

$$\theta(z) = \Sigma H\left(\frac{\alpha_i z + \beta_i}{\gamma_i z + \delta_i}\right)(\gamma_i z + \delta_i)^{-2m}$$

where $H(z)$ is an arbitrary rational function of z having no pole on the fundamental circle, and m is an integer greater than 1. If H has poles at $a_1, \ldots, a_p$ inside the unit circle then $\theta(z)$ is infinite at all the points $\frac{\alpha_i a_k + \beta_i}{\gamma_i a_k + \delta_i}$, but the series converges everywhere else. In Section V of the paper Poincaré investigated whether H could be chosen that $\theta(z)$ vanished identically, and gave examples to show that this could happen. E.g. (no. 4) if α_r is a vertex of the polygon R_0 and α_r' its image in the fundamental circle, then the Fuchsian group contains the substitution $\left(\frac{z - \alpha_r}{z - \alpha_r'}, e^{2\pi i/\beta_r}\left(\frac{z - \alpha_r}{z - \alpha_r'}\right)\right)$ for some integer β_r. If $H(z) := \left(\frac{z - \alpha_r}{z - \alpha_r'}\right)^p \frac{1}{(z - \alpha_r')^{2m}}$ and $p + m \equiv 0 \mod \beta_r$ then $\theta(z) = 0$ for all z. For the function θ may be written as

$$\theta(z) = \Sigma_i \frac{(f_i - \alpha_i)^p}{(f_1 - \alpha_r')^{p+2m}}\left(\frac{df_i}{dz}\right)^m$$ and if the summation is taken first over

the rotations around α_r and then over each coset the factor $e^{2n(p+m)\pi i/\beta_r}$ appears, which is identically zero. Poincaré showed how one example automatically generates infinitely many more, but gave no characterisation of the H for which θ vanishes.

The bulk of the paper was given over to establishing the theorem that every theta-Fuchsian can be written as $\left(\frac{dx}{dz}\right)^m F(x,y)$, where F is a rational function and x and y are two Fuchsian functions in terms of which every other Fuchsian function can be written rationally and between which there exists an algebraic relation $\psi(x,y) = 0$. Conversely, every function of the form $\left(\frac{dx}{dz}\right)^m F(x,y)$ can be written as

a theta-Fuchsian function provided it vanishes whenever z is a vertex
of R_0 which lies on the fundamental circle. The proof distinguished
between finite and infinite R_0 and the cases where the genus does and
does not equal zero. It follows from the theorem that every
Fuchsian function can be written infinitely many ways as a quotient
of two theta-Fuchsian functions. Furthermore, the theta-Fuchsian
functions which do not have poles in the fundamental circle form a
finite dimensional space, and Poincaré showed how various linear
relationships between such functions might be found. The dimension
is q = (2m - 1)(n - 1), where m is as above and 2n is a number of sides
of R_0.

Poincaré's papers of 1883 and 1884

Mention should be made of Poincaré's other three big papers in
Acta Mathematics: [1883] on Kleinian groups, [1884a] on the groups
associated to linear differential equations, and [1884b] on
Zeta-Fuchsian functions. The difficulties involved in studying
Kleinian groups and Kleinian functions are much greater than those of
the Fuchsian case, and are still not properly resolved today.
Poincaré does little more than indicate how much of the analogy goes
over, and how three-dimensional non-Euclidean geometry might help.
The papers on differential equations and Zeta-Fuchsians are more
substantial. Poincaré raised two questions: given a linear equation
with algebraic coefficients, find its monodromy group; and, given a
second-order linear equation containing accessory parameters, choose
them in such a way that the group is Fuchsian. His conclusions, based
on the method of continuity (which he admitted was not at all
obviously true) were that every equation

$$\frac{d^2 v}{dx^2} = \phi(x,y)v,$$

where $\theta(x,y) = 0$, ϕ and θ are rational in x and y, and the exponent differences of the solutions at the singular points are zero or aliquot parts of unit is such that, if z is a quotient of two independent solutions, then x := x(z)

  (i)   will be a Fuchsian function existing only inside a circle for exactly one choice of the accessory parameters;

 (ii)   will be a Fuchsian or Kleinian function existing for all z for exactly one choice of the accessory parameters;

(iii)   will be a Kleinian function existing only in a subregion of $\mathbb{C}$ for infinitely many choices of the accessory parameters.

The zeta-Fuchsian paper likewise confined itself to equations all of whose solutions are regular. These equations whose coefficients may be uniformized by Fuchsian functions, are called Fuchsian equations. At the end of this series of papers, and 390 pages of Acta Mathematica, Poincaré modestly bade farewell to these subjects for a while with the words: "This will suffice to make it clear that, in the five memoirs of Acta Mathematica which I have dedicated to the study of Fuchsian and Kleinian transcendents I have only skimmed a vast subject which without doubt will furnish geometers with the occasion for numerous important discoveries."

Conclusion

With Poincaré's papers a certain process is completed. Gauss's insights into the hypergeometric equation, the theory of modular transformations, the centrality of the theory of functions of a complex variable, are here combined and in that way raised to a new level. It would be possible to insist on the neatness of this synthesis, to

stress the unity underlying the apparently varied techniques of
non-Euclidean geometry, group theory, and ordinary differential
equations. But it is surely preferable to close by stressing the
fortuitous nature of these developments; it was after all the
generously inventive Poincaré and not the more learned Klein who
achieved this consumation. Tradition, even historical momentum,
cannot guarantee progress. The unities of mathematics naturally
reflect the unities of a mathematician's training, the shared
frameworks within which questions are posed. But they are
maintained and re-formed by the original perceptions of fortunate
mathematicians. They are real, but fragile like all living things.
They are questions as well as answers, and so one task for
historians is to capture this aspect of mathematical work. Poincaré
drew together questions asked by different mathematicians, and his
answers showed that these concerns belonged together, but his
strangely patchy education makes this achievement the more
remarkable. Confronted with this fact, the historian may also feel,
and feel pleased, that there are things that cannot be fully
explained.

APPENDICES, NOTES, AND BIBLIOGRAPHY.

Appendix 1: Riemann, Schottky, and Schwarz on conformal representation.

In his Thesis [1851], Riemann sought to prove that if a function u is given which satisfies certain conditions on the boundary of a surface then it is the real part of a unique complex function which can be defined on the whole of T. He was able to show that any two simply connected surfaces (other than the complex plane itself – Riemann was considering bounded surfaces) can be mapped conformally onto one another, and that the map is unique once the images of one boundary point and one interior point are specified; he claimed analogous results for any two surfaces of the same connectivity [§19]. To prove that any two simply connected surfaces are conformally equivalent he observed that it is enough to take for one surface the unit disc $K = \{z : |z| \leq 1\}$ and he gave [§ 21] an account of how this result could be proved by means of what he later, [1857c, 103], called Dirichlet's principle. He considered [§ 16]:

(i)   the class of functions, λ, defined on a surface T and vanishing on the boundary of T, which are continuous, except at some isolated points of T, for which the integral

$$L = \int_T \left(\left(\frac{\partial \lambda}{\partial x} \right)^2 + \left(\frac{\partial \lambda}{\partial y} \right)^2 \right) dt$$

is finite; and

(ii)   functions $\alpha + \lambda = \omega$, say, satisfying

$$\int_T \left(\frac{\partial \omega}{\partial x} - \frac{\partial \beta}{\partial y} \right)^2 + \left(\frac{\partial \omega}{\partial y} + \frac{\partial \beta}{\partial x} \right)^2 \, dt = \Omega < \infty$$

for fixed but arbitrary continuous functions α and β.

He claimed that Ω and L vary continuously with varying λ but cannot be zero, and so Ω takes a minimum value for some ω. The claim that this value is attained for some λ in the first class of functions had earlier been made by Green [1883, first pub. 1835] and Gauss [1839/40] but it was questioned in another context by Weierstrass [1870], who showed by means of a counter-example that there is no general theorem of the kind: 'a set of functions bounded below attains its bound'. The use of Dirichlet's principle became contentious, and several mathematicians, notably Schwarz, sought to avoid it (Schwarz's work is described below). It was finally vindicated, under slightly restricted conditions, by Hilbert [1900a] in a beautifully simple paper.

In his next major paper on complex function theory, "Theorie der Abel'schen Functionen" [1857c], published just after his paper on the hypergeometric equation, Riemann recapitulated most of the above analysis. A complex function ω of $z = x+iy$ is one which satisfies $i\frac{\partial\omega}{\partial x} = \frac{\partial\omega}{\partial y}$, and hence can be written uniquely as a power series in $z-a$ where a is any point near which ω is continuous and single valued. If ω is defined on a subset of the complex plane it may be continued analytically along strips of finite width. The continuation is unique at each stage, but if the path crosses itself the function may take different values on the overlap. If it does not it is said to be single-valued or monodromic (einwerthig), otherwise multivalued (mehrwerthig). Multivalued functions possess branch points; e.g. $\log(z-a)$ has a branch point at a. The different determinations of the function according to the path of the continuation he now called its branches or leaves [Zweige, Blättern].

Riemann extended his analysis of surfaces to include those without boundary by the simple trick of calling an arbitrary point of an unbounded surface its boundary. He concluded the elementary part of this paper by repeating his argument, based on what he now explicitly called Dirichlet's principle, that a complex function can be defined on a connected surface T, which, on the simply connected surface T' obtainable from T by boundary cuts,

(i) has arbitrary singularities, in the sense of becoming infinite at finitely many points in the manner of a rational function; and

(ii) has a real part which takes arbitrary values on the boundary.

The conformal representation of multiply connected regions was studied by Schottky [1877]. He considered the integrals of rational functions defined on such a region, A, and showed (§4) by looking at their periods that every single-valued function on A, which behaves like a rational function on the interior of A and is real and finite on the boundary of A, can be written as a rational function of functions like

$$u = \frac{1}{x - a} + u_0 + \sum_{i=1}^{\infty} u_i (x - a)^i$$

$$v = \frac{i}{x - a} + v_0 + \sum_{i=1}^{\infty} v_i (x - a)^i.$$

He defined the genus of A(§16) as Weierstrass had done (Schottky referred to Weierstrass's lectures of 1873-74), noted that it agreed with Riemann's definition and that genus was preserved by conformal maps. He concluded the paper (§16) by taking the special case when A is bounded by a circle L_0 which encloses n-1 circles which are disjoint and do not enclose one another. This region has genus n-1 and there is a group of conformal self-transformations of the disc bounded by L_0 obtained by inverting A in each $L_1, \ldots, L_{n-1}$ and then in the circle which bounds the images, and so on indefinitely. (The limit point set can be quite dramatic; Fricke in Fricke-Klein Automorphe Fuchtionen, I, 104, gave an example when A is bounded by three circles and the limit set is a Cantor set). Such regions had already been considered by Riemann, and the fragmentary notes he left on the question edited into a brief coherent text by Weber (Riemann [1953f], and published in 1876). Like Schottky, Riemann showed that a conformal map of the region can always be found which maps the boundary circles onto the real axis and takes every value in the upper half plane, H, n times. Repeated inversion then produces a function invariant under the group and this function is the inverse of the quotient of a second-order differential equation with algebraic coefficients, that is to say, the differential equation is defined on the above covering of H. Conversely, a conformal representation of this covering on A is obtained by solving such a differential equation, provided the coefficients take conjugate imaginary values at conjugate points. Schottky's work is essentially his thesis of 1875, and consequently was done independently of Riemann.

Schwarz wrote several papers on the conformal representation of one region upon another between 1868-1870 which should be mentioned, although they do not immediately relate to differential equations.

In the first of them, [1869a], Schwarz remarked that Mertens, a fellow student of his in Weierstrass's 1863-64 class on analytic function theory, had observed to him that Riemann's conformal representation of a rectilineal triangle on a circle raised a problem: the precise determination of such a function seemed to be beyond one's power to analyse because of the discontinuities at the corners. Discontinuity at that time meant any form of singularity, in this case the map is not analytic. That had set Schwarz thinking, for he knew no examples of the conformal representation of prescribed regions, and in this paper be proposed to map a square onto a circle. He said [1869a=Abh,II,66]

"For this and many other representation problems this fruitful theorem leads to the solution:

"If, to a continuous succession of real values of a complex argument of an analytic function, there corresponds a continuous succession of real values of the function then to any two conjugate values of the argument correspond conjugate values of the function."

This result is now called the Schwarz reflection principle. In symbols, it asserts that if f is analytic and $f(x)$ is real for real x, then $f(x + yi) = \overline{f(x-yi)}$. To represent the square on a circle by a function $t = f(u)$ Schwarz said it was simpler to replace the circle by a half-plane, which can be done by inversion ("transformation by reciprocal radii"). The singular points would be

$t = \infty$, -1, 0, and 1 and the upper half plane is to correspond to the inside of the square. The boundary of the square will be mapped by t onto the real axis. As the variable u crosses a side of the square $t(u)$ will cross into the lower half plane, and by the reflection principle t is thus defined on the four squares adjacent to the original one, so on iterating this process t is seen to be a doubly periodic function defined on a square lattice, thus a lemniscatic function.

Indeed, Schwarz went on, the function $v = u^2$ converts the wedge-shaped region within an angle of $\pi/2$ to a half plane, and any analytic function with non-vanishing derivative at $v = 0$ will then produce an conformal copy of the wedge. One might speak of such a function straightening out the corner. But even so, this u is not sufficiently general, for an everywhere analytic map sending lines to lines, namely $u' = C_1 u + C_2$, where C_1 and C_2 are constants, is surely equivalent to it. To eliminate the constants Schwarz put forward the equation

$$\frac{d}{dt} \log \frac{du'}{dt} = \frac{d}{dt} \log \frac{du}{dt},$$

which he said was an important step, for $\frac{d}{dt} \log \frac{du}{dt}$ is infinite whenever $\frac{du}{dt}$ is zero or infinite, which happens at each singular point. So

$$\frac{d}{dt} \log \frac{du}{dt} = -\frac{1}{2} t^{-1} + d_1 + d_2 t + \ldots ,$$

and by considering the vertices $t = -1$, 0 and 1, Schwarz showed that the expression

$$\frac{d}{dt} \log \frac{du}{dt} + \frac{1}{2}(\frac{1}{t+1} + \frac{1}{t} + \frac{1}{t-1}) = *$$

is to be analytic for all finite t. The transformation $t' = \frac{1}{t}$

enabled him to consider t infinite, and it turned out $\frac{d}{dt} \log \frac{du}{dt}$ was

zero at infinity, so * is constant and indeed zero, which implies

that

$$\frac{d}{dt} \log \frac{du}{dt} = -\frac{1}{2}(\frac{1}{t+1} + \frac{1}{t} + \frac{1}{t-1}).$$

This led Schwarz to his result:

$$u = C_1 \int_0^t \frac{dt}{(4t(1-t^2))^{\frac{1}{2}}} + C_2$$

To represent a regular n-gon on a circle, a simple generalization

showed that

$$u = \int_0^s \frac{ds}{(1-s^n)^{n/2}}$$

would suffice, a result which Schwarz said he had presented for his

promotion at Berlin University in 1864. A continuity argument which

Schwarz attributed to Weierstrass enabled him also to deal with an

irregular n-gon.

To consider figures bounded by "the simplest curved lines", curves,

Schwarz applied an inversion

$$u' = \frac{C_1 u + C_2}{C_3 u + C_4}$$

where C_1, C_2, C_3 and C_4 are arbitrary constants, to obtain the fullest

generality. He eliminated the arbitrary constants and obtained

$$\frac{d^2}{dt^2} \log \frac{du}{dt} - \frac{1}{2}(\frac{d}{dt} \log \frac{du}{dt})^2,$$ which he denoted $\Psi(u,t)$

[it is equal to $\dfrac{\dfrac{d^3u}{dt^3}\dfrac{du}{dt} - \dfrac{3}{2}(\dfrac{d^2u}{dt^2})^2}{(\dfrac{du}{dt})^2}$].

If t has no winding point inside the circular-arc polygon, then $\Psi(u,t)$ is analytic inside the polygon, and so, if the polygon is mapped onto the upper half plane, $\Psi(u,t)$ will be a rational function $F(t)$. Furthermore, the general solution of the differential equation $\Psi(u,t) = F(t)$ will be the quotient of two solutions of a second order linear differential equation with rational coefficients. Schwarz thanked Weierstrass for drawing this observation to his attention. We have seen that this observation was a standard move in the theory of elliptic functions.

Schwarz then considered the function

$$u = \int_{t_0}^{t} (t-a)^{\alpha-1}(t-b)^{\beta-1}(t-c)^{\gamma-1} \, dt$$

for real constants, a, b, c, α, β, and γ, where α, β, and γ are positive and $\alpha + \beta + \gamma = 1$. It maps the upper and lower half planes onto two rectlilinear triangles with angles $\pi\alpha$, $\pi\beta$, and $\pi\gamma$. By reflection an infinitely many valued function is obtained unless α, β, γ take one of these four sets of values

$$\frac{1}{2}, \frac{1}{4}, \frac{1}{4}; \quad \frac{1}{2}, \frac{1}{3}, \frac{1}{6}; \quad \frac{1}{3}, \frac{1}{3}, \frac{1}{3}; \quad \frac{2}{3}, \frac{1}{6}, \frac{1}{6};$$

in which case the triangles tessellate the plane, as Christoffel [1867] and Briot and Bouquet had shown [1856a, 306-308].

During 1869 Schwarz's views on the Dirichlet principle hardened,
one supposes in discussions with Weierstrass, who indeed published his
counter-example to a related Dirichlet-type argument the next year.
In his paper [1869c] Schwarz observed that Dirichlet's principle still
lacked a proof (it was becoming Dirichlet's problem), and he supplied
one for simply-connected polygonal regions, or more generally for convex
regions. The proof consisted of showing that one could deform a
conformal map of a small convex region onto a circle analytically so
that the domain was a slightly larger convex region, which was also
mapped conformally into a circle. The deformation of the domain went
from one convex boundary curve to another through a series of (non-
convex) rectangular approximations. One started trivially with a
circle inside the given region and finished with a map of the whole
region onto a circle.

By the next year Dirichlet's principle had become quite
problematic. In his [1870a] Schwarz brought forward an approach based,
he said, on discussion he had had with Kronecker and other mathematicians in November 1869 on the partial differential equation $\Delta u = 0$.
This was his "alternating method" ("alternirendes Verfahren"), and it
applied to regions bounded by curves with a finite radius of curvature
everywhere, and only finitely many vertices, although cusps were not
excluded. The idea is an attractive one. He considered two overlapping regions T_1 and T_2 of the kind he could already solve Dirichlet's
problem for, called the boundary of T_1 L_0 outside T_2 and L_2 inside,
and likewise the boundary of T_2 L_3 outside T_1 and L_1 inside, and called
the overlap T^*. He sought a map u such that $\Delta u = 0$ and u took

prescribed values on L_0 and L_3.

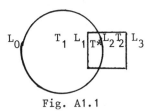

Fig. A1.1

Invoking the metaphor of an air pump he regarded T_1-T* and T_2-T* as air cylinders, and L_1 and L_2 as valves. The first stroke of the first cylinder maps T_1 by u_1 onto a half-plane, taking arbitrary prescribed values on L_0 and a fixed value on L_2 equal to the least value to be taken by u on L_0 and L_3. The first stroke of the second cylinder did the same for T_2, with a function u_2 which always took the maximum value of u on L_1. This is possible because of his assumptions about T_1 and T_2. The second stroke of the first cylinder converted u_1 into u_3, a function agreeing with u on L_0 and u_2 on L_2. The second stroke of the second cylinder likewise turned u_2 into u_4, which agreed with u on L_3 and u_3 on L_1. The pump works smoothly, and Schwarz found the sequences

$$u' = u_1 + (u_3 - u_1) + (u_5 - u_3) + \ldots + (u_{2n+1} - u_{2n-1}) + \ldots$$

$$u'' = u_2 + (u_4 - u_2) + (u_6 - u_4) + \ldots + (u_{2n+2} - u_{2n}) + \ldots$$

converge to the same, sought-for function u.

He then gave examples of what could be done, remarking at one point [<u>Abh</u>. II, 142]:

"Let a circular arc triangle be given. The conformal representation of the surface of a circle onto the inside or the outside of a

circular arc triangle can be carried out without difficulty by means
of hypergeometric series."

An extended version of this paper was published in [1870b].
Schwarz took the opportunity to point out that the analytic functions
he was constructing might have a natural boundary beyond which they
could not be analytically continued, a circumstance of importance
for function theory which, he said, Weierstrass had remarked upon in
general some years ago. But he gave only a rather contrived example.
Finally he again presented his proof of Dirichlet's principle for
simply connnected regions with, so to speak, 'rectifiable' boundaries.

Twelve years later, on February 1, 1882, Schwarz wrote a letter
to Klein, [1882] in which he showed how to extend the alternating
method to multiply-connected regions. A two-fold connected region
with boundary, such as an annulus, T, becomes a simply-connected
region once a boundary cut is drawn. Let the cut Q produce the region

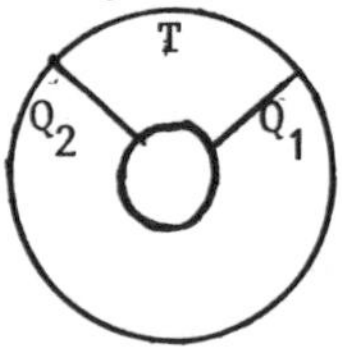

Fig. A1.2

T_1, and Q_2 produce T_2. The alternating method enables one to equalize
the two solutions to Dirichlet's Problem for T_1 and T_2, and find a
function which jumped by a prescribed constant amount upon crossing
a cut. The argument readily extended by induction to regions of
higher connectivity, but Schwarz admitted he had found it much harder
to deal with unbounded, closed, Riemann surfaces. The difficulty, he
said, was overcome by removing two concentric circular patches R_2
containing R_1 from the surface, which lay in the same leaf and

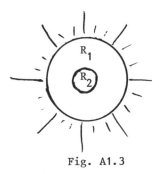

Fig. A1.3

enclosed no singular point. The problem could be solved for the Riemann surface without the smaller patch, as it had a boundary, and one could indeed assume the solution function, u_1, took a constant value on the boundary of R_1 and prescribed moduli of periodicity along the cuts. The function did not extend into R_1, but one could solve the Dirichlet problem for the surface patch R_2 by a function u_2 which took the values of u_1 on the boundary of R_2. One could then solve the problem outside R_2 and find a function which agreed with u_2 on R_1 and so on. In this way one could solve the problem.

One supposes this letter was a reply to one of Klein's about the validity of the Dirichlet principle, perhaps to the publication of Klein's essay on Riemann, to which Schwarz referred. Schwarz still regarded the Riemann surface as spread out above the complex plane. When Klein shortly came to a different view he wrote again to Schwarz, as is described in Chapter VI. I have examined the collection of letters from Schwarz to Klein in the Klein Archive in Göttingen, [N.S.U. Bib. Göttingen, Klein XI, 934-938] but they add nothing to our knowledge of Schwarz's idea. 937 (8 April) looks forward to Klein's visit, 938 (2 July) talks about different matters - models of Kummer's and other surfaces.

Appendix 2: Riemann's lectures on differential equations.

Much of Riemann's work was not published until 1876 and some extracts from his notebooks and lectures were not published until 1902 in the Nachträge added to the second edition of his Werke. By then much of what he had found had been independently rediscovered, and these ideas of Riemann's will only be briefly described here. The publication of 1902 excited much interest and Schlesinger [1904] and Wirtinger [1904] reported on it to the Heidelberg International Mathematical Congress.

Riemann had lectured on differential equations at Göttingen in the Winter semester 1856/57, the year before he published his paper on the hypergeometric equation. In those lectures [Nachträge 67-68] he analysed the behavior of two independent solutions of a second order linear differential equation near a branch point, b. He showed that there are two different cases which can arise. The two independent solutions, y_1 and y_2, are each transformed under analytic continuation around b, into $ty_1 + uy_2$ and $ry_1 + sy_2$ respectively. If there is a combination $y_1 + \varepsilon y_2$ which returns as $(y_1 + \varepsilon y_2)$ constant then ε must satisfy

$$\varepsilon(t + \varepsilon r) = u + \varepsilon s. \qquad (A2.1)$$

Let ε satisfy this equation, and choose α so that $t + \varepsilon r = e^{2\pi i\alpha}$, then $(y_1 + \varepsilon y_2).(z-b)^{-\alpha}$ is unaltered on analytic continuation around b, and so Riemann wrote $y_1 + \varepsilon y_2 = (z-b)^{\alpha} \sum_{n=-\infty}^{\infty} a_n(z-b)^n$. The two cases arise according as (2.1) has one root or two. If it has two roots, say ε and ε', then the corresponding α and α' are different and two independent solutions can be found which are invariant (up to multiplication by a scalar) upon analytic continuation around b. But

if, and this is the second case, (A2.1) has repeated root, ε, then

the second solution must have the form y,

$$y = (z-b)^\alpha \log(z-b) \sum_{-\infty}^{\infty} a_n(z-b)^n + (z-b)^\alpha \sum_{-\infty}^{\infty} b_n(z-b)^n.$$ That is

to say the second solution contains one term which is log(z-b) times

the first solution, and one term having the form of the first solution.

These two cases correspond to the cases where the monodromy matrix

$\left(\begin{smallmatrix} t & u \\ r & s \end{smallmatrix}\right)$ can be diagonalized and has distinct eigenvalues and where it

cannot. They can be read off from the quadratic equation for the

eigenvalues of the matrix, so it is quite possible to pass directly

from that algebraic equation at the branch points to the analytic form

of the solutions, without the intervening geometry being apparent.

This has been the method usually adopted subsequently, perhaps at a

cost in intelligibility. The connection between the occurrence of

matrices in non-diagonal form and the presence of logarithmic terms

in the solution, as described by Fuchs, Jordan, and Hamburger, is

discussed in chapter II. From von Bezold's summary of these lectures

[Nachträge, 108] and the extract [Werke, 379-385] it appears that

Riemann, after deriving what is essentially the Fuchsian form of a

differential equation, considered the nth order case only when the

monodromy matrices can be put in diagonal form [1953a, 381] but gave

a complete analysis of the 2x2 case neglecting the trivial diagonal

case u=r=0, t=s.

Riemann lectured on the hypergeometric equation [Nachträge, 67-

94] again in the Winter semester 1858-59. In Section A(67-75) he gave a treatment of his P-function in terms of definite integrals of the form $\int s^a(1-s)^b(1-xs)^c dx$. The theme of differential equations emerges more strongly in Section B(76-94), which teems with ideas which were not to be developed by other mathematicians until much later, and then without direct knowledge of Riemann's pioneering insights.

Riemann began by considering the equation $a_0 y'' + a_1 y' + a_2 y = 0$, where a_0, a_1, a_2 are function of z, and $y' = \frac{dy}{dz}$, $y'' = \frac{d^2 y}{dz^2}$. If Y_1 and Y_2 are independent solutions, and $Y = Y_1/Y_2$, then under analytic continuation around a closed path in the z domain Y is transformed into $\frac{\alpha Y + \beta}{\gamma Y + \delta}$, where α, β, γ and δ are constants. The inverse function to Y, say $z = f(Y)$, has the attractive property that $f(\frac{\alpha Y + \beta}{\gamma Y + \delta}) = f(Y) = z$, so it is invariant under all the substitutions corresponding to circuits in z.

Conversely, Riemann claimed that the inverse of any function f of Y invariant under a set of substitutions $Y \to \frac{\alpha Y + \beta}{\gamma Y + \delta} = Y'$ satisfies a second order linear ordinary differential equation. For $\frac{dY}{dz}$ is transformed by the substitution

$$Y \to \frac{\alpha Y + \beta}{\gamma Y + \delta} \text{ into } \frac{dY'}{dz} = \frac{d}{dz}(\frac{\alpha Y + \beta}{\gamma Y + \delta}) = \frac{\alpha \delta - \beta \gamma}{(\gamma Y + \delta)^2}\frac{dY}{dz} = \frac{1}{(\gamma Y + \delta)^2}\frac{dY}{dz},$$

assuming, as Riemann did, that $\alpha \delta - \beta \gamma = 1$,

so
$$(\frac{dY'}{dz})^{-\frac{1}{2}} = (\frac{dY}{dz})^{-\frac{1}{2}}(\gamma Y + \delta)$$

and
$$Y'(\frac{dY'}{dz})^{-\frac{1}{2}} = (\frac{dY}{dz})^{-\frac{1}{2}}(\alpha Y + \beta).$$

So, setting $Y_1 = (\frac{dY}{dz})^{-\frac{1}{2}}$, $Y_2 = Y (\frac{dY}{dz})^{-\frac{1}{2}}$,

Y_1 and Y_2 are two particular solutions of the equation $y'' + a_2 y = 0$,

where $a_2 \doteq - \frac{dY}{dz}^{-\frac{1}{2}} \frac{d^2}{dz^2} \frac{dY}{dz}^{-\frac{1}{2}}$ is, moreover, an algebraic function (it

is $\frac{d^2 Y}{dz^2} . \frac{dY_1}{dz} - \frac{d^2 Y}{dz^2} . \frac{dY_2}{dz}$ and so a constant multiple of the Schwarzian

of Y). Riemann observed that this analysis is very important for the

study of algebraic solutions of a differential equation. I cannot

find that the idea of specifying the monodromy group in advance of

the differential equation was taken up by anyone before Poincaré in

1880, who undoubtedly came to it independently (see Chapter VI). It

is the origin of Hilbert's 21st problem, see Hilbert [1900b]. Riemann

then turned to the hypergeometric equation and took the constants

$\alpha, \alpha', \beta, \beta', \gamma, \gamma'$ to be real. Then the real axis is mapped by Y, the

quotient $P^\alpha / P^{\alpha'}$, onto a triangle with sides the real interval $[0, Y(1)]$,

a straight line from 0, and circular arc from $Y(1)$. The angles of

this triangle are $\pi(\alpha-\alpha')$ at 0, $\pi(\gamma-\gamma')$ at $Y(1)$, and $\pi(\beta-\beta')$ where

the two other sides meet . The function takes every value inside this

triangle exactly once. Conversely, the inverse function $Z = Z(Y)$ is

single valued inside the triangle, since $\frac{dY}{dz}$ never vanishes, and maps

it onto the upper half plane.

This triangle will be a spherical triangle if the angles

$\lambda\pi = (\alpha-\alpha')\pi$ at $Z = 0$, $\mu\pi = (\beta-\beta')\pi$ at $Z = \infty$, $\nu\pi = (\gamma-\gamma')\pi$ at

$Z = 1$ are such that $\lambda + \mu + \nu > 1$. Riemann studied the particular

cases when this is so which are connected to the quadratic and cubic transformations that can sometimes be made to a P-function, so this is a convenient place to record what he had earlier found in [1875a, §5]. Let $P(z_1)$ and $\widetilde{P}(z_2)$ denote P-functions of their respective arguments, possibly with different exponents. A quadratic transformation replaces z by $\sqrt{z}$, and one wants to define $\widetilde{P}(\sqrt{z})$:= $P(z)$. This clearly forces $\widetilde{P}(z) = \widetilde{P}(-z)$ $(= P(z^2))$, so the exponent difference at 0 in the $\sqrt{z}$-domain must be −1. There are singular points at $\sqrt{z} = +1$ and $\sqrt{z} = -1$ which must have the same exponent difference, say $\gamma - \gamma'$, and the exponents at $\sqrt{z} = \infty$ are, say 2β and $2\beta'$. The map $\sqrt{z} \to z$ therefore maps the quadrilateral

to

and one has the P-function

$$
P \left\{ \begin{array}{ccc} 0 & \infty & 1 \\ 0 & \beta & \gamma \\ \tfrac{1}{2} & \beta' & \gamma' \end{array} z \right\} = \widetilde{P} \left\{ \begin{array}{ccc} -1 & \infty & 1 \\ \gamma & 2\beta & \gamma \\ \gamma' & 2\beta & \gamma' \end{array} z \right\} .
$$

Entirely similar considerations of the transformation $\sqrt[3]{z} \mapsto z$ show that

$$
P \left\{ \begin{array}{ccc} 0 & \infty & 1 \\ 0 & 0 & \gamma \\ \tfrac{1}{3} & \tfrac{1}{3} & \gamma' \end{array} z \right\} = P \left\{ \begin{array}{ccc} 1 & \rho & \rho^2 \\ \gamma & \gamma & \gamma \\ \gamma' & \gamma' & \gamma' \end{array} \right\} , \qquad \rho^3 = 1
$$

is also possible. So quadratic and cubic transformations can be made if the exponents are suitable.

Riemann wrote $P(\mu, \nu, \tfrac{1}{2}, x)$, $P(\mu, 2\nu, \mu, x_1)$, and $P(\nu, 2\mu, \nu, x_2)$

when he wanted to display the exponent differences, where $x = 4x_1(1-x_1)$ $= 1/4x_2(1-x_2)$. These transformations he discussed in the Nachträge extract. If x lies in triangle AOB, then x_1 lies in ADB and x_2 in ACB. Conversely, it was straight forward for him to show that the rational function of x which transforms AOB into ADB is necessarily of the form $x = cx_1(1-x_1)$ for some constant c, which can be normalized to c = 4. This gave him a geometric interpretation of the quadratic transformations of the P function. He also knew that the regular solids provide further examples in which the function mapping the triangle onto the half-plane is algebraic. There is no suggestion, however, that Riemann knew these are essentially the only example of such functions. The recognition of that fact had to wait for Klein (see Chapter III). Riemann did not consider the problem of analytically continuing Y around several circuits of the branch points, when the image triangles may overlap (unless $\lambda\pi$, $\mu\pi$, and $\nu\pi$ are the angles of a regular solid with triangular faces), an idea Schwarz was the first to develop (see Chapter III).

Finally, Riemann considered the periods K and iK' of an elliptic integral as functions of the modulus k^2, and considered a branch of the function $\frac{K'}{K}$. As k^2 goes from 0 to 1, $\frac{K'}{K}$ is real and goes from ∞ to 0. The part of the real axis $(-\infty,0)$ is mapped onto the right half line through $-i$ parallel to the real axis, the part $(1,\infty)$ is mapped onto a semicircle centre $\frac{-i}{2}$ radius $\frac{1}{2}$ (see figure p.338). So one branch of $\frac{K'}{K}$ maps the values of k^2 on the real axis onto the figure made up of two horizontal straight lines and a semicircular arc. As k^2 ranges over the upper half plane $\frac{K'}{K}$ takes every value

338

exactly once within this domain. When k^2 was allowed to vary over
the plane, Riemann found that each branch of $\frac{K'}{K}$ mapped a half plane
into a similar circular arc triangle in the right-hand half plane,
and $\frac{K'}{K}$ takes every value exactly once in that space, whatever circuit
k^2 performs about 0, 1, ∞. So, given an arbitrary function, $Y(z)$ say,
with branch points at 0, 1, and ∞, setting $z' = \phi(z)$, where $k^2 = \phi(\frac{K'}{K})$,
produces $Y(z)$ which is single valued in z.

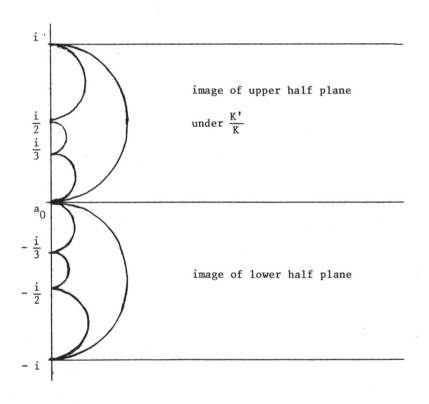

Riemann has shown that a multiple-valued function Y defined on C with branch points at 0, 1, ∞ can be lifted to a single valued function on H, a half plane, by means of the modular function k^2.

Thus Y has been globally parameterized by k^2, for which the term is 'uniformized'.

$$H \searrow \\ \downarrow \\ \mathbb{C} - \{0,1\}\ldots\searrow \mathbb{C} \\ Y$$

Interestingly, the following appears in the first edition of volume III of Gauss's Werke [1866, 477]:

"If the imaginary part of t, $\frac{1}{t}$ has between −i, +i then the Real Part of $(\frac{Qt}{Pt})^2$ is positive

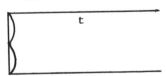

Space for t and $\frac{1}{t}$

"The equation $(\frac{Qt}{Pt})^2$ = A has exactly 1 solution in the

t

space $\alpha\delta - \beta\gamma = 1, \alpha \equiv \delta \equiv 1 \mod 4, \beta, \gamma$ even

$$t' = i \left(\frac{\alpha t + \beta i}{\gamma t + \delta i}\right) \text{ then } \left(\frac{Qt'}{Pt'}\right)^2 = i^\gamma \left(\frac{Qt}{Pt}\right)^2 . "$$

P and Q play the role of theta-functions in Gauss's theory of elliptic functions, and Gauss here described the fundamental domain for the modular function t but the editor (Schering) missed the point and rendered the arcs as mere doodles. In the second edition the diagrams have been fattened up to represent semicircles, as they should. Had Riemann seen that part of the Nachlass? Certainly he would not have failed to see the import of Gauss's drawings.

The question of the boundary values of $\frac{K'}{K}$ was also considered by Riemann and the fragment he left [Werke 455-465] was discussed by Dedekind [Riemann Werke 466-479]. The matter is a subtle one with number-theoretic and topological implications, and Dedekind was able to develop it to advantage when Fuchs, on Hermite's instigation, raised it again in 1876. This was discussed in Chapter IV.

Appendix 3: Fuchs's analysis of the nth order equation.

I shall now run through Fuchs's argument as it applies to the nth order equation (2.1.1).

It has been seen that such an equation always has n linearly independent solutions in the neighbourhood of any non-singular point in the domain of definition of the coefficients $p_1, \ldots, p_n$; Fuchs took this domain to be

$$\mathbb{C} \smallsetminus \{a_1, a_2, \ldots, a_{\rho+1} = \infty\} = T',$$

Suppose $y_1, \ldots, y_n$ is a system of solutions which, upon analytic continuation around a_k, return as $\overset{\smile}{y}_1, \ldots, \overset{\smile}{y}_n$ respectively. Then there are linear relations between each $\overset{\smile}{y}_i$ and the $y_1, \ldots, y_n$:

$$\overset{\smile}{y}_i = \sum_j \alpha_{ij} y_j,$$

and the matrix $A = (\alpha_{ij})$ has eigenvalues which determine the properties of the fundamental system near a_k. (Of course, the matrix depends upon which a_k has been chosen.) The equation for the eigenvalues

$$|A - wI| = 0$$

was called by Fuchs the fundamental equation (Fundamentalgleichung)
at the singular point a_k [1866, §3].

A change of basis permits one solution, u, to be found near a_k
for which u = w.u, where w is a solution of the fundamental equation.
If an r is chosen so that $w = e^{2\pi i r}$ (Fuchs called r an exponent at a_k)
then $u.(x-a_k)^{-r}$ is single-valued near a_k.

If all the eigenvalues w_1, ..., w_n are distinct then a fundamental system of solutions can be made up of n elements each with the
property that $\tilde{y}_i = w_{ik} y_i$ and $(x-a_k)^{r_{ik}} y_i$ is single-valued near
$x = a_k$, and $w_{ik} = e^{2\pi i r_{ik}}$. All solutions near a_k are therefore sums of
the form

$$\Sigma c_i (x - a_k)^{r_{ik}} \phi_{ik} (x - a_k)$$

for single-valued functions ϕ_{ik}. If, however, the eigenvalues are
not all distinct, and w is an ℓ-times repeated root, then Fuchs
showed [1866, 136 = 1904, 175] a fundamental system of solutions exists
in blocks of the form u_1, u_2, ..., u_ℓ which transform under analytic
continuation around a_k as

$$u_1 = wu_1$$

$$u_2 = w_{21} u_1 + wu_2$$

$$u_\ell = w_{\ell 1} u_1 + w_{\ell 2} u_2 + ... + wu_\ell$$

where the w_{ij} are constants. So Fuchs presented the solutions in
the form

$$u_1 = (x - a_k)^r \phi_{11}$$

$$u_2 = (x - a_k)^r \phi_{21} + (x - a_k)^r \phi_{22} \log (x - a_k)$$

$$\cdots\cdots\cdots\cdots$$

$$u_\ell = (x - a_k)^r \phi_{\ell 1} + (x - a_k)^r \phi_{\ell 2} \log (x - a_k) + \cdots$$

$$+ (x - a_k)^r \phi_{\ell\ell} [\log (x - a_k)]^{\ell-1}$$

where r again satisfies $w = e^{2\pi i r}$, and each ϕ_{ij} is a linear conbination of ϕ_{11}, ϕ_{21}, ..., $\phi_{\ell 1}$. This is a mildly confusing presentation, but blocks of solutions were found for each repeated eigenvalue w, and the resulting solution then involved logarithmic terms. Fuchs did not write his solutions in the simplest possible way, which would have amounted to putting the monodromy matrix in its Jordan canonical form; this was first done by Jordan [1871] and Hamburger [1873] (see Hawkins [1877]), as was discussed in Chapter II. On the other hand Fuchs's presentation is two years earlier than Weierstrass's theory of elementary divisors, which gives a canonical presentation of a matrix equivalent to Jordan's but one couched in a more forbidding study of the minors of the matrix. Weierstrass's theory was designed to explain the simultaneous diagonalization of two matrices and Fuchs's simpler approach benefits from needing to consider only one. Fuchs's minor modifications of his presentation in the solution in his [1868, §1], but it is possible that Fuchs's paper joined with others to re-awaken Weierstrass's interest in the problem of canonical forms.

To relate the solution valid near a_1 to those valid near a_k,

Fuchs extended them analytically in T". He found relations of the form

$$\underline{y}^1 = B_k \underline{y}^k,$$

and $\widetilde{\underline{y}}^k = B_k R_1 B_k^{-1} \underline{y}^k = R_k \underline{y}^k,$

extending the earlier notation in the obvious way, and as before

$$\Pi_i \det R_i = \Pi_i (w_{i1} \dots w_{in}) = \Pi_i e^{2ki \sum_k r_{ik}} = 1, \text{ so}$$

$$\sum_{i=1}^{\rho+1} \sum_{k=1}^{n} r_{ik} = K, \text{ an integer, i.e. the sum of the exponents}$$

is an integer. To find K, Fuchs formed the determinants

$$\Delta_0^K = |(d_{pq})| = \left| \frac{d^{n-q}}{dx^{n-q}} y_p^k \right| ,$$

and Δ_k^i, which is obtained from Δ_0^i by replacing the kth row of Δ_0^i with the transpose of

$$\left(\frac{d^n y^i}{dx^n} , \dots , \frac{d^n y^i}{dx^n} \right) , \quad k = 1, \dots, n.$$

So $\quad \Delta_0^i P_k = -\Delta_k^i$, and

$$\Delta_k^1 = (\det B_i) \Delta_k^i \text{ for } k = 1, \dots, n.$$

Again, $\widetilde{\Delta}_k^i = \det R_i \cdot \det \Delta_k^i$, so

$$\Delta_k^i = (x - a_i)^\varepsilon \psi, \text{ where } \psi \text{ is a single-valued function of x near}$$

a_i, and ε satisfies $e^{2\pi i \varepsilon} = \det R_i$. Therefore

$\Delta_0^i \ (x - a_i)^{-\Gamma}i$ is finite, single-valued, and continuous

near a_i, where $\Gamma_i = \Sigma_k \ r_{ik} - \dfrac{n(n-1)}{2}$, for

Δ_0^i has a pole of order $\Sigma_k \ r_{ik} - \dfrac{n(n-1)}{2}$, being made up of

products of y_{ik} and its derivatives, and $y_{ik} \ (x - a_i)^{-r_{ik}}$ has only

a logarithmic infinity at a_i, as does

$\dfrac{d^s y}{dx^s}_{ik} \ (x - a_i)^{-r_{ik} + s}$. Similarly $\Delta_0^{\rho+1}$ has a Pole of order

$\Sigma \ r_{\rho+1, k} \quad + \dfrac{1}{2}n \ (n - 1)$ at ∞.

The expression $K_0 = \Delta_0^1 \ (x - a_1)^{-\Gamma}1 \ \dots \ (x - a_\rho)^{-\Gamma_\rho}$ is finite,

continuous, and single-valued everywhere in $\mathcal{C}$. It is infinite at ∞

in a way which can be determined by looking at the solutions near ∞,

and the behaviour of $\Delta_0^{\rho+1}$. Indeed K_0 has a pole at ∞ of order at

most

$$- \Sigma_{i=1}^\rho \ \Sigma_{k=1}^n \ r_{ik} + (\rho - 1) \ \dfrac{n(n-1)}{2} = -K + (\rho - 1) \ \dfrac{n(n-1)}{2} \ .$$

Fuchs took as a basis of solutions:

$$y_1^i, \ y_2^i = y_1^i \int z_1 dx, \ y_3^i = y_1^i \int dx z_1 \int u_1 dx, \ \dots$$

$$y_n^i = y_{n-1}^i \int w_1 dx, \ \text{so that}$$

$$\Delta_0^i = C(y_1^i)^n \ z_1^{n-1} \ \dots \ w_1 .$$

$\Delta_0^i \ (x - a_i)^{-\Gamma}i$ does not vanish at a_i, and

$$\Delta_0^i = C. \ e^{\int p_1 dx} \ , \ \text{so}$$

$e^{\int p_1 dx} (x - a_i)^{-\Gamma_i}$ is regular a_i, i.e. $\lim\limits_{x \to a_i} (x - a_i)p_1$ is finite,

and Γ_1 is the coefficient of $(x - a_i)^{-1}$ in the power series for p_1

near a_i. Likewise near $a_{\rho+1} = \infty$, $\Gamma_{\rho+1} = \sum\limits_k r_{\rho+1,k} + \dfrac{n(n-1)}{2}$ is the

coefficient of x^{-1} in the power series for p_1 near ∞.

So $\sum\limits_{i=1}^{\rho+1} \Gamma_i = K - (\rho - 1) \dfrac{n(n-1)}{2}$.

But $\lim\limits_{x \to \infty} x \, p_1$ being finite entails firstly that p_1 is rational,

secondly that the degree of the denominator of p_1 exceeds that of the

numerator by at most 1, so $\sum\limits_{i=1}^{\rho+1} \Gamma_i = 0$ and Fuchs has obtained the

following equation for K, the sum of the exponents:

$K = (\rho-1) \dfrac{n(n-1)}{2}$. (A3.1) [1865, §4, equation 8,

1866, §4 equation 10].

This reduces K_0 to a constant, which cannot be zero because Δ_0^1

cannot vanish.

$K_k := \Delta_k^1 (x - a_1)^{-\Gamma_1} \ldots (x - a_\rho)^{-\Gamma_\rho}$ for $k = 1, \ldots, n$ is

likewise related to p_k, and turns out to have a degree at most

$-K + (\rho - 1) \dfrac{n(n-1)}{2} + (\rho - 1)k = (\rho - 1)k$, so it is of the form

$F_{(\rho-1)k}(x) / [(x - a_1) \ldots (x - a_\rho)]^k$, where $F_s(x)$ is a

polynomial in x of degree at most s.

The general linear ordinary differential equation of the Fuchsian

class is therefore of the form:

$$\frac{d^n y}{dx^n} + \frac{F_{\rho-1}(x)}{\psi}\frac{d^{n-1}y}{dx^{n-1}} + \frac{F_{(\rho-1).2}(x)}{\psi^2}\frac{d^{n-2}y}{dx^{n-2}} + \ldots + \frac{F_{(\rho-1)n}(x)}{\psi^n}y = 0.$$

It can be put in various alternative forms. Near $x = a_i$ for example, setting $F_{k(\rho-1)}(x)(x-a_i)^k \psi^{-k} = P_{ik}(x)$, it takes the form

$$\frac{d^n y}{dx^n} + \frac{P_{i1}(x)}{x-a_1}\cdot\frac{d^{n-1}y}{dx^{n-1}} + \frac{P_{i2}(x)}{(x-a_i)^2}\frac{d^{n-2}y}{dx^{n-2}}\ldots + \frac{P_{in}(x)}{(x-a_i)^n}\, y = 0$$

Upon further setting

$$y = (x-a_i)^r u \text{ it becomes}$$

$$\frac{Q_{i0}(x)}{x-a_i}\cdot\frac{d^n u}{dx^n} + \frac{Q_{i1}(x)}{x-a_i}\frac{d^{n-1}u}{dx^{n-1}} + \ldots + \frac{Q_{in}(x).u}{(x-a_i)^n} \doteq 0$$

where $Q_{ij}(x)$ is a linear combination of the P_{ik}'s, $k = 1, \ldots, j$ and r is one of the exponents $(r_{i1},\ldots,r_{in})$ at a_i and so satisfies

$$r(r-1)\ldots(r-n+1) - r(r-1)\ldots(r-n+2)P_{i1}(a_i)$$

$$-r(r-1)\ldots(r-n+3)P_{i2}(a_i) - \ldots - P_{in}(a_i) = 0.$$

The advantage of this form is that it can easily be shown to have solutions of the form

$$\sum_0^\infty C_k (x-a_i)^k.$$

To show this Fuchs cleared the denominator $(x-a_i)^n$ from the equation and rearranged it in a form suitable for calculation of the constants C_k. A comparison with the more familiar method of Frobenius is given in Section 2.3 above.

Appendix 4: On the History of non-Euclidean Geometry.

Throughout the nineteenth century geometries were developed
which differed from Euclid's. The most prominent of these was
projective geometry, whether real or complex; n-dimensional geometries
were also increasingly introduced. However, these geometries were
considered to be mathematical constructs generalizing the idea of
real existing Space which could still be regarded as a priori
Euclidean. The overthrow of Euclid is due to the successful development of what may strictly be called non-Euclidean geometry: a
system of ideas having as much claim as Euclid's to be a valid
description of Space, but presenting a different theory of parallels.
In non-Euclidean geometry, given any line ℓ and and point P not on ℓ
there are infinitely many lines through P coplanar with ℓ but not
meeting it. The consequences of this new postulate (together with
the other Euclidean postulates, which are taken over unchanged)
include: the angle sum of a triangle is always less than π, by an
amount proportional to the area - so all triangles have finite area;
and there is an absolute measure of length. Moreover, a distinction
must be made between parallel lines (which are not equidistant) and
the curve equidistant to a straight line (which is not itself straight).

As is well-known, non-Euclidean geometry was first successfully
described by Lobachevskii [1829] and J. Bolyai [1831] independently,
although Gauss had earlier come to most of the same ideas. The story
is well told in Bonola [1912, 1955], and Klein [1972, Chs. 36, 38].
I have argued elsewhere [Gray 1979a, b] that the crucial step taken
by Lobachevskii and Bolyai was the use of hyperbolic trigonometry.

Their descriptions of non-Euclidean geometry are in this sense analytic, it is the power and wealth of their formulae which convinces the reader that non-Euclidean geometry must exist, for its existence is in fact taken for granted. Hyperbolic trigonometry, in turn, was successful because it allowed for a covert use of the differential geometric concepts of length and angle in a more flexible way than earlier formulations of the problem would permit. Moreover, the problem of parallels was not taken to be one of the mere logical consistency of a set of postulates. Saccheri, Lambert, and Taurinus, three of its most vigorous investigators, knew that spherical geometry differed from Euclid's, but regarded it as irrelevant. For them the problem had to do with the nature of the straight line and the plane (which they could not really define), and so broke into two parts: could there be a system of geometry with many parallels? and, if so, could it describe Space? The Lobachevskii-Bolyai system seemed to answer both questions affirmatively, and thus to reduce the nature of Space to an empirical question, one Lobachevskii sought unsuccessfully to resolve by measuring the parallax of stars (which is bounded away from zero in non-Euclidean geometry). However, not everyone was convinced, and for a generation little progress was made. It was not until the hyperbolic trigonometry was given an explicit grounding in differential geometry and non-Euclidean geometry based explicitly on the new, intrinsic, metrical ideas that it could be said to be rigorously established. The central figures in this development are Riemann [1854, 1867] and Beltrami [1868]. From 1868 on, non-Euclidean geometry met with increasing acceptance from mathematicians, if not, predictably, amongst philosophers.

The growth of non-Euclidean geometry from, say, 1854 to 1880, has seldom been described. I shall concentrate on the more purely mathematical developments, chiefly its formulation in projective terms, for that line leads to Klein and Poincaré, and leave aside the more philosophical aspects[1], which are well discussed in Radner [1979], Scholz [1980], and Torretti [1978].

In his Habilitationsvortrag [1854] "On the hypotheses which lie at the foundations of geometry" Riemann does not explicitly mention non-Euclidean geometry, and he never mentions the names Bolyai and Lobachevskii. He refers only to Legendre when describing the darkness which he says has covered the foundations since the time of Euclid, and later (Sections II, 5 and III, I) he describes 2 homogeneous geometries in which the angle sum of all triangles is determined once it is in one triangle, as occuring on surfaces of constant non-zero curvature. This oblique glance at the characterization of 3 homogeneous geometries is typical of the formulation of the 'problem of parallels' since Saccheri, it can be found in many editions of Legendre's Elémens de Géometrié, e.g. [1823], and would have alerted any mathematician to the implications for non-Euclidean geometry without its being mentioned by name. On the other hand, not naming it explicitly would avoid philosophical misapprehensions about what he had to say, so one may perhaps ascribe the omission to a Gaussian prudence. It is less certain, however, that Riemann had read Lobachevskii, and very unlikely he had read Bolyai. Only Gauss appreciated them in Göttingen at that time, and it is not known if he discussed these matters with Riemann. Whatever might be the resolution of that small question of Riemann's sources, the crucial idea in Riemann's paper in his presentation of geometry as intrinsic, grounded in the free mobility of infinitesimal measuring rods, and to be expressed mathematically in terms of curvature. This is an immense generalization of Gauss's idea of the intrinsic curvature of a surface [1827], itself a profound novelty. It has the effect of basing any geometry on specific metrical considerations, and so removes Euclid's geometry from its paramount position as the geometry of space and the source of geometrical concepts which are induced onto embedded surfaces. Following Minding [1839], he gave an informal local description of a surface of constant negative curvature as a piece of a curved narrowing cylinder but left the global existence question unresolved, and he did not discuss the connection with non-Euclidean geometry.

Riemann did not seek to publish these ideas. They were somewhat further developed in his Paris Prize entry of 1861, also unpublished until 1876, and first appeared[2] in 1867, after his death. By then Beltrami had independently discovered the import of Gauss's ideas for non-Euclidean geometry. Beltrami's famous 'Saggio' [1868] begins with an account of geometry as based upon the notion of superposability of figures and thus existing on a surface of constant curvature. Beltrami modified the metric on the sphere to obtain a metric with negative curvature $K = -1/R^2$ and a coordinate system in which straight lines had linear equations. This enabled him to map the geometry, which Beltrami said resided on the surface of a pseudosphere (p. 290) onto the interior of a disc with radius a; in this map a point with coordinates (r, θ) is, in the non-Euclidean metric, $\rho = \frac{a}{2} \log \frac{a + r}{a - r}$ away from the (Euclidean) centre of the disc. This metric allows him to derive the formulae of non-Euclidean trigonometry, and he observed that they agreed with the trigonometric formulae of Minding [1839] and Codazzi [1857] for the surface of constant negative curvature, as well as those of Lobachevskii. Beltrami also sketched an account of non-Euclidean three-dimensional geometry – the original starting point of Lobachevskii and Bolyai's descriptions – but his grasp of higher dimensional differential geometry was not yet sufficiently certain.

Beltrami described the relationship between non-Euclidean geometry and its image in the disc as one in which approximate or qualitative Euclidean pictures are given of non-Euclidean figures. He discussed the way in which such pictures, necessarily distorted, may mislead the mind if features proper to the one geometry (such as points at infinity) are illicitly transposed to the other. He did not state that the non-metrical projective properties of non-Euclidean geometry are represented exactly in the disc – a point made soon afterwards by Klein – and he did not indicate that this gave a projective proof of the existence of the new geometry – for him the new geometry was grounded in differential geometric considerations[3]. The crucial point, clearly grasped by Beltrami, is that his description is the first global description of the space of non-Euclidean geometry. By contrast, Riemann's model has unacceptable features globally: self-intersecting geodesics and a boundary of singular points.

In a paper published soon afterwards, [1868b], Beltrami was able to extend his reasoning to n dimensions and give a thorough

account of three dimensional non-Euclidean space. This paper, but
not the earlier one, makes extensive reference to Riemann's recently
published Habilitationsvortrag, so it seems that Beltrami only
learned of them after his first paper was finished. Scholz [1980, 107]
confirms that this is so, but that a delay in the publication of the
[1868a] enabled Beltrami to read Riemann's work before his own was
published, and to add a note linking it so the forthcoming [1868b].
His source for Lobachevskii's work was Houël's translation of some
of it [1866], but he does not seem to have known of Houël's French
translation of Bolyai's Tentamen [1867].

The Italians had become very interested in non-Euclidean
geometry, Battaglini and Forti published reports on it in 1867 and
Battaglini also translated Bolyai into Italian in 1868. The French,
led by Houël, were discovering the new geometry too, and in 1866
Germany interest awakened by the publication of the Gauss-Schumacher
correspondence[4] [Gauss 1860], was decisively quickened by Helmholtz's
essays [1866] and, more important, [1868]. Christoffel's first works
on differential geometry [1868, 1869a,b] also appeared at this time,
on the subject of transformations of differential expressions like
$\Sigma g_{ij} dx_i dx_j$ (which represent Riemannian metrics). The English,
notably Cayley, were an anomalous case and require separate discussion.

The generally accepted view of these developments is that only
ignorant philosophers and elderly mathematicians refused to accept
the new geometry, but this view has been claimed to be historically
inexact by Toth [1977, 144]. However, Toth's criticism of philosophers
seems to be that they were not ignorant (and so had no excuse), and

the mathematicians he assembles are a mixed bunch. In Russia, Ostrogradskii and Buniakovskii; in England Dodgson (= Lewis Carroll), de Morgan, and Cayley; and in Hungary the astonishing case of Wolfgang Bolyai, the father of Janos. The Russians' furious criticism of Lobachevskii dates from the 1830's and 1840's (when Gauss taught himself Russian so as to be able to follow it) and so predates the acceptance of differential geometric concepts as suitable foundation for geometry. Without it Lobachevskii's work is indeed open to criticism, and in other works [1829, 1837] Lobachevskii made unsuccessful attempts to ground his geometry in something more basic than his intuitive ideas of lines and surfaces. As for Wolfgang Bolyai, Toth cites no evidence for his claim, and I can find none to support it in the two volumes written by Stäckel on the Bolyai's [Stäckel, 1913], so it must be set aside.

Dodgson's and de Morgan's importance should not be greatly stressed, so Toth's argument rests on only one major mathematician , Cayley, and cannot really be accepted. Once a grave weakness in the Bolyai-Lobachevskii approach had been repaired, non-Euclidean geometry commanded widespread acceptance among mathematicians, as can be seen from the annual reviews in Fortschritte, if depressingly little amongst philosophers[5] (Frege amongst them). The case of Cayley is interesting, however, and has recently been ably treated by Richards [1979]. She argues that Cayley accepted non-Euclidean geometry as a piece of mathematics, but refused to accept differential geometry as the foundations of geometry because of its empiricist philosphical undertones. Cayley preferred to regard the new geometry

as an intellectual construct based ultimately on projective or even
Euclidean geometry given a priori. Richards also cites Jevons [1871]
who gave a related argument, endorsed by Cayley, that the geometry
on the tangent space to the surface of constant negative curvature is
Euclidean. Their absolutism was opposed by Clifford precisely
because Clifford wanted to entertain empiricism elsewhere, in religion,
morality, and politics, whereas conservatives saw mathematics as an
example of a priori reasoning which endorsed such reasoning on those
other issues. So the English debate on non-Euclidean geometry became
embroiled in other contemporary debates which gradually advanced
science at the expense of religious orthodoxy.

Cayley's purely mathematical papers made a more secure contribution. In his "A Sixth Memoir upon Quantics"[1859a] he described how
a metrical geometry may be defined with respect to a fixed conic
which he called the 'Absolute' in (complex) projective space. He
first developed the projective geometry of such a space, taking as
the points of the geometry all the points of projective space. He
then introduced a metric by considering two points (x,y) and (x',y')
as defining a line which met the Absolute in two points, and
defining equidistance in terms of cross-ratio (making it a projective
invariant). The additivity of distance led him to define the
distance between the given points as

$$\frac{\cos^{-1}(a,b,c\!\!\;(x,y)(x',y')}{\sqrt{(a,b,c(x,y)^2}\;\sqrt{(a,b,c(x',y')^2}} \qquad\qquad (\S211)$$

His notation is:

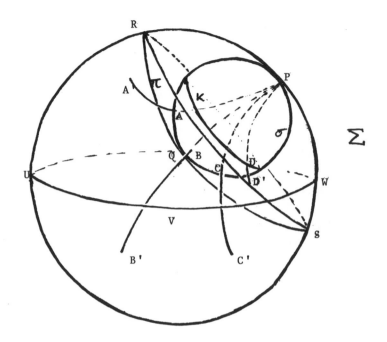

Non-Euclidean 3-space in the Poincaré model. The 2-sphere,
Σ , PRUVWS, bounds a ball in Euclidean 3-space but its interior
an be made to carry a metric which makes it a model of non-Euclidean
-space. The non-Euclidean lines PAA', . . . PDD' are represented
y arcs of circles perpendicular to Σ . The smaller sphere, σ ,
ABCD, touches Σ internally at P, so it represents what Lobachevskii
alled a horosphere. The spherical shell, π , RQS, meets Σ at
ight angles, so it represents a non-Euclidean plane. It touches π
t Q. The non-Euclidean metric inside Σ induces, remarkably, the
uclidean metric on σ . Lines like PDD' which meet π only on points
f Σ are asymptotic to π . They cut σ in a circle which represents
ayley's absolute conic, κ , which bounds a model of non-Euclidean
-space in σ just as Σ (Cayley's absolute quadric) bounds a model
n 3-space.

$$(a,b,c)(x,y)(x',y') := axx' + b(xy' + x'y) + cyy'$$

$$(a,b,c)(x,y)^2 \quad := ax^2 + 2bxy + cy^2$$

and the equation of the Absolute is $(a,b,c)(x,y)^2$. In the simplest general case (§225) the absolute is $x^2 + y^2 + z^2 = 0$ (in homogeneous form), and the distance between (x,y,z) and (x',y',z') is

$$\frac{\cos^{-1} xx' + yy' + zz'}{\sqrt{(x^2 + y^2 + z^2)} \ \sqrt{(x'^2 + y'^2 + z'^2)}}$$

He defined angles dually. What Cayley considered most important was the distinction between a proper conic as the Absolute and a point-pair as Absolute. The choice of proper conic led to spherical geometry, the choice of the point-pair to Euclidean geometry (§225).

Cayley did not consider non-Euclidean geometry at all in this paper. Subsequently, in [1865], he quoted Lobachevskii's observation that the formulae of spherical trigonometry become those of hyperbolic or non-Euclidean trigonometry on replacing the sides a,b,c of the spherical triangle by ai, bi, and ci respectively [Lobachevskii 1837], and went on to say "I do not understand this; but it would be very interesting to find a *real* geometrical interpretation [of the equations of hyperbolic trigonometry]" (Cayley's italics). Such an interpretation was soon to be given by Felix Klein[6].

In early 1870 Felix Klein was in Berlin with his new friend Sophus Lie, attending the seminars of Kummer and Weierstrass on geometry. Weierstrass was interested in the Cayley metrics, and

Klein raised the question of encompassing non-Euclidean geometry
within this framework, but Weierstrass was sceptical, preferring to
base geometry on the idea of distance and to define a line as the
curve of shortest length between its points[7]. Klein was later to
claim that during seminars he and Lie had had to shout their view
from the back, and he was not deterred by Weierstrass's opinion.
Spurred on by his friends, notably Stolz, who had read Lobachevskii,
Bolyai, and von Staudt (which Klein admitted he could never master
[1921, 51-52]) he continued to work on his view of geometry. It
became crucial to his unification of mathematics behind the group
concept, and his Erlanger Programm [1872] reflects his satisfaction
at succeeding. His first paper on non-Euclidean geometry [1871]
presents the new theory. The space for the geometry is the interior
of a fixed conic (possibly complex) in the real projective plane.
Klein called the geometries he obtained hyperbolic, elliptic, or
parabolic according as the line joining any two points inside the
conic meets the conic in real, imaginary, or coincident points (§2),
Non-Euclidean geometry coincides with hyperbolic geometry, Euclidean
geometry is Cayley's special case of a highly degenerate conic. In
the hyperbolic case the distance between two points is defined as the
log of the cross-ratio of the points and the two points obtained by
extending the line between them to meet the conic (divided by a
suitable real constant c (§3))[8]. This function is additive, since
cross-ratio is multiplicative, and puts the conic at infinity.
Angles were again defined dually. The Cayley metric yields a
geometry for which the infinitesimal element of length can be found,
and in this way Klein showed that the geometry on the interior of

the conic was in fact non-Euclidean (§7). Klein showed that his
definition of distance gave Cayley's formulae when c = i/2 and the
conic is purely imaginary. Klein then considered the group of
projective transformations which map the conic to itself, showed
that these motions were isometries, and interpreted them as translations or rotations (§9). He also considered elliptic and parabolic
geometry, and extended the analysis to geometries of three dimensions.
In a subsequent paper [1873] he developed non-Euclidean geometry in
n dimensions.

Klein's view of non-Euclidean geometry was that it was a species
of Projective geometry, and this was a conceptual gain in that it
brought harmony to the proliferating spread of new geometries which
Klein felt was in danger of breaking geometry into several separate
disciplines (see the Erlanger Programm, p4). In particular it made
clear what the group of non-Euclidean transformations was, and Klein
regarded the group idea as the key to classifying geometries. But
this view was perhaps less well adapted to understanding the new
geometry in its own right than was the more traditional standpoint
of differential geometry.

The most brilliant exponent of the traditional point of view
was Poincaré. It would be interesting to know how he learned of
non-Euclidean geometry, but we are as ignorant of this as we are of
the details of most of his early career. He might have read Beltrami's
[1868], perhaps in Houël's translation [Beltrami 1869] published in
the Annals of the Ecole Normale Supérieure, or any of Houël's articles,
or the papers [J. Tannery 1876, 1877] or [J.M. de Tilly 1877]. He

might have learned of it from Hermite, or have been introduced to the

works of Helmholtz and Klein (he could read German, and Helmholtz had

been translated into French), but in any case did not say. The

description he gave of non-Euclidean geometry is a novel one. In it

geodesics are represented as arcs of circles perpendicular to the

boundary of a circle or half plane. The group of non-Euclidean proper

motions is then obtained as all matrices $(\begin{smallmatrix} a & b \\ \bar{b} & \bar{a} \end{smallmatrix})$, $a\bar{a}-b\bar{b} = 1$, if the

circle is $|z| = 1$, or as all matrices $(\begin{smallmatrix} a & b \\ c & d \end{smallmatrix})$, $ad-bc = 1$, a, b, c, d $\in \mathbb{R}$,

in the half-plane case. In each case a single reflection ($z \rightarrow \bar{z}$ or

$z \rightarrow -\bar{z}$, respectively) will then generate the improper motions.

Poincaré's representation of non-Euclidean geometry is conformal, and

may be obtained from Beltrami's as follows. Place a sphere, having

the same radius as the Beltrami disc, with its South pole at the

centre of the disc and project vertically from the disc to the

Southern hemisphere. Now project that image stereographically from

the North pole back onto the plane tangent to the sphere at the south

pole. The first projection maps lines onto arcs of circles perpen-

dicular to the equator, and since stereographic projection is conformal

these arcs are then mapped onto circular arcs perpendicular to the

image of the equator. To see that the groups correspond it is

enough to show that they are triply transitive on boundary points and

that the map sending any given triple to any other is unique, which

it is. The metric may be pulled back from the one representation to

the other, or calculated directly from the group. (It is particularly

simple in the half-plane model:

$$ds^2 = \frac{dx^2 + dy^2}{y^2} \;.)$$

Since the elements of the group are all compositions of Euclidean inversions in non-Euclidean lines they are automatically conformal.

The main text of this chapter describes how Poincaré published his interpretation of non-Euclidean geometry. To my knowledge, there is no account of how he came to discover it. I have argued elsewhere [1982] that Poincaré did not know of Klein's Erlanger Program in 1880 and seems only to have known of Beltrami's work on non-Euclidean geometry. This belief is confirmed by a letter Poincaré wrote to Lie in 1890 in which he explicitly disclaimed any knowledge of Klein's tract, which was surely little known in 1880. I am indebted to T. Hawkins for informing me of this letter, which he discusses in passing in his convincing paper arguing that the Erlanger Programm had much less influence than is generally believed [Hawkins 1984]. D. Rowe [1983] has also pointed out that the written Programm is quite different from the address Klein gave at his inauguration, which was on mathematics and education.

Notes on Chapter I

1.   The most thorough recent treatment of all these topics is
Houzel's essay in [Dieudonné, 1978, II]. Gauss's work on elliptic
functions and differential equations is treated in the essays by
Schlesinger [Schlesinger 1909a,b] and in his Handbuch [Schlesinger
1898, Vol. II, 2]. Among many accounts of the mathematics, two
which contain valuable historical remarks are Klein's Vorlesungen
über die hypergeometrische Function [1894] and, more recently,
Hille's Ordinary Differential Equations in the Complex Domain [1976].

2.   An integral form of the solution was also to prove of interest
to later mathematicians:

$$f(x) = \int_0^1 u^{b-1}(1-u)^{c-b-1}(1-xu)^{-a} du$$

where the integral is taken as a function of the parameter. It is
assumed that $b > 0$, $c-b > 0$, $x < 1$, so that the integral will
converge. It is easy to see, by differentiating under the integral,
that $f(x)$ satisfies (1.1.1). To obtain f as a power series
expand $(1-xu)^{-a}$ by the binomial theorem and replace each term in
the resulting infinite series of integrals by the Eulerian Beta
functions they represent. An Euler Beta function is defined by

$$B(b+r,c-b) = \int_0^1 u^{b+r}(1-u)^{c-b-1} du;$$

it satisfies the function equation or recurrence relation
$B(b+r+1, c-b) = \frac{b+r}{c+r} B(b+r, c-b)$, which yields precisely the
relationship between successive terms of the hypergeometric series,
and so $f(x)$ is, up to a constant factor, represented by (1.1.2).
This argument is given in Klein [1894, 11], see also [Whittaker

and Watson, 1973, Ch. XII].

3.   Biographical accounts of Gauss can be found in Sartorius von
Waltershausen, Gauss zum Gedächtnis [1856], G. W. Dunnington [1955],
and H. Wussing [1979]. Accounts of most aspects of his scientific
work are contained in Materialien für eine wissenschaftlichen
Biographie von Gauss, edited by Brendel, Klein, and Schlesinger
1911-1920 and mostly reprinted in [Gauss, Werke X.2, 1922-1933].
The best introductions in English are the article on Gauss in the
Dictionary of Scientific Biography [May, 1972], which is weak
mathematically, Dieudonné [1978], and W. K. Bühler [1981].

4.   There are detailed discussions in Schlesinger [1898] and
[Gauss, Werke, X.2, 1922-1933], [Krazer, 1909], [Geppert, 1927], and
[Houzel in Dieudonné 1978 vol. II]. In particular Geppert supplies
the interpretation of the fragments in [Gauss, Werke III, 1866,
361-490] on which this account is based.

5.   Later Gauss calculated M(1, √2) to twenty decimal places, his
usual level of detail, and found it to be 1.19814 02347 35592 20744
[Werke III, 364].

6.   This is not contiguous in the sense of Arbogast [1791], which
more or less means continuous.

7.   The lemniscate has equation $r^2 = \cos 2\theta$ in polar coordinates,
and arc-length $\int_0^t \frac{dx}{(1-x^4)^{\frac{1}{2}}}$. Introduced by Jacob Bernoulli in 1694
it had been studied by Fagnano (1716, 1750) who established a formula
for the duplication of arc, and by Euler (1752) who gave a formula

for increasing the arc n times, n an integer. Euler

considered the problem an intriguing one because it showed that

the differential equation $\dfrac{dx}{(1-x^4)^{\frac{1}{2}}} = \dfrac{dy}{(1-y^4)^{\frac{1}{2}}}$ had algebraic

solutions: $x^2 + y^2 = c^2 + 2xy(1-c^4)^{\frac{1}{2}} - c^2 x^2 y^2$. The comparison

with the equation $\dfrac{dx}{(1-x^2)^{\frac{1}{2}}} = \dfrac{dy}{(1-y^2)^{\frac{1}{2}}}$, which leads to

$x^2 + y^2 = c^2 + 2xy(1-c^2)^{\frac{1}{2}}$, guided his researches. (It provides

the algebraic duplication formula for sine and cosine.) Gauss

observed that, whereas in the trigonometric case the formula

for multiplication by n leads to an equation of degree n,

for the lemniscatic case the formula is of degree n^2, and

invented double periodicity on March 19, 1797, see Diary entry

#60, to cope with the extra roots. He was fond of this discovery,

and hints dropped about it in his Disquisitiones Arithmeticae

(§335) inspired Abel to make his own discovery of elliptic

functions. An English translation of Gauss's diary is available

in [Gray, 1984b and corrigenda].

8.   Gauss had discussed the multiple-valued function log x in

a letter to Bessel the previous year [Werke II, 108] and also

stated the residue theorem for integrals of complex functions around

closed curves. From the discussion in Kline [1972, 632-642] it

seems that Gauss was well ahead of the much younger Cauchy on

this topic; see also Freudenthal [1971], and Bottazzini [1981, 133].

9.   For a comparison of the work of Abel and Jacobi see [Krazer

1909] or [Houzel, Dieudonné, 1978, II]. It seems that Abel was

ahead of Jacobi in discovering elliptic functions; there is no

doubt he was the first to study the general problem of inverting integrals of all algebraic functions. Jacobi's work was perhaps more influential because of his efforts as a teacher and an organizer of research, and also because Abel died in 1829 at the age of 26.

10.   The expression $\dfrac{\dfrac{d\lambda}{dk}\dfrac{d^3\lambda}{dk^3} - \dfrac{3}{2}(\dfrac{d^2\lambda}{dk^2})^2}{(\dfrac{d\lambda}{dk})^2}$ has come to be known,

following Cayley [1883], as the Schwarzian derivative of λ with respect to k. Schwarz's use of it is discussed in detail in Chapter III.

11.   See [Edwards, 1977] for a discussion of Kummer's number theory, and [Lampe 1892-3] and [Biermann 1973] for further biographical details about Kummer.

12.   Kummer wrote this equation as

$$2\,\frac{d^3 z}{dxdx^2} - 3(\frac{d^2 z}{dzdx})^2 = (2\,\frac{dP}{dz} + P^2 - 4Q)(\frac{dz}{dx})^2 + (2\,\frac{dp}{dz} + p^2 - 4q)$$

using the then customary notation for higher derivatives.

13.   In section III Kummer studied the special cases when γ depends linearly on α and β and quadratic changes of variable produce other hypergeometric series. Typical of his results is his equation 53:

$$F(\alpha,\beta,\ \alpha-\beta+1,\ x) = (1-x)^{-\alpha}\ F(\frac{\alpha}{2},\ \frac{\alpha-2}{2}\,\frac{\beta+1}{2},\ \alpha-\beta+1,\ \frac{-4x}{(1-x)^2}\ ).$$

His results were incomplete and were extended by Riemann [1857a, §5].

14.   Neuenschwander [1979, 1981b] discusses the limited use Riemann
had for the theory of analytic continuation, of which he was certainly
aware, see e.g. [1857c, 88-89].

15.   For a discussion of Riemann's topological ideas and their
implications for analysis see Pont [1974 Ch. II]. Pont does not make
as much as he should have done of the distinction between homotopy-
and homology-theoretic ideas. A better discussion will be found in
[Scholz, 1980, Chapter 2].

16.   For a history of Cauchy's Theorem see Brill and Noether [1892-93,
Chapter II]. They also discuss Gauss's independent discovery of it.

17.   Riemann wrote (A), (B), and (C) for A, B, and C respectively.

18.   Riemann wrote (b) for B' and (c) for C'. I have introduced vector
notation purely for brevity.

19.   E. Scholz tells me there is evidence in the Riemann Nachlass to
show that Riemann had read Puiseux. See also Neuenschwander, [1979, 7].

Notes on Chapter II

1.   These biographical details are taken from Biermann, [1973b, 68, 94, 103].

2.   Weierstrass [1841]. These series are called Laurent series after Laurent's work, reported on in Cauchy, [1843]. Laurent's paper is reprinted in Peiffer [1978].

3.   Singuläre Punkte = singular points, in this case poles of finite order.

4.   In his [1868, §8] Fuchs investigated the singularities more closely, and found that there are exceptional cases when a singular point of a coefficient function does not give rise to a singular point of the solutions. Such a point he called an accidental singular point (ausserwesentlich singulärer Punkt) in contradistinction to the other singular points which he called essential (wesentlich), a term he attributed to Weierstrass. Since 'essential' when applied to singularities now means something different I shall use the word 'actual' for them instead.

      A singular point $x = a$ is accidental if the determinant of a fundamental system vanishes, for, as has been seen (p. 341) $p_i = \dfrac{-\Delta_i}{\Delta_0}$, so if $p_i(a) = \infty$ and no solution is infinite at $x = a$, then $\Delta_0(a) = 0$. Fuchs showed that for $x = a$ to be accidental it is necessary and sufficient:

(i) that the equation have the form

$$(x-a)^n \frac{d^n y}{dx^n} + (x-a)^{n-1} p_1(x) \frac{d^{n-1} y}{dx^{n-1}} + \ldots + p_n(x)y = 0,$$

where the p_i's are analytic near $x = a$;

(ii) that $p_i(a)$ is a negative integer;

(iii) that the roots of the indicial equation at $x = a$ all be different and are positive numbers or zero; and finally

(iv) that none of the solutions contain logarithmic terms.

5.   See Hawkins [1977].

6.   Fuchs did not call the w Eigenwerthe (eigenvalues), and had no special term for them other than 'roots of the fundamental equation'.

7.   I introduce the vector notation purely to abbreviate what Fuchs wrote in coordinate form.

8.   Fuchs did not specify the sense in which a circuit of a point is to be taken, but it must be taken in an agreed way each time.

9.   See Biermann [1973b, 69-70].

10.   Strictly, Fuchs has made a mistake: (2.1.9) admits solutions $n=1$, ρ arbitrary or $n=2$, $\rho=2$.

11.   See Biermann [1973b, 77].

12.   If η_λ is the period taken along the λth part, $\eta_\lambda = \frac{1}{2}\displaystyle\int_{(\lambda)} y\,dx$,

then

$$\frac{d^i \eta_\lambda}{du^i} = \frac{1}{2}\int_{(\lambda)} \frac{\partial^i u}{\partial y^i}\,dx,$$

where (λ) is any appropriate curve. The solutions $\eta_1, \ldots, \eta_{n-1}$ of (2.2.2) are regular. Indeed, if Π is the product of the difference of the roots of y^2, then for each λ, $\Pi\eta_\lambda$ is everywhere finite and non-zero away from $k_1, \ldots, k_n$.

$$\tilde{\Pi} = \prod_{i<j} (k_i - k_j) = \prod_{i<j} [(k_i - x) - (k_j - x)]$$

$$= \Sigma c_\ell (k_1 - x)^{\ell_1}(k_2 - x)^{\ell_2}\ldots(k_{n-1} - x)^{\ell_n}.$$

where c_ℓ is an integer, the numbers $\ell_1, \ell_2, \ldots, \ell_n$ are all taken from the set $\{0, 1, \ldots, n - 1\}$, and $\ell_1 + \ell_2 + \ldots + \ell_n = \frac{1}{2}n(n - 1)$. The quotient $\tilde{\Pi}/y$ is a sum of terms of the form

$$(-1)^{n/2} c_\ell (k_1 - x)^{\ell_1 - \frac{1}{2}} (k_2 - x)^{\ell_2 - \frac{1}{2}} \ldots (k_{n-1} - x)^{\ell_n - \frac{1}{2}}$$

is finite even at $k_1, \ldots, k_{n-1}$. Near k_i, say, $\tilde{\Pi}$ has the form $(u-k_i)^m \tilde{\Pi}'$ where m is a positive integer and $\tilde{\Pi}'$ is a finite, continuous, single-valued function of n vanishing at $u = k_i$. This makes the functions η_λ regular.

13.   Legendre's relation had been generalized by Weierstrass [1848/49] to a system of $2n^2 - n$ relations between periods of a hyper-elliptic integral of order n, and, as a determinant, by Haedenkamp [1841], who, however, considered the periods only over real paths. Fuchs was easily able to extend Haedenkamp's work, and he found (Fuchs [1870b]) that if

$$\eta_{ij} = \int_{\gamma_j} \frac{x^i dx}{[\phi(x)]^{\frac{1}{2}}} ,$$

where

$$\phi(x) = (x-u)(x-k_1) \ldots (x-k_{n-1})$$

and γ_j is a path from k_{j-1} to k_j $(1 \leq j < n-1)$

and γ_1 connects u and k_1,

then the $(n-1) \times (n-1)$ determinant

$$H: = \begin{vmatrix} \eta_{01} & \eta_{02} & \cdots & \eta_{0,n-1} \\ \eta_{11} & \eta_{12} & \cdots & \eta_{1,n-1} \\ \cdot & \cdot & \cdots & \cdot \\ \eta_{n-2,1} & \eta_{n-2,2} & \cdots & \eta_{n-2,n-1} \end{vmatrix}$$

is independent of u, as Haedenkamp had claimed. So it can be evaluated when $\phi(x) = x^n - 1$, and Fuchs found

$$H = \frac{(-1)^{\frac{n-1}{4}} (2\pi)^{\frac{n}{2} - 1}}{(n-2)(n-4)\ldots 3.1} , \qquad n \text{ odd,}$$

$$H = \frac{(-1)^{\frac{n}{4}} (2\pi)^{\frac{n}{2} - 1} \pi}{(n-2)(n-4)\ldots\ldots 2} , \qquad n \text{ even. } [1870, 135 = 1904, 291]$$

Legendre's relation is the special case n = 2.

14.   The first generalization of the hypergeometric series is due to Heine [1848, 1847], who studied the series

$$\phi(\alpha, \beta, \gamma, q, x) = 1 + \frac{(1-q^\alpha)(1-q^\beta)}{(1-q)(1-q^\gamma)} x + \frac{(1-q^\alpha)(1-q^{\alpha+1})(1-q^\beta)(1-q^{\beta+1})x^2}{(1-q)(1-q^2)(1-q^\gamma)(1-q^{\gamma+1})} + .$$

for α, β, γ, x real or imaginary, and q real. He showed that, as $\frac{1-q^\varepsilon}{1-q} \to \varepsilon$ as $q \to 1$, $\phi(\alpha, \beta, \gamma, 1, x) = F(\alpha, \beta, \gamma, x)$. For particular values of α, β and particular q, Heine's series can represent any of Jacobi's θ-functions, so it stands in the same relation to elliptic functions as Gauss's series to the trigonometric functions. However, Heine's series satisfies not a differential equation but a difference equation with respect to x. This difference equation was later studied by Thomae [1869] from a Riemannian point of view. Thomae also sought to generalize the hypergeometric equation. Thomae [1870] studied the power series

$$y = 1 + \frac{a_0 \cdot a_1 \cdot a_2}{1 \cdot b_1 b_2} x + \frac{a_0(a_0+1)a_1(a_1+1)a_2(a_2+1)x^2}{1 \cdot 2 \cdot b_1(b_1 + 1) b_2(b_2 + 1)} + \ldots \text{ and showed}$$

that many of the properties of the hypergeometric series carried over to the higher hypergeometric series, as he called them. For example, the series converges inside $|x| = 1$, it satisfies an nth order linear ordinary differential equation, it can be written as an (n-1)-fold integral, contiguous functions can be defined in the obvious way, and any n+1 contiguous functions satisfy linear relationships with rational coefficients. Thomae was much influenced by Riemann's [1857a] and presented his solutions as generalized P functions, which he denoted $F\begin{pmatrix} \alpha & \alpha' \ldots \alpha^{(h-1)} \\ \beta & \beta' & \beta^{(h-1)} & x \end{pmatrix}$.

For computational reasons he concentrated in his [1870] on the

3rd order case, and deduced a host of relationships for the

monodromy coefficients, his generalizations of Riemann's α_β's.

In his [1874], Thomae generalized the P-function to a function

$P\begin{pmatrix} \alpha, & \beta, & \gamma, & \delta \\ \alpha', & \beta', & \gamma', & \delta' \end{pmatrix} k, x$ which had branch points at 0, ∞, 1, and $1/k$

and for which the exponent pairs were α, α', at 0, etc. (and

$\alpha + \alpha' + \beta + , \ldots + \delta' = 2$). He showed that much of Riemann's

theory went over to the new functions with inessential changes,

for example P satisfies a second order differential equation, but

that P depended in an essential and complicated way on a parameter

t (which, in later terminology, is an accessory parameter).

Thomae showed that if $\delta = \delta' = 0$ then t can be chosen

arbitrarily without affecting the differential equation satisfied

by P and that if $\delta = 1$, $\delta' = 0$ the equation reduced to Riemann's

form of the hypergeometric equation.

15.   New readers can start with Thomé's own summary [1884].

Notes on Chapter III

1.   There are several alternative forms for the Schwarzian
which can be of use. For example

$$\Psi(s,x) = \frac{2s's''' - 3s''^2}{2s'^2} = \frac{s'''}{s'} - \frac{3}{2}\left(\frac{s''}{s'}\right)^2 \text{ (where } s' = \frac{ds}{dx} \text{ , etc.)}$$

and this can be written in terms of the logarithmic derivative as

$$\Psi(s,x) = \frac{d^2}{dx^2}\left(\log \frac{ds}{dx}\right) - \frac{1}{2}\left(\frac{d}{dx}\log \frac{ds}{dx}\right)^2.$$

2.   There is a change of variable which can simplify the
original differential equation and help to make the argument
clearer. Set $y = gu$ where $g = e^{-\frac{1}{2}\int pdx}$, then the differential
equation for y becomes the following equation for u:

$$\frac{d^2u}{dx^2} = Pu, \text{ where } P = \frac{1}{4}p^2 + \frac{1}{2}\frac{dp}{dx} - q.$$

P is a rational function whose only singularities are those of
the original p and q, and for such equations, if the quotient of
two linearly independent solutions is algebraic, then so are all
solutions. It was this form of the general second order
differential equation that was later studied by Fuchs.
Set $\eta = \frac{f}{g}$, the quotient of two independent solutions of
$y'' + Py = 0$, then $\frac{\eta'''}{\eta'} - \frac{3}{2}\left(\frac{\eta''}{\eta'}\right)^2 = 2P$.

3.   Kronecker's example is discussed in Neuenschwander,
[1977, 7] which is based on Casorati's notes of a conversation
with Kronecker, 16 October 1864.

4.   See Kiernan [1972], Wussing [1969], Purkert [1971, 1973].

5. $A_m (ju_1)^m + A_{m-1}(ju_1)^{m-1} + \ldots + A_0 = 0$, and (3.2.2) is

irreducible, so $\dfrac{A_{m-i}}{A_m} = \dfrac{A_{m-i}}{A_m} j^{-i}$, which implies, since $A_m \neq 0$,

$A_{m-i} = 0$ $1 \leq i \leq m$, or $j^{-i} = 1$.

The irreducibility of (3.2.2) precludes $A_{m-i} = 0$ for all i, so j

is a primitive root of unity.

6.   The most general expression which is the root of a rational

function is $y = (z-a_1)^{\alpha_1} (z-a_2)^{\alpha_2} \ldots (z-a_\rho)^{\alpha_\rho} g(z)$, where g(z) is

a polynomial, and $\alpha_1, \ldots, \alpha_n$ are rational numbers which can be

assumed not to be positive integers. If y is to satisfy an nth

order differential equation, then the singular points of the

equation must be $a_1, \ldots, a_\rho$ and the exponents must be

$\alpha_1, \ldots, \alpha_n$. When y is substituted into the given differential

equation an equation is obtained for the coefficients of g(z).

If this equation has solutions then y is a solution of the

differential equation, and if not, not.

7.   Suppose $g_1, g_2, \ldots, g_N$ are the elements of G, then

$\Pi(x) = (x-g_1 a)(x-g_2 a) \ldots (x-g_N a)$ is a polynomial of order N which

vanishes on the orbit of G containing the point a, and any other

polynomial with those zeros is a linear multiple of Π. In

particular $\Pi(gx)$ has the same zeros, is monic, and so coincides

with $\Pi(x)$, and $\Pi(x)$ is said to be G-invariant. Let

$\Pi'(x) = (x-g_1 a') \ldots (x-g_N a')$ be the monic polynomial defining

the orbit containing a' and Q(x) a polynomial defining the orbit

of a point b. To show $Q(x) = \kappa\Pi(x) + \kappa'\Pi'(x)$, consider $Q(x) - Q(a)$.

It vanishes when x = a and is G-invariant, so it is some multiple of $\Pi(x)$, say $Q(x) - Q(a) = \beta\Pi(x)$. Likewise $Q(x) - Q(a') = \beta'\Pi'(x)$, so

$$Q(x)[Q(a') - Q(a)] = Q(a')\beta\Pi(x) - Q(a)\beta'\Pi'(x)$$

or $\quad Q(x) = \dfrac{Q(a')\beta\Pi(x) - Q(a)\beta'\Pi'(x)}{(Q(a') - Q(a))}$

8.   Transvectants are a somewhat mysterious collection of invariants of given forms. Let $f(x_1, x_2)$ and $\phi(y_1, y_2)$ be two binary forms, of degrees m and n respectively, and introduce the symbolic operator

$$\Omega = \frac{\partial^2}{\partial x_1 \partial y_2} - \frac{\partial^2}{\partial x_2 \partial y_1}$$

so, e.g. $\Omega(f.\phi) = \dfrac{\partial f}{\partial x_1} \dfrac{\partial \phi}{\partial y_2} - \dfrac{\partial f}{\partial x_2} \dfrac{\partial \phi}{\partial y_1}$

Then the form

$$\frac{1}{m.n} \Omega(f.\phi)$$

is the <u>first transvectant</u> of f and ϕ.

The rth transvectant of f and ϕ is similarly defined as

$\dfrac{(m-r)!}{m!} \dfrac{(n-r)!}{n!} \Omega^r (f.\phi).$

To obtain the rth transvectant of f with itself, one calculates with $f(x_1, x_2)$ and $f(y_1, y_2)$ and then sets $y_1 = x_1$, $y_2 = x_2$ after the differentiation is over. So, for example, the second transvectant of f with itself is obtained by first calculating

$$(\frac{1}{m(m-1)})^2 \Omega^2 (f.f) = (\frac{1}{m(m-1)})^2 \{\frac{\partial^2 f}{\partial x_1^2} \cdot \frac{\partial^2 f}{\partial y_2^2} - 2 \frac{\partial^2 f}{\partial x_1 \partial x_2} \cdot \frac{\partial^2 f}{\partial y_1 \partial y_2} + \frac{\partial^2 f}{\partial x_2^2}\frac{\partial^2 f}{\partial y_2^2}\}$$

and then substituting x_1 and x_2 for y_1 and y_2 respectively, obtaining

$$(\frac{1}{m(m-1)})^2 . 2 . \{ \frac{\partial^2 f}{\partial x_1^2} \frac{\partial^2 f}{\partial x_2^2} - (\frac{\partial^2 f}{\partial x_1 \partial x_2})^2 \}.$$

This, up to a constant factor, is the Hessian of f. The odd order transvectants of a form with itself all vanish. A modern account of transvectants is Dieudonné and Carrell [1971], and light is also shed on them by the aptly-named umbral calculus of Kung and Rota, [1984].

9.   This problem has recently been studied in great generality in Baldassarri and Dwork, [1979] and Baldassarri, [1980]. They regard Klein's method as getting close to solving the recognition problem, although insufficient to deal with the dihedral case, but as not well adapted for the construction of all second order linear differential equations with algebraic solutions and prescribed singular parts.

10.   H(f) is the Hessian up to a constant factor. For

$$f_6 = y_1^5 y_2 + y_1 y_2^5, \quad H(f_6) = y_1^8 - 14 y_1^4 y_2^4 + y_2^8. \quad \text{For}$$

$$f_{12} = y_1^{11} y_2 + 11 y_1^6 y_2^6 + y_1 y_2^{11}, \quad H(f_{12}) = y_1^{20} - 228 y_1^{15} y_2^5$$

$$+ 494 y_1^{10} y_2^{10} + 228 y_1^5 y_2^{15} + y_2^{20}.$$

11.   Sylow presented his discovery that subgroups of order p^r exist in a group of order n, whenever p is a prime and p^r divides n, in [1872]. Jordan was much impressed with this result, and wrote to Sylow about it; Sylow's letters are preserved in the collection of Jordan's correspondence at the Ecole Polytechnique (catalogue numbers I, 19–22). Sylow's proofs were permutation theoretic, and the history of their reformulation in abstract terms is well traced in Waterhouse [1980]. Jordan himself never published a proof of Sylow's theorems.

12.   Jordan called the linear transformations 'substitutions' and the name attached itself to the elements of groups as his theory of groups became progressively more abstract. In matrix form this transformation would be written

$$\begin{pmatrix} u_1 \\ u_2 \end{pmatrix} \rightarrow \begin{pmatrix} \alpha & \beta \\ \gamma & \delta \end{pmatrix} \begin{pmatrix} u_1 \\ u_2 \end{pmatrix}$$

13.   They are as follows, where Jordan's notation for a typical element has been replaced by matrix notation:

1. Groups generated by

$$\begin{pmatrix} a & 0 & 0 \\ 0 & b & 0 \\ 0 & 0 & c \end{pmatrix} \quad \text{and} \quad \begin{pmatrix} 0 & a' & 0 \\ 0 & 0 & b' \\ c' & 0 & 0 \end{pmatrix} , \ a, \ b, \ c, \ a', \ b', \ c'$$

all roots of unity, and their subgroups [no. 198].

2. Extensions of such groups by the group generated by

$$\begin{pmatrix} 0 & a'' & 0 \\ b'' & 0 & 0 \\ 0 & 0 & c'' \end{pmatrix} , \ a'', \ b'', \ c'' \text{ roots of unity [no. 204].}$$

3. A group generated by mI, $\begin{pmatrix} \tau & 0 & 0 \\ 0 & \tau^{-1} & 0 \\ 0 & 0 & 1 \end{pmatrix}$, $\begin{pmatrix} 0 & 1 & 0 \\ 1 & 0 & 0 \\ 0 & 0 & -1 \end{pmatrix}$, and

$$C = \begin{pmatrix} a & -(1 + a) & -2a^2 \\ -(1 + a) & a & 2a^2 \\ 1 & -1 & -(1 + 2a) \end{pmatrix} , \text{ where m is a root of unity,}$$

$\tau^5 = 1$, and a is defined by $a(\tau + \tau^{-1} - 2) = 1$. If $m = 1$ this group reduces to the group of proper motions of an icosahedron [no. 208].

4. A group generated by mI,

$$A = \begin{pmatrix} 1 & 0 & 0 \\ 0 & \theta & 0 \\ 0 & 0 & \theta^2 \end{pmatrix}, \quad B = \begin{pmatrix} 0 & 1 & 0 \\ 0 & 0 & 1 \\ 1 & 0 & 0 \end{pmatrix},$$

$$rD = \begin{pmatrix} rj & 0 & 0 \\ 0 & rj\theta^2 & 0 \\ 0 & 0 & rj \end{pmatrix}, \quad \text{and } r^2E = r^2a \begin{pmatrix} 1 & 1 & \theta \\ 1 & \theta & 1 \\ 1 & \theta^2 & \theta^2 \end{pmatrix}, \quad \text{where}$$

$$\theta^3 = 1, \quad j^3 = \theta, \quad a^3 = \frac{1}{3(1 - \theta^2)}, \quad m^{3\rho} = 1, \quad r^3 = m^\mu,$$

(ρ arbitrary, $\mu = 0, 1,$ or 2). If $r = 1 = \rho$ this group has order 27.24 [nos. 201-3, 209].

5. A subgroup of the previous one of order 27.2.2 generated by mI, A, B, and sDE, where $s^4 = 1$, m, m^2, or m^3.

6. A subgroup of order 27.2.4 generated by mI, A, B, sDE, and tED where $t^2 = s^2$ or ms^2, which Jordan called Hesse's group.

Hesse's group is a subgroup of the Galois group of the equation of the 9 inflection points on a general cubic. Jordan had first considered the larger group in his Traité, 302-305, where he showed it had order 432 and was the symmetry group of the configuration of the lines joining the inflection points, by

treating it as the symmetries of the affine plane over the field
of 3 elements. Hesse's group consists of those matrices having
determinant 1. The group in (5) above, of order $27.2.2 = 108$ is
the symmetry group of the Pappus configuration. The group is
discussed in Miller, Blichfeldt, and Dickson, Theory and
Applications of Finite Groups, Chapter XVIII.

14.   In a letter from Jordan to Klein, 11 October 1878, [N.S.U.
Bibliothek zu Göttingen, Klein, cod Ms. F. Klein, 10, number
11] Jordan wrote:

"Mon cher ami

"Vous avez parfaitement raison. En énumérant les groupes
linéaires à trois variables, j'ai laissé échapper celui
d'ordre 168 que vous me signalez".

Jordan corrected his mistake, and went on to discuss the
group of the modular equation at the prime 11 (PSL (2; 11), of
order 660) which Klein had presumably raised in his letter to
Jordan. Jordan observed that the group has an element, say A',
of order 11, and one, B', of order 5. If it is to be a linear
group in three variables, which he doubted, then, he wrote:

"En supposant qu'il n'y ait que 3[?] variables, il faudra
trouver une substitution C' telle que A', B', C' combinées entre
elles ne donnent que 12.11.5 substitutions. Il ne serait pas
difficile sans doute de vérifier si la chose est possible, mais

je suis trop occupé en ce moment pour pouvoir exécuter ce
calcule."

Jordan went on to discuss the Hessian group of order 216,
and then congratulated Klein on his good fortune in working with
Gordan.

"Je ne me sens pas aucune en état de le suivre sur le
terrain des formes ternaires; mais j'ai longuement approfondi
sa belle démonstration de l'Endlichkeit der Gründformen des
formes binaires..."

Jordan's intuition about PSL (2; 11) was correct, it cannot
be represented as a group in three variables; however, both he
and Klein missed a group of order 360 which can be represented
by ternary collineations. This is Valentiner's group
(Valentiner, [1889]) which is discussed in Wiman [1896], where it
is shown to be abstractly isomorphic to A_6, the group of even
permutations of 6 objects, and, in Klein, [1905 = 1922, 481-502.].

15.   See Hermite's letter to P.du Bois-Reymond, 3 September 1877,
in Hermite [1916], referred to in Hawkins, [1975, 93n].

16.   Jordan was somewhat imprecise. Poincaré showed [1881x =
Oeuvres III, 95-97] that to each finite monodromy group in three
or more variables there correspond infinitely many differential
equations having rational coefficients and algebraic solutions.

Notes on Chapter IV

1.   Complex multiplication and singular moduli. Let Λ be a lattice in $\mathbb{C}$. The only holomorphic maps $f: \mathbb{C}/\Lambda \to \mathbb{C}/\Lambda$ are those lifting to a map $f: \mathbb{C} \to \mathbb{C}$ satisfying $f(z + \omega) - f(z) \in \Lambda$ whenever $\omega \in \Lambda$. This implies that $f(z + \omega) - f(z)$ is a constant independent of z and so $\dfrac{df}{dz}$ is constant and $f(z) = az + b$. By suitably renormalizing one can always rewrite f as $z \to az$, so the only maps of a lattice Λ to itself are of the form

$$\left. \begin{array}{l} a\omega = \alpha\omega + \beta\omega' \\[2mm] a\omega' = \gamma\omega + \delta\omega' \end{array} \right\} \quad \alpha, \ \beta, \ \gamma, \ \delta \in \mathbb{Z},$$

where ω and ω' are generators of the lattice. Either $a = \alpha = \delta$, $\beta = \gamma = 0$, or, setting $\tau = \omega'/\omega$, $\beta\tau^2 + (\alpha - \delta)\tau - \gamma = 0$, so τ is a quadratic imaginary. In this case the lattice is said to possess a complex multiplication, by a. The theory of singular moduli was first developed by Abel [1828 2nd ed. p. 426] in terms of elliptic integrals. Each transformation of the lattice gives a holomorphic map from $\mathbb{C}/\Lambda$ to $\mathbb{C}/\Lambda$, Abel sought values of a for which the differential equation

$$\frac{dy}{\sqrt{[(1 - y^2)(1 + \mu y^2)]}} = a \frac{dx}{\sqrt{[(1 - x^2)(1 + \mu x^2)]}}$$

has algebraic solutions, and found that a is either rational or a quadratic imaginary $m + \sqrt{n}i$, m, n rational. This occurs only for certain μ which he called the singular moduli, and which he conjectured satisfied an algebraic equation solvable by radicals.

The conjecture was proved by Kronecker [1857].

2.   Hadamard [1954, 109] said of Hermite: "Methods always seemed to be born in his mind in some mysterious way".

3.   See also Dugac [1976], Mehrtens [1979].

4.   Compare Lang [1976], Serre [1973], or Schoeneberg [1974].

5.   I cannot find that he ever took an opportunity to do this.

6.   The group of matrices $\begin{pmatrix} a & b \\ c & d \end{pmatrix} \equiv \begin{pmatrix} 1 & 0 \\ 0 & 1 \end{pmatrix}$ mod 2 was connected with modular equations at arbitrary N, not merely the primes, by H. J. S. Smith [1877], in a paper much admired by Klein. Smith wanted to connect the theory of binary quadratic forms $ax^2 + 2bx + cy^2$, $N = b^2 - ac > 0$, with the modular equation at N. Results of Hermite and Kronecker concerning the case when N is negative were well known, he said, but the positive case was more difficult and little had been done beyond Kronecker [1863] which discussed the solution of Pell's equation by elliptic functions.

Smith's paper dealt with the study of the reduced forms equivalent to a given form $Q(x,y) = ax^2 + 2bxy + cy^2$, a, b, and c integers. Reduced forms are those $a'x^2 + 2b'xy + c'y^2$ for which the quadratic equation $a' + 2b't + c't^2 = 0$ has real roots of opposite sign such that the absolute values of one root is greater, and the other less, than unity. They are equivalent to $Q(x,y)$ if there is $g = \begin{pmatrix} \alpha & \beta \\ \gamma & \delta \end{pmatrix} \in SL(2;\mathbb{Z})$ such that

$Q(\alpha x + \beta y, \gamma x + \delta y) = a'x^2 + 2b'xy + c'y^2$. They are associated

with the continued fraction expansion of $\sqrt{N}$ in a way first

described by Dirichlet [1854] (for the history of this matter

see Smith [1861, III, §93]). Smith found it convenient to work

with the weaker notion of equivalence where g satisifies

$\begin{pmatrix} \alpha & \beta \\ \gamma & \delta \end{pmatrix} \equiv \begin{pmatrix} 1 & 0 \\ 0 & 1 \end{pmatrix}$ mod 2, $\alpha \equiv \delta \equiv 1$ mod 4, presumably to avoid the

fixed points i, ρ of the Γ-action. He only remarked of his

restriction that "c'est uniquement pour abréger le discours

que nous l'admettons ici". This group acts on the upper half

plane and a set of inequivalent points, Σ, is obtained by taking

the region bounded by $P = x - 1 = 0$, $P^{-1} = x + 1 = 0$, and the

two semicircles $Q = x^2 + y^2 - x = 0$, $Q^{-1} = x^2 + y^2 + x = 0$,

including P and Q but not P^{-1} and Q^{-1}. To each quadratic form

$[a, b, c]: = ax^2 + 2bxy + cy^2$ he associated a circle

$(a, b, c): = a + 2bx + c(x^2 + y^2) = 0$, and he studied the image

of this circle reduced mod $\Gamma(2)$, a set of arcs in Σ "en

apparence sera brisée, mais dont on mettra en evidence la

continuité". The way in which these arcs went from one boundary

of Σ to another was, he showed, described by the periodic

continued fraction expansion of $\sqrt{N}$, and thus to the production of

the reduced forms equivalent to (a, b, c) (see Smith [1861, Part

IV, §93]).

The connection with the modular equation was established by

introducing Hermite's functions $\phi^8(\omega) = k^2 = \frac{1}{2} + X + iY$,

$\psi^8(\omega) = k'^2 = \frac{1}{2} + X - iY$ and mapping from the upper half ω-plane

to the X,Y plane. Under this map $\omega = i$ goes to $X = 0$, $Y = 0$, the imaginary axis of Σ is mapped onto the X-axis (Smith

incorrectly said the real points of Σ are sent to points with

Y = 0) and the map is conformal even at $\omega = 0$, $\omega = 1 \equiv -1$, and

$\omega = i\infty$. Each circle (a, b, c) is sent to an algebraic curve in

the X, Y plane whose equation is $F(\phi^8(\omega), \psi^8(\omega)) = 0$, where

F = 0 is the modular equation at N between k^2 and λ^2. If

$k^2 = \phi^8(\omega)$ then $\lambda^2 = \phi^8 (\frac{\gamma\omega + 2\delta}{\gamma'})$, where $\gamma\gamma' = N$, $\delta \equiv 0, 1, \ldots, \gamma'-1$

mod γ and $(\gamma, \gamma', \delta) = 1$. F is symmetric with respect to k^2 and

λ^2 and is invariant under the six permutations of k^2 (and λ^2)

which form the cross-ratio group $(k^2, 1-k^2, \frac{1}{k^2}$ etc.$)$, so there

are 6 modular curves $F(k^2, \lambda^2) = 0$, $F(k^2, 1-\lambda^2) = 0$, etc. These

curves spiral around the images $A_1 = (+ \frac{1}{2}, 0)$ of $\omega = 0$ and

$A_2 = (- \frac{1}{2}, 0)$ of $\omega = i\infty$ like interlacing lemniscates

symmetrically situated about the X axis. If the XY plane is cut

along the X axis from A_1 to $+ \infty$ and A_2 to $- \infty$ the spirals are cut

into whorls ('spires') which correspond to the individual arcs of

the reduced image of (a, b, c) in Σ, and consequently to the

continued fraction expansion of N.

7.   The implicit isomorphism between PSL(2, $\mathbb{Z}/5\mathbb{Z}$) and A_5 was

made explicit by Hermite [1866 = Oeuvres, II, 386-387].

8.   See Dugac [1976, 73] for a discussion of priorities, and

Dauben [1978, 142] for Kronecker's tardiness. The matter is fully

discussed in H. M. Edwards [1980, 370-371].

9.   Klein [1926, I, 366] - strictly, a description of work on automorphic function theory.

10.   This highly transitive group inspired Mathieu to look for others, and so led to his discovery of the simple, sporadic groups which bear his name, Mathieu [1860, 1861].

11.   The multiplier equation associated to a modular transformation describes M (see Chapter I p.19) as a function of k.

12.   Bottazzini is editing some unpublished correspondence between Italian and other mathematicians on this topic.

13.   Gordan made a study of quintics in his [1878], discussed in Klein [1922, 380-4].

14.   Notes of Weierstrass's lectures were presumably available, see Chapter VI n. 4. Klein based his approach to the modulus of elliptic functions on the treatments in Müller [1867, 1872 a, b] which I have not seen. Hamburger's report on them in Fortschritte, V, 1873, 256-257, indicates that they are based on Weierstrass's theory of elliptic functions and the invariants of biquadratic forms.

Notes on Chapter V

1.   The plane is generally the complex rather than the real
plane, but early writers e.g. Cayley are ambiguous. Later
writers, like Klein, are more careful.

2.   If 4 or more inflection points were real they would all
have to be, since the line joining two inflection points meets
the curve again in a third. But then they would all lie on the
same line, which is absurd. The configuration of inflection
points on a cubic coincides with the points and lines of the
affine plane over the field of 3 elements, an observation first
made, in its essentials, by Jordan, Traité 302.

3.   H. J. S. Smith [1877] pointed out that Eisenstein used the
Hessian earlier, in 1844. Eisenstein's 'Hessian' is the Hessian
of the cubic form $ax^3 + 3bx^2y + 3cxy^2 + dy^3$, i.e. $(b^2 - ac)x^2 +$
$(bc - ad)xy + (c^2 - bd)y^2$, see Eisenstein [1975, I, 10].

4.   It takes two to tango.

5.   In a subsequent paper on this topic, [1853], Steiner was led
to invent the Steiner triple system. However, these configurations
had already been discovered and published by Kirkman [1847], as
Steiner may have known, see the sympathetic remarks of Klein,
Entwicklung, 129.

6.   It has proved impossible to survey the literature of these
wonderful configurations. For some historical comments see
Henderson [1911, 1972], who also describes how models of the 27
lines may be constructed. For modern mathematical treatments
see Griffiths and Harris [1978], Hartshorne [1977] and
Mumford [1976]. For their connection with the Weyl group of E_7
see Manin [1974]. The existence of a finite number of lines on
a cubic surface was discovered by Cayley [1849], and their
enumeration is due to Salmon [1849]. The American mathematician
A. B. Coble [1908] and [1913] seems to have been the first to
illuminate the 27 lines and 28 bitangents with the elementary
theory of geometries over finite fields. I am grateful to
J. W. P. Hirschfeld for drawing his work to my attention.

7.   See Weierstrass Werke, IV, 9, quoted and discussed in
Neuenschwander [1981b, 94].

8.   Riemann tacitly assumed that the genus was finite. His
proof that it was well-defined was imprecise and was improved
by Tonelli [1875].

9.   Jacobi [1829 = 1969, I, 249-275] had introduced, in
connection with the multiplier equation at the prime p, a class
of equations of degree p + 1 whose roots depend linearly on
$\frac{p+1}{2}$ quantities $A_0, \ldots, A_{\frac{p+1}{2}}$. The p+1 roots take the
form

$$\sqrt{z}_{\infty} = (-1)^{\frac{p-1}{2}} \cdot p \cdot A_0$$

$$\sqrt{z}_i = A_0 + \varepsilon^{\nu} A_1 + \varepsilon^{4\nu} A_2 + \ldots + \varepsilon^{(\frac{p-1}{2})^2 \nu} A_{\frac{p-1}{2}}$$

where $\varepsilon = e^{2\pi i/p}$. Brioschi [1858] was able to show what the general form of an equation of degree 6 is if it is a Jacobian equation, and to exhibit explicitly the corresponding quintic. The multiplier equation was, he showed, a special Jacobian equation for which the corresponding quintic took Bring's form, so the solution of the quintic by elliptic functions was again accomplished. Kronecker [1858] showed that the general quintic yields rational resolvents which are Jacobian equations of degree 6, and conjectured that equations of degree 7 which were solvable by radicals would likewise be associated with Jacobian equations of degree 8.

Klein [1879 = 1922, 390-425] argued that, since the group of the modular equation of the prime 7 is $G_{168} = PSL(2; \mathbb{Z}/7\mathbb{Z})$ it is only possible to solve those equations of degree 7 whose Galois group is a quotient of G_{168}, i.e. is G_{168} itself. The general Jacobian equation of degree 8 has Galois group $SL(2:\mathbb{Z}/7\mathbb{Z})$ so that task of solving every equation of degree 7 by modular functions is impossible. In fact $SL(2:\mathbb{Z}/7\mathbb{Z})$ can also send each $\sqrt{z}_i$ to $-\sqrt{z}_i$, and each $\sqrt{z}_i$ ($i = 0, 1,\ldots,6, \infty$) is linear in the A's. Klein took $[A_0:A_1:A_3:A_4]$ as coordinates in $\mathbb{C}P^3$, so each $\sqrt{z_i} = 0$ defines a plane and (dually) a point. These eight points define a net of quadrics (after Hesse). If the Jacobian equation has Galois group G_{168} then the net contains a degenerate family of conics whose vertices lie on a space sextic (K, above, p. 243) which is the Hessian of a quartic [1879, §9]. Klein also showed that any polynomial equation whose Galois group was G_{168} could be reduced to the modular equation at the prime 7, thus answering affirmatively Kronecker's conjecture about the 'Affect' of such equations.

10.   The simplicity of the group emerges from Klein's description of its subgroups without Klein remarking on it explicitly. Silvestri's comments [1979, 336] in this connection are misleading, and he is wrong to say the group is the transformation group of a seventh order modular function: G_{168} is the quotient of PSL(2; $\mathbb{Z}$) by such a group.

11.   In particular, the inflection points of this curve are its Weierstrass points. For example, when the equation of the curve is obtained in the form $\lambda^3\mu + \mu^3\nu + \nu^3\lambda = 0$ the line $z = 0$ is an inflection tangent at $[0,1,0]$ and meets the curve again at $[1,0,0]$. The function $g(\lambda,\mu,\nu) = \mu/\nu$ is regular at $[1,0,0]$ but has a triple pole at $[0,1,0]$.

12.   Gordan [1880a,b] derived a complete system of covariants for $f := x_1^3 x_2 + x_2^3 x_3 + x_3^3 x_1 = 0$ from a systematic application of transvection and convolution (Faltung). Convolution replaces a

product $\alpha_x \beta_y$ in a covariant by the factor $(\alpha\beta)$, where

$(\alpha\beta) := \alpha_1\beta_2 - \alpha_2\beta_1$, and $\alpha_x = (\alpha_1 x_1 + \alpha_2 x_2)^n = \Sigma\binom{n}{r} a_r x_1^{n-r} x_2^r$.

These papers and other related ones by Gordan were summarized

by Klein in [1922], 426-438, where it is shown that they lead to

a solution of the 'form problem' for $PSL(2, \mathbb{Z}/7\mathbb{Z})$ just as

Klein's work had solved the 'form problem' for the

Icosahedron. Klein also showed how Gordan's work enables one to

resolve Kronecker's conjecture. The surviving correspondence

in Göttingen between Klein and Gordan throws no extra light on

the working relationship between the two men in this period,

and Professor K. Jacobs has kindly informed me that there is no

correspondence between them kept in Erlangen.

13.   Halphen's solution appeared as a letter to Klein in

[Halphen 1884 = Oeuvres IV 112 - 5]. Hurwitz's more general

solution was published as [Hurwitz 1886]. He argued that three

functions λ, μ, ν of J such that $\lambda^3\mu + \mu^3\nu + \nu^3\lambda = 0$ could be

found which are the solutions of a 3rd order linear differential

equation as follows. Pick 3 linear independent everywhere

finite integrals J_1, J_2, J_3 on $\lambda^3\mu + \mu^3\nu + \nu^3\lambda = 0$ such that

$dJ_1 : dJ_2 : dJ_3 = \lambda : \mu : \nu$. Then Fuchs's methods [Fuchs 1866,

esp. 139 - 148] imply that, if $y_i = \dfrac{dJ_i}{dJ}$ $1 \leq i \leq 3$, then

y_1, y_2, and y_3 are the solutions of an equation of the form

$$\frac{d^3y}{dJ^3} + \frac{aJ + b}{J(J - 1)} \frac{d^2y}{dJ^2} + \frac{a'J^2 + b'J + c'}{J^2(J - 1)^2} \frac{dy}{dJ} + \frac{a''J^3 + b''J^2 + c''J + d''}{J^3(J - 1)^3} y = 0$$

since the only possible branch points are J = 0, 1, and ∞. Klein's

[1878/79] establishes that the branching is 7-fold at J = ∞,

3-fold at J = 0, and 2-fold at J = 1, whence Fuchs's equation for the sum of the exponents yields values for most of a, b,...,d". Since the exponent differences are integers at J = 1, log terms could occur in the solution but they do not, so the differential equation can be found explicitly. It is

$$J^2(J - 1)^2 \frac{d^3y}{dJ^3} + (7J - 4)J(J - 1)\frac{d^2y}{dJ^2} + [\frac{72}{7}J(J - 1) - \frac{20}{9}(J - 1)$$

$$+ \frac{3J}{4}] \frac{dy}{dJ} + [\frac{72.11}{73}(J - 1) + \frac{5}{8} + \frac{2}{63}]y = 0.$$

A study of the invariants associated to the curve $\lambda^3\mu + \mu^3\nu + \nu^3\lambda = 0$ enabled Hurwitz to show that his differential equation took Halphen's form when the substitution $y = (J(J - 1))^{-2/3}y'$ is made.

14.   Two of Klein's students at this time deserve particular mention: Walther Dyck, and Adolf Hurwitz. Dyck presented his thesis at Munich in 1879 on the theory of regularly branched Riemann surfaces, and published two papers on this topic in the Mathematische Annalen the next year [1880a, b]. He called a surface regularly branched if each leaf was mapped by the self-transformations of the surface into every other leaf, and was particularly interested in the case where the group (supposed to be of order N) has a normal ("ausgezeichneten") subgroup of order N_1. He then interested himself in the surfaces corresponding to the subgroup and the quotient group, and in this way analysed surfaces until simple groups were reached. The map of a subgroup into a bigger group corresponds to a map of one Riemann surface onto another, and he gave explicit forms for these in the cases he looked at. In the example which most

attracted him, the surface was composed of triangles with angles
$\pi/2$, $\pi/3$, and $\pi/8$, and its genus is 3. The principal group has order
96, so Dyck regarded the surface as having 96 leaves. There is a
normal subgroup of order 48, and so the surface is also made up of
triangles with angles $\pi/3$, $\pi/3$, and $\pi/4$, in this case. This in turn
sits over a 16-leaved surface made up of 2-sided leaves with vertical
angles of $\frac{\pi}{3}$ and $\frac{\pi}{3}$, and so on until a double cover of the sphere is
reached. He also noted that other decompositions are possible and
showed that the corresponding algebraic curve is the 'Fermat' curve
$x^4 + y^4 + z^4 = 0$.

Dyck was in some ways very like Klein as a mathematician,
attracted to "anschauliche" geometry and the role of groups in
describing the symmetry of figures. He also became very
interested in constructing models of various mathematical objects,
and collaborated with Klein in the design of these for the firm,
owned by A. Brill's brother in Munich, which made them.

Klein's other, and younger, student was a more independent
character. Hurwitz's Inauguraldissertation at Leipzig was
published in the Mathematische Annalen for 1881 [Hurwitz 1881]. It
recalls both Klein and Dedekind's work on modular functions, but it
develops the theory entirely independently of the theory of elliptic
functions. This had been Dedekind's aim as well, but he had not been
able to carry it through completely. Hurwitz succeeded because he
had read and understood Eisenstein, a debt he fully acknowledged.
He did not say he was the first person to have read Eisenstein since
the latter's death in 1852, although he might well have been. Weil points
out [Weil 1976, 4] that only Hurwitz and Kronecker (in 1891) seem
to mention their

illustrious predecessor in the whole of the later nineteenth
century. In Hurwitz's case the reference to Eisenstein is a
little more substantial that Weil's comment might suggest, but
even so it was not sufficient to rescue him from the shades.
With some reluctance I have chosen to suppress the details of
Hurwitz's important treatment of modular forms and modular
quotients and the invariance group for $\Delta^{\frac{1}{12}}$, although it well
illustrates how Hurwitz grasped the connections between group
theory, geometry, and function theory which I am tracing
historically.

Notes on Chapter VI

1.   The Lamé equation is discussed in Whittaker and Watson, Ch. XXIII.

2.   The applications of elliptic functions are discussed in Houzel [1978, §15].

3.   Halphen defined a single-valued function (fonction uniforme) thus (p4n): "Par ce mot fonction uniforme, j'entends ici une fonction ayant l'aspect d'une fonction rationnelle, c'est-à-dire développable sous la forme

$$(\alpha - \alpha_0)^n [A + B(\alpha - \alpha_0) + C(\alpha - \alpha_0)^2 + \ldots],$$

dans laquelle n est un nombre entier, positif on négatif."

4.   I have not seen [Mittag-Leffler, 1876] which is in Swedish. A propos of the Weierstrassian theory Halphen commented [69n]: "Il existe des tableaux de formules lithographiées qui ont été rédigées d'après les leçons de M. Weierstrass, et qui sont entre les mains de presque tous les géomètres allemands."

5.   This information is taken from Darboux, Éloge Historique d'Henri Poincaré, printed in Poincaré, Oeuvres, II, vii-lxxi.

6.   7 letters between Poincare and Fuchs were printed in Acta Mathematica (38) 1921 175-187, and reprinted in Poincaré Oeuvres, XI, 13-25. An eighth letter is given in photographs in Oeuvres, XI, 275-276.

7.   When the essay was published in 1923 this figure was incorrectly printed as figure 6, before the one depicting the situation which Fuchs had shown to be impossible. The text refers to the annular case as the second one, which it is in Poincaré's original essay as deposited in the Académie.

8.   Accounts of the supplements appear in Gray [1982a,b] and Dieudonné [1983]. I would like to thank J. Dieudonné for his help during my work on these supplements.

9.   Poincaré presented his ideas on non-Euclidean geometry and arithmetic to l'Association Française pour l'avancement des Sciences at its 10th session, Algiers, 16 April 1881, [Oeuvres V, 267-274]. The ternary indefinite form $\xi^2 + \eta^2 - \zeta^2 = -1$ is preserved by a group of 3×3 matrices (SO(2,1)). If new variables $X := \frac{\xi}{\zeta+1}$ and $Y = \frac{\eta}{\zeta+1}$ are introduced and ζ is taken to be positive (a condition Poincaré forgot to state) then the point (X,Y) must lie inside the unit circle. The linear transformations in SO(2,1) induce transformations of the unit disc, planes through $(\xi,\eta,\zeta) = (0,0,0)$ cut the upper half of the hyperboloid of 2 sheets $(\xi^2 + \eta^2 - \zeta^2 = -1)$ in lines which correspond to arcs of circles in the (X,Y) domain meeting the unit circle at right angles, and the induced transformations map these arcs to other similar arcs. Poincaré showed that the concepts of non-Euclidean distance and angle may be imposed on the disc to yield a conformal map of the geometry and that the induced transformations are then non-Euclidean isometries. His interest in the group SO(2,1) derived from Hermite's number –

theoretical work [1854]. I presume that Poincaré's Comptes
Rendus paper [1881, 11 July] appeared in print before the
account of this talk. There is no reason to suppose Poincaré
knew of the Weierstrass-Killing interpretation of
non-Euclidean geometry in terms of the hyperboloid of 2 sheets
[Hawkins, 1980, 297].

10.   See [Bottazzini 1977, 32].

11.   c.f. Riemann, "Vorlesungen über die hypergeometrische
Reihe", Nachträge, 77.

12.   The correspondence is printed in Klein, Gesammelte
Mathematische Abhandlungen, III, 1923, 587-626, in Acta
Mathematica (38) 1922, and again in Poincaré Oeuvres, XI.
The simplest way to refer to the letters is to give their
date.

13.   Published somewhat accidentally in Poincaré's Oeuvres XI,
275-276.

14.   The same approach was taken by Picard in proving his
celebrated 'little' theorem that an entire function which fails
to take two values is constant, Picard [1879a]. He argued that,
if there is an entire function which is never equal to a, b, or
∞ (being entire), then a Möbius transformation produces an
entire function, G, say, which is never 0, 1, or ∞. So G never
maps a loop in the complex plane onto a loop enclosing 0, 1, or ∞.
Now, G maps a small patch into a region of, say, the upper half

plane, and if k^2 : $H \to \mathbb{C}$ is the familiar modular function, then composing G with a branch of the inverse of k^2 maps the patch into half a fundamental domain for k^2. Analytic continuation of this composite function cannot proceed on loops enclosing 0, 1, or ∞ (by the remark just made) so the composite $(k^2)^{-1} \circ G$ is single-valued. But it is also bounded, so, by Liouville's theorem, it is constant, and the 'little' theorem is proved. To see that the function is bounded replace the domain of k^2 by the conformally equivalent unit disc; Picard missed that trick and proved it directly. For a nearly complete English translation see Birkhoff and Merzbach, 79-80.

15.   Many years later, Mittag-Leffler wrote in an editorial in Acta Mathematica ((39) 1923, iii) that Poincaré's work was not at first appreciated. "Kronecker, for example, expressed his regret to me via a mutual friend that the journal seemed bound to fail without help on the publication of a work so incomplete, so immature and so obscure."

16.   The uniformization theorem was first proved by Koebe and Poincaré independently in 1907, after an incomplete attempt by Schlesinger. The validity of the continuity method was a central question in the new theory of the topology of manifolds, and was decisively studied by Brouwer in 1911-12. For a full discussion and references see Scholz [1980, 198-222]. Hilbert raised uniformization as a problem as the 22nd in his famous list, and one should also consult Bers' article on it in Mathematical developments arising from Hilbert Problems, AMS, PSPUM, 28, 1976, 559-609, and D. M. Johnson [1982].

Notes on Appendix 4

1.　　　This presentation therefore contributes to the misrepresentation of Riemann's ideas against which Radner, Scholz and others have fought.　There is much more to this paper of Riemann's than a sketch of a theory of differential geometry, and the treatment of non-Euclidean geometry is but a speck.　The influence of Herbart's ideas on Riemann has been given an interesting and surprising examination by Scholz [1982].

2.　　　17 July, 1867 is the date when Dedekind handed the paper over to Weber for final editing [Dugac 1976, 171], and it is the date given in the French edition.　The bound edition of the Göttingen Abhandlungen, volume 13, covers the years 1866 and 1867 and was published in 1868.　Sommerville [1911, 38] gives 1866 and Fortschritte 1868.

3.　　　I wish this account to replace the misleading one in Gray [1979a].　In particular the reference (p. 247) to geodesics appearing as circular arcs in the disc is a totally wrong summary of Beltrami's ideas.

4.　　　A brief reference was also made to Gauss's views on non-Euclidean geometry in the biography [von Waltershausen, 1856, 80-81].　Gauss is a difficult case to understand, for he never brought his ideas together in one place, nor did he seem to like the idea.　For a detailed discussion see [Reichardt 1976].

5.   For a truly stupid piece of philosophizing in a book well received at the time, consider e.g. "The foregoing discussion has brought us to the point where the reader is in a condition, I hope, to realize the great fundamental absurdity of Riemann's endeavour to draw inferences respecting the nature of space and the extension of its concept from algebraic representations of "multiplicities". An algebraic multiple and a spatial magnitude are totally disparate. That no conclusion about forms of extension or spatial mangitudes are derivable from the forms of algebraic functions is evident upon the most elementary considerations". [Stallo, 1888 - 1960, 278]. On Frege, see Toth [1984].

6.   Cayley solved his problem himself in [1872].

7.   Weierstrass's seminar of 1872 seems to be the source of Killing's interest in non-Euclidean geometry. See [Hawkins 1979, Ch. 2].

8.   Both Klein and Cayley made an attempt to define distance by (i) saying that if the cross ratio of A,B,C, and D is −1 and D is at infinity, then B is the midpoint of AC, and (ii) somehow passing to the infinitesimal case. The passage (ii) is not clear, nor was it generally accepted. The whole process recalls von Staudt's attempts [1856 - 1860] to define cross-ratio without the concept of distance and then to define distance in terms of cross-ratio. The difficulty is the introduction of continuity into the projective space.

Bibliography

The following abbreviations have been used:

Abh. Deutsch. Akad. Wiss. Berlin = Abhandlungen den Deutschen Akademie
der Wissenschaften zu Berlin.

Acta Math = Acta Mathematica.

A. H. E. S. = Archive for the History of the Exact Sciences.

Ann. di Mat. = Annali di matematiche pura e applicata.

Ann. Sci. Ec. Norm. = Annales des Sciences de l'École Normale
Supérieure.

Ann. Sci. Mat. Fis. = Annali di scienze matematiche e fisiche,
(ed. Tortolini).

Astr. Nachr. = Astronomische Nachrichten.

Bull. A. M. S. = Bulletin of the American Mathematical Society.

Comm. Soc. Reg. Gött. = Commentationes (recentiores) societatis regiae
scientiarum Göttingensis.

C. R. = Comptes Rendus de l'Académie des Sciences, Paris.

D. S. B. = Dictionary of Scientific Biography, New York.

Enc. Math. Wiss. = Encyklopädie der Mathematischen Wissenschaften.

Fortschritte = Jahrbuch über die Fortschritte der Mathematik.

Giornale di Mat. = Giornale di Matematiche (ed. Battaglini).

H. M. = Historia Mathematica.

J. D. M. V. = Jahrsbericht den Deutschen mathematiker Vereinigung.

J. de Math = Journal des mathématiques pures et appliquées (ed.
Liouville).

J. Ec. Poly = Journal de l'École Polytechnique, Paris.

J. F. M. = Journal für die reine und angewandte Mathematik (ed.
Crelle, Borchardt,...).

Math. Ann. = Mathematische Annalen.

Mem. Caen = Mémoires de l'Académie nationale des Sciences, Arts et
Belles - Lettres de Caen.

Monatsber. Königl. Preuss. Akad. der Wiss. zu Berlin = Monatsbericht
der Königlich Preussichen Akademie der Wissenschaften zu Berlin.

Nachr. König. Ges. Wiss. zu Gött = Nachrichten Königlichen
Gesellschaft der Wissenschaften zu Göttingen.

Nouv. Ann. Math = Nouvelles Annales des Mathématiques.

Quetelet Corr. Math = Correspondence mathématique et physique (ed.
Quetelet).

Rep. f. Math. = Repertorium für die reine und angewandte Mathematik.

Rev. phil = Revue philosophique.

Schlömilch's Zeitschrift = Zeitschrift für Mathematik und Physik ...

Tortolini's Ann = Ann. Sci. Mat. Fis.

402

N. H. Abel (1802-1829)

1828a   Ueber einige bestimmte Integrale, J. F. M. 2, 22-30,

translated as Sur quelques intégrales définies, in

Oeuvres, 2nd ed. (no. 15) I, 251-262.

1828b   Solution d'un problème général concernant la transformation

des Fonctions elliptiques, Astr. Nachr., VI, col. 365-388,

and Addition au Mémoire précédent, ibid VII, 147, =

Oeuvres 2nd ed., (nos. 19, 20) I, 403-428 and 429-443.

1881   Oeuvres complètes de Niels Hendrik Abel, 2 vols, 2nd ed.

edited L. Sylow and S. Lie, Christiania.

W. Abikoff

1981   The Uniformization Theorem, American Mathematical Monthly,

88.8, 574-592.

L-F. A. Arbogast (1759-1803)

1791   Mémoire sur la nature des fonctions arbitraires... St.

Petersburg.

S. H. Aronhold (1819-1884)

1872   Partial French translation of: Ueber den gegenseitigen

Zusammenhang der 28 Doppeltangenten einer allgemeinen

Curve 4-ten Grades (Berlin Monatsber. 1864, 499-523), in

Nouv. Ann. Math. XI, 438-443.

F. Baldassarri

1980   On second order linear differential equations with
        algebraic solutions on algebraic curves, Seminario
        Matematico, Università di Padova, Preprint.

F. Baldassarri and B. Dwork

1979   On second order linear differential equations with
        algebraic solutions, American Journal of Mathematics,
        101, 42-76.

E. Beltrami (1835-1900)

1868   Saggio di interpretazione della geometria non-Euclidea,
        Giornale di Mat, VI, 284-312.

1869   Essai d'interpretation de la géométrie non-Euclidienne,
        Annales École Norm. Sup. VI, 251-288 (translation of
        previous by Höuel).

Jakob Bernoulli (1654-1705)

1694   Curvatura laminee elasticae... Acta erud, 262-276 =
        Opera, II, 576-600.

E. Betti (1823-1892)

1853   Sopra l'abbassamento delle equazioni modulari delle
        funzioni ellittiche, Tortolini, Annali, IV, 81-100.

404

K. R. Biermann

1971a   (Julius Wilhelm) Richard Dedekind, D. S. B., IV, 1-5.

1971b   Ferdinand Gotthold Max Eisenstein D. S. B., IV, 340-343.

1973a   Ernst Eduard Kummer, D. S. B., V, 521-524.

1973b   Die Mathematik und ihre Dozenten an der Berliner
           Universität 1810 - 1920, Akademie Verlag, Berlin.

1976   Karl Theodor Wilhelm Weierstrass, D. S. B., XIV, 219-224.

G. Birkhoff and U. Merzbach

1973   A Source Book in Classical Analysis, Harvard U. P.

J. Bolyai (1802-1860)

1832   Science Absolute of Space, trans. G. B. Halsted,
           published as an Appendix in Bonola [1912].

R. Bonola (1874-1911)

1912   Non-Euclidean Geometry, trans. by H. S. Carslaw of
           original Italian edition of 1906, Dover reprint 1955.

U. Bottazzini

1981   Il calcolo sublime: storia dell'analisi matematica da
           Euler a Weierstrass, Editore Boringhieri, Turin.

A. Brill (1842-1935) and M. Noether (1844-1921)

1874   Ueber die algebraischer Functionen und ihre Anwendung in
           der Geometrie, Math. Ann. 7, 269-310.

1892-93   Die Entwicklung der Theorie der algebraischen Functionen in alterer und neuerer Zeit, J. D. M. V., 3, 107–566.

E. S. Bring (1736–1798)

1786   Promotionschrift, Lund.

F. Brioschi (1824–1897)

1858a   Sulla equazioni del moltiplicatore per la transformazione

            delle Funzioni ellittiche Ann. di Mat. (I), I, 175–177 =

            Opere, I, (no. XLIX), 321–324.

1858b   Sulla risoluzione delle equazioni del quinto grado Ann.

            di Mat (I), I, 256–259 and 326–328 = Opere, I, (no. LII),

            335–341.

1877a   La théorie des formes dans l'intégration des équations

            différentielles linéaires du second ordre, M. Ann. 11,

            401–411 = Opere, II, (no. CCXXXIV), 211–223.

1877b   Sopra una classe di forme binarie, Ann. di. Mat., (II),

            VIII, 24–42 = Opere, II (no. LXII), 157–175.

1878/79a   Sopra una classe di equazioni differenziali lineari del

            secondo ordine, Ann. di Mat. (II), IX, 11–20 = Opere,

            II (no. LXXIII) 177–187.

1878/79b   Nota alla memoria del sig. L. Kiepert "Uber die Auflösung

            der Gleichungen fünften Grades", Ann. di Mat. (II) IX,

            124–125 = Opere II (no. LXXXIV) 189–191.

1877c   Extrait d'une lettre de M. F. Brioschi à M. F. Klein,

            M. Ann, 11, 111–114.

1883/84   Sulla Teoria delle funzioni ellittiche Ann. di Mat. (II)

            XII 49–72 = Opere II (no. LXXXVI), 295–318.

1901   <u>Opere Matematiche</u>, 5 vols.

C. Briot (1817-1882) and J. Bouquet (1814-1885)

1856a   Étude des fonctions d'une variable imaginaire, J. Ec.
            Poly, vol 21, 85-132.

1856b   Recherches sur les propriétés des fonctions définies par
            des équations différentielles, ibid, 133-198.

1856c   Mémoire sur l'intégration des équations différentielles
            au moyen des fonctions elliptiques, ibid, 199-254.

1859   <u>Théorie des fonctions doublement périodiques et, en
            particulier, des fonctions elliptiques</u>, Paris.

W. K. Bühler

1981   <u>Gauss, a biographical study</u>, Springer.

A. L. Cauchy (1789-1857)

1841   Note sur le nature des problèmes que présente le calcul
            intégrale, in <u>Exercises d'Analyse et de physique
            mathématiques</u>, I, Bachelier, Paris = <u>Oeuvres complètes</u>
            (2) 12, 263-271.

1840   Mémoire sur l'intégration des équations différentielles
            in <u>Exercises d'Analyse et de physique mathématiques</u>,
            II, Bachelier, Paris = <u>Oeuvres complètes</u> (2), 11,
            399-465.

1843   Rapport sur un Mémoire de M. Laurent qui a pour titre:
            Extension du théorème de M. Cauchy relatif à la
            convergence du développment d'une fonction suivant les
            puissances ascendantes de la variable x, C.R. 17,

938-940 = <u>Oeuvres</u> (I) 8 (no. 233) 115-117.

1845   Sur le nombre des valeurs égales ou inégales que peut

acquérir une fonction, etc C. R., 21, 779-797 =

<u>Oeuvres complètes</u> (1), 9, 323-341.

1851   Rapport sur un Mémoire présenté à l'Académie par

M. Puiseux CR, pp 32, 276, 493 = <u>Oeuvres complètes</u> (1),

11, 380-382.

A. Cayley (1821-1895)

1859   A sixth memoir upon quantics, Phil. Trans. Roy. Soc.

London, 149, 61-90 = <u>Coll. Math. Papers</u>, 2, (no. 158)

561-592.

1859b   On the conic of 5 point contact at any point of a plane

curve, Phil. Trans. Roy. Soc. CXLIX 371-400 = <u>Coll.</u>

<u>Math. Papers</u>, 4, (no. 261) 207-239.

1865   Note on Lobatchewsky's imaginary geometry, Phil. Mag., 29,

231-233 = <u>Coll. Math. Papers</u>, 5, (no. 362) 471-472.

1883   On the Schwarzian derivative and the polyhedral functions,

Trans. Camb. Phil. Soc., 13.1, 5-6 = <u>Coll. Math. Papers</u>.

11, (no. 745) 148-216.

M. Chasles (1793-1880)

1837   Apercu historique sur l'origine et le développement des

méthodes en Géométrie, Bruxelles, Mem. Couronn. XI.

E. Christoffel (1829-1900)

1867   Über einige allgemeine Eigenschaften der Minimumsflächen,

J. F. M., 67, 218-228.

1868   Allgemeine Theorie der geodetischen Dreiecke, Abh. Deutsch. Akad. Wiss. Berlin 119-176.

1869a   Über die Transformation der homogenen Differentialausdrücke Zweiten Grades, J. F. M., 70, 46-70.

1869b   Über ein die Transformation homogener Differentialausdrücke Zweiten Grades betreffendes Theorem, J. F. M. 70, 241-245.

R. F. A. Clebsch (1833-1872)

1864   Ueber die Anwendung der Abelschen Functionen in der Geometrie, J. F. M. 63, 189-243.

1876   Vorlesungen über Geometrie, ed., F. Lindemann, Teubner, Leipzig.

R. F. A. Clebsch and P. Gordan (1837-1912)

1866   Theorie der Abelschen Functionen. Teubner, Leipzig.

C. H. Clemens

1980   A Scrapbook of Complex Curve Theory, Plenum, New York.

A. Coble (1878-   )

1908   A configuration in finite geometry isomorphic with that of the 27 lines on a cubic surface, Johns Hopkins University Circular, no. 7, 80-88, 736-744.

1913   An application of finite geometry to the characteristic theory of the odd and even theta functions, Trans. A. M. S., 14, 241-276.

G. Codazzi (1816-1877)

1857   Intorno alle superficie le quali hanno costante il prodotto

de' due raggi di curvatura, Ann. Sci. Mat. Fis, 8,

346-355.

R. Dedekind (1831-1916)

1872   Stetigkeit und irrationale Zahlen, trans. Beman as

Continuity and irrational numbers, Dover 1963.

1877   Schreiben an Herrn Borchardt über die Theorie der

elliptischen Modulfunctionen J. F. M. 83, 265-292 = Werke

I, 174-201.

1882   (with H. Weber) Theorie der algebraischen Functionen einer

Veränderlichen J. F. M. 92, 181-290 = Werke I 238-350.

1969   Gesammelte mathematische Werke, I, New York.

L. E. Dickson (1874-1954)

1900   Linear Groups, reprinted Dover 1958.

J. Dieudonné

1974   Cours de géométrie algébrique. I, (Collection SUP) Presses

Univ. de France, Paris.

1978   Abrégé d'histoire des mathématiques, (editor) Hermann,

Paris.

1983   La découverte des fonctions fuchsiennes, Académie des

Sciences, Paris.

J. Dieudonne, J. Carroll

410

1971   Invariant Theory, old and new, New York.

P. G. L. Dirichlet (1805-1859)

1829   Sur la convergence des séries trigonométriques, J. f. M.
            4, 157-169 = Werke, I, 117-132.

1850   Die reduction der positiven quadratischen Formen mit
            drei unbestimmten ganzen Zahlen J. f. M. 40, 213-220,
            = Werke, II, 34-41.

1889-1897   Gesammelte Werke, 2 vols, ed. L. Fuchs and L. Kronecker,
            Berlin.

P. Dugac

1976   Richard Dedekind et les fondements des mathématiques
            Vrin, Paris.

G. W. Dunnington

1955   Carl Friedrich Gauss, Titan of Science, New York.

W. von Dyck (1856-1934)

1880a   Ueber Ausstellung und Untersuchung von Gruppe und
            Irrationalität regular Riemann'scher Flächen, Math. Ann.
            17, 473-509.

1880b   Notiz uber eine reguläre Riemann'scher Flächen vom
            Geschlechte drei... Math. Ann. 17, 510-516.

H. M. Edwards

1977   Fermat's Last Theorem, Springer.

1978   On the Kronecker Nachlass, H. M. 5. 4, 419-426.

1980   The Genesis of Ideal Theory, A. H. E. S., 23, 321-378.

F. G. M. Eisenstein (1823–1852)

1847   Genaue Untersuchungen der unendlichen Doppelproducte...

J. f. M. 35, 137–274 = <u>Mathematische Werke</u>, I, 1975,

(no. 28) 357–478.

L. Euler (1707–1783)

1769   <u>Institutiones Calculi integralis</u>, II = <u>Opera Omnia</u>, ser.

1, 11–13.

1794   Specimen Transformationis singularis serierum, Nova Acta

Acad Sci Petrop, 58–70 = <u>Opera Omnia</u> ser. 2, 16, 41–55

(E710).

H. M. Farkas and I. Kra

1980   <u>Riemann Surfaces</u>, Springer.

C. Fisher

1966   The death of a mathematical theory A. H. E. S. 3,

137–159.

L. R. Ford (1886–   )

1929   <u>Automorphic functions</u>, reprinted Chelsea, 1972.

H. Freudenthal

1954   Poincaré et les fonctions automorphes, in Poincaré Oeuvres,

XI, 213–219.

1971   Augustin – Louis Cauchy, <u>D. S. B.</u> vol III, 131–148.

1975   Bernhard Riemann, <u>D. S. B.</u> vol XI, 447–456.

412

1970   L'Algèbre Topologique en particulier les groupes

topologiques et de Lie, XIIe Congrès International d'Histoire

des Sciences, Paris, Actes, I, 223-243.

R. Fricke (1861-1930)

1913   Automorphe Funktionen mit Einschluss der elliptischen

Modulfunktionen, Enc. Math. Wiss. II B 4.

A. R. Forsyth (1858-1942)

1900-1906 Theory of Differential Equations 6 vols.

F. G. Frobenius (1849-1917)

1875a   Über algebraisch integrirbare lineare Differentialgleichungen

J. f. M., 80, 183-193 = Ges. Abh I, 221-231.

1873a   Über die Integration der linearen Differentialgleichungen

durch Reihen J. f. M., 76, 214-235 = Ges. Abh., I, 84-105.

1873b   Über den Begriff der Irreducibilität in der Theorie der

linearen Differentialgleichungen J. f. M., 76, 236-270 =

Ges. Abh. I, 106-140.

1875b   Über die reguläre Integrale der linearen

Differentialgleichungen J. f. M., 80, 317-333 = Ges. Abh.

I, 232-248.

1968   Gesammelte Abhandlungen ed. J. P. Serre, Springer Verlag.

L. I. Fuchs (1833-1902)

1865   Zur Theorie der linearen Differentialgleichungen mit

veränderlichen Coefficienten, Jahrsber. Gewerbeschule,

Berlin, Ostern = <u>Werke</u>, I, 111-158.

1866   ibid, J. f. M., 66, 121-160 = <u>Werke</u>, I, 159-204.

1868   ibid, Ergänzungen zu der im 66-sten Bande dieses Journal enthaltenen Abhandlung, J. f. M. 68, 354-385 = <u>Werke</u>, I, 205-240.

1870a   Die Periodicitätsmoduln der hyperelliptischen Integrale als Functionen eines Parameters aufgefasst, J. f. M. 71, 91-127 = <u>Werke</u>, I, 241-282.

1870b   Über eine rationale Verbindung der Periodicitätsmoduln der hyperelliptischen Integrale, J. f. M., 71, 128-136 = <u>Werke</u>, I, 283-294.

1870c   Bemerkungen zu der Abhandlungen "Über hypergeometrische Functionen unter Ordnung" in diesem Journal, 71, 316 = J. f. M. 72, 255-262 = <u>Werke</u>, I, 311-320.

1871a   Über die Form der Argumente der Thetafunctionen und über die Bestimmung von $\theta(0, 0,\ldots,0)$ als function der Klassenmoduln, J. f. M., 73, 305-323 = <u>Werke</u>, I, 321-342.

1871b   Über die linearen Differentialgleichungen, welchen die Periodicitätsmoduln der Abelschen Integrale genügnen, und über verschiedene Arten von Differentialgleichungen für $\theta(0, 0,\ldots,0)$, J. f. M., 73, 324-339 = <u>Werke</u>, I, 343-360.

1875   Über die linearen Differentialgleichungen zweiter Ordnung, welche algebraische Integrale besitzen, und eine neue Anwendung der Invariantentheorie.... Nachr. Königl. Ges. der Wissenschaften, Göttingen 568-581, 612-613 = <u>Werke</u>, II, 1-10.

1876a   Über die linearen Differentialgleichungen zweiter Ordnung, welche algebraische Integrale besitzen, und eine neue Anwendung der Invariantentheorie. J. f. M., 81, 97-142 = <u>Werke</u>, II, 11-62.

1876b   Extrait d'une lettre adressée à M. Hermite J. de Math, (3), 2, 158-160 = <u>Werke</u>, II, 63-66.

1876c   Sur les équations différentielles linéaires du second ordre C. R. 82, 1494-1497 and 83, 46-47 = <u>Werke</u> II, 67-72.

1877a   Selbstanzeige der Abhandlung: "Über die linearen Differentialgleichungen zweiter Ordnung, welche algebraische Integrale besitzen und eine neue Anwendung der Invariantentheorie". Borchardts Journal, Bd. 81, p. 97 sqq. Rep. f. Math, I, 1-9 = <u>Werke</u>, II, 73-86.

1877b   Sur quelques propriétés des intégrales des équations différentielles, auxquelles satisfont les modules de périodicité des intégrales elliptiques des deux premières espèces. Extrait d'une lettre adressée à M. Hermite. J. f. M. 83, 13-37 = <u>Werke</u>, II, 87-114.

1878a   Über die linearen Differentialgleichungen zweiter Ordnung, welche algebraische Integrale besitzen. Zweite Abhandlung, J. f. M. 85, 1-25 = <u>Werke</u>, II, 115-144.

1877c   Extrait d'une lettre adressée à M. Hermite, C. R. 85, 947-950 = <u>Werke</u>, II, 145-150.

1878b   Über eine Klasse von Differentialgleichungen, welche durch Abelsche oder elliptische Functionen integrirbar sind, Nachrichten von der Königl. Gesellschaft der Wissenschaften, Göttingen 19-32; Ann di mat (II), 9, 25-35 = <u>Werke</u>, II, 151-160.

1878c   Sur les équations différentielles linéaires qui admettent des intégrales dont les différentielles logarithmiques sont des fonctions doublement périodiques. Extrait d'une lettre à M. Hermite J. de Math. (3),

4, 125–140 = <u>Werke</u>, II, 161–176.

1879a   Selbstanzeige der Abhandlung: "Sur quelques propriétés
des intégrales des équations différentielles, auxquelles
satisfont les modules de périodicité des intégrales
elliptiques des deux premières espèces. Extrait d'une
lettre adressée à M. Hermite". Borchardts Journal,
83, 13, Rep. Math 2. 235–240 = <u>Werke</u>, II, 177–184.

1880a   Über eine Klasse von Functionen mehrerer Variabeln,
welche durch Umkehrung der Integrale von Lösungen der
linearen Differentialgleichungen mit rationalen
Coefficienten entstehen. Nachr Königl. Ges. der
Wissenschaften, Göttingen 170–176 = <u>Werke</u>, II, 185–190.

1880b   Über eine Klasse von Functionen mehrerer Variabeln,
welche durch Umkehrung der Integrale von Lösungen der
linearen Differentialgleichungen mit rationalen
Coefficienten entstehen, J. f. M. 89 151–169 = <u>Werke</u>, II,
191–212.

1880c   Sur une classe de fonctions de plusieurs variables tirées
de l'inversion des intégrales de solutions des équations
différentielles linéaires dont les coefficients sont des
fonctions rationnelles. Extrait d'une lettre adressée
à M. Hermite C. R. 90, 678–680 and 735–736, = <u>Werke</u>, II,
213–218.

1880d   Über die Functionen, welche durch Umkehrung der Integrale
von Lösungen der linearen Differentialgleichungen
entstehen. Nachr Königl Ges. der Wissenschaften,
Göttingen 445–453 = <u>Werke</u>, II, 219–224.

1881a   Auszug aus einem Schreiben des Herrn L. Fuchs an
C. W. Borchardt J. f. M. 90, 71–73 = <u>Werke</u>, II, 225–228.

1880e   Sur les fonctions provenant de l'inversion des
intégrales des solutions des équations différentielles
linéaires... Bulletin des Sciences mathématiques et
astronomiques. (2), 4, 328-336 = <u>Werke</u>, II, 229-238.

1881b   Über Functionen zweier Variabeln, welche durch Umkehrung
der Integrale zweier gegebener Functionen entstehen...
Abh. König. Gesellschaft der Wissenschaften zu
Göttingen, 27, 1-39 = <u>Werke</u>, II, 239-274.

1881c   Sur les fonctions de deux variables qui naissent de
l'inversion des intégrales de deux fonctions données.
Extrait d'une lettre adressée à M. Hermite C. R. 92,
1330-1331, 1401-1403 = <u>Werke</u>, II, 275-289.

1881d   Sur une équation différentielle de la forme f $(u, \frac{du}{dz}) = 0$.
Extrait d'une lettre adressée à M. Hermite, C. R. 93,
1063-1065 = <u>Werke</u>, II, 283-284.

1882a   Über Functionen, welche durch lineare Substitutionen
unverändert bleiben. Nachr. Konigl. Gesellschaft der
Wissenschaften, Göttingen, 81-84 = <u>Werke</u>, II, 285-288.

1882b   Über lineare homogene Differentialgleichungen, zwischen
deren Integralen homogene Relationen höheren als ersten
Grades bestehen, Sitzungsberichte der K. preuss
Akademie der Wissenschaften zu Berlin, 703-710 = <u>Werke</u>
II, 289-298.

1882c   Über lineare homogene Differentialgleichungen, zwischen
deren Integralen homogene Relation höheren als ersten
Grades bestehen, Acta Mathematica, I, 321-362 = <u>Werke</u>,
II, 299-340.

1884a   Über Differentialgleichungen, deren Integrale feste
Verzweigungspunkte besitzen., Sitzungsberichte der

K. Preuss. Akademie der Wissenschaften zu Berlin, 699–710, = <u>Werke</u>, II, 355–368.

1884b   Antrittsrede gehalten am 3 Juli 1884 in der öffentlichen Sitzung zur Feier des Leibniztages der Königl. Preuss Akademie der Wissenschaften zu Berlin, Sitzungsberichte, 744–747 = <u>Werke</u>, II, 369–372.

1904   <u>Gesammelte Mathematische Werke</u>, ed. R. Fuchs and L. Schlesinger, vol I, Berlin.

1906   ibid, II.

1909   ibid, III.

E. Galois (1811–1832)

1846a   Lettre à M. A. Chevalier sur la théorie des équations et les fonctions intégrales, J. de Math. 11, 408–416 = <u>Oeuvres</u>, 25–32.

1846b   Mémoire sur les conditions de la résolubilité des équations par radicaux J. de Math. 11, 417–433.

1897   <u>Oeuvres Mathématiques</u>, ed. E. Picard.

C. F. Gauss (1777–1855)

1801   <u>Disquistiones Arithmeticae</u> = <u>Werke</u>, I.

1812a   Disquisitiones generales circa seriem infinitam, Pars prior, Comm. Soc. reg Gött. II = <u>Werke</u>, III, 123–162.

1812b   Determinatio seriei nostrae per aequationem differentialem secundi ordinis, MS. = <u>Werke</u>, 207–230.

1822   Allgemeine Auflösung etc. Astron. Abh. III, 1 = <u>Werke</u>, IV, 189–216.

1827   Disquisitiones generales circa superficies curvas, Comm. Soc. reg. Gött. VI = <u>Werke</u>, IV, 217–258.

418

1839/40	Allgemeine Lehrsatz ein Beziehung auf die im verkehrten Verhältnisse des Quadrats der Entfernung, Resultate aus den Beobachtungen des magnetischen Vereins im Jahre 1839. Herausg. v. Gauss u. Weber, Leipzig = <u>Werke</u>, V, 195–240.
1860–65	<u>Briefwechsel zwischen C. F. Gauss and H. C. Schumacher</u>, 6 vols, Altona.
1863–1867	<u>Werke</u>, vols 1–V.

C. F. Geiser (1843–1930)

1869	Ueber die Doppeltangenten einer ebenen Curve Vierten Grades, Math. Ann. 1, 129–138.

H. Geppert (1902–1945)

1927	Bestimmung der Anziehung eines elliptischen Ringes, Nachlass zur Theorie des arithmetisch – geometrischen Mittels und der Modulfunktion von C. F. Gauss. Ostwald Klassiker no. 225, Leipzig.

C. Gilain

1977	La théorie géométrique des équations différentielles de Poincaré et l'histoire de l'analyse, Thesis, Paris.

A. Göpel (1812–1847)

1847	Theorie transcendentium Abelianarum primi ordinis adumbratio levis, J. f. M. 35, 277–312.

P. Gordan (1837-1912)

1868   Beweis, dass jede Covariante und Invariante einer
        binären Form eine ganze Function mit numerischen
        Coefficienten einer endlichen Anzahl solcher Formen
        ist, J. f. M. 69, 323-354.

1877a   Über endliche Gruppen linearer Transformationen einer
        Veränderlichen Math. Ann. 12, 23-46.

1877b   Binäre Formen mit verschwinden Covarianten Math. Ann.
        12, 147-166.

1878   Über die Auflösung der Gleichungen vom fünften Grade,
        Math. Ann, 13, 375-404.

1880   Ueber das volle Formen system der terneren
        biquadratischen Form $f = x_1^3 x_2 + x_2^3 x_3 + x_3^3 x_1$, Math.
        Ann. 17, 217-233.

J. J. Gray

1979a   Non-Euclidean Geometry - a re-interpretation, H. M. 6,
        236-258.

1979b   Ideas of Space, Oxford U. P.

1981   Les trois suppléments au Mémoire de Poincaré, écrit en
        1880, sur les fonctions fuchsiennes et les équations
        différentielles. C. R. 293, Vie Académique, 87-90.

1982   The three supplements to Poincaré's prize essay of 1880
        on Fuchsian functions and differential equations.
        Arch. Int. Hist. Sci. 32, No. 109, 221-235

1984a   Fuchs and the theory of differential equations, Bull.
        A. M. S., 10.1, 1-26, corrigendum to appear.

1984b   A commentary on Gauss's mathematical diary, 1796-1814,

with an English translation, Expositiones Mathematicae,
2, 97-130, with corrigenda to appear.

1984c   The Riemann-Roch theorem: acceptance and rejection of
geometric ideas. Astérisque, N,

G. Green (1793-1841)

·1835   On the Determination of the Exterior and Interior
Attractions of Ellipsoids of Variable Densities. Trans
Camb Phil Soc 5 (3), 395-430 = <u>Mathematical Papers</u> 1883,
187-222.

J. Hadamard (1865-1963)

1920   The Early Scientific Work of Poincaré, Rice Institute
Pamphlet, no. 9.3, 111-183.

1954   <u>The Psychology of Invention in the Mathematical Field</u>,
Dover.

H. Haedenkamp

1841   Ueber Transformation vielfacher Integrale, J. f. M.
22, 184-192.

G. H. Halphen (1844-1889)

1881a   Sur des fonctions qui proviennent de l'équation de Gauss,
C. R. 92, 856 = <u>Oeuvres</u>, II, 471-474.

1881b   Sur un systeme d'équations différentielles, C. R. 92,
1101, = <u>Oeuvres</u>, II, 475-7.

1884a   Sur une Équation différentielle linéaires du troisième ordre, Math. Ann. 21, 461 = <u>Oeuvres</u>, IV, 112-115.

1884b Sur le réduction des équations différentielles linéarires

aux formes intégrables, Mem. présentes par divers savants

à l'Académie des Sciences, 28, 1-260 = <u>Oeuvres</u>, 3,

1-260.

1921 <u>Oeuvres</u>, 4 vols, ed. C. Jordan, H. Poincaré, E. Picard,

E. Vessiot, Gauthier-Villars, Paris.

M. Hamburger (1838-1903)

1873 Bemerkungen über die Form der Integrale der linearen

Differentialgleichungen mit veränderlichen

Coefficienten J. f. M., 76, 113-125.

J. Harkness, W. Wirtinger, R. Fricke

1913 Elliptische functionen, Enc. Math. Wiss. II B3.

R. Hartshorne

1977 <u>Algebraic Geometry</u>, Springer.

M. W. Haskell (1863-)

1890 Über die zu der Curve $\lambda^3\mu + \mu^3\nu + \nu^3\lambda = 0$ ... American

Journal of Mathematics 13, 1-52.

T. Hawkins

1975 <u>Lebesgue's Theory of Integration</u> Chelsea (2nd edition).

1977   Weierstrass and the Theory of Matrices, A. H. E. S. 17. 2, 119-163.

1980   Non-Euclidean Geometry and Weierstrassian Mathematics: The background to Killing's work on Lie Algebras, H. M. 7.3, 289-342.

H. E. Heine (1821-1881)

1845   Beitrag zur Theorie der Anziehung und der Wärme, J. f. M. 29, 185-208.

1846   Über die Reihe $1 + \dfrac{(q^{\alpha}-1)(q^{\beta}-1)}{(q-1)(q^{\gamma}-1)} x + $ etc., J. f. M. 32, 210-212.

1847   Untersuchungen über die Reihe $1 + \dfrac{(1-q^{\alpha})(1-q^{\beta})}{(1-q)(1-q^{\gamma})} x + $ etc., J. f. M. 34, 285-328.

H. Helmholtz (1821-1894)

1868   Über die tatsachlichen Grundlagen der Geometrie, Nachr. König. Ges. Wiss. zu Göttingen 15, 193-221 = <u>Abh.</u> 2, 1883, 618-639.

C. Hermite (1822-1901)

1851   Sur les fonctions algébriques, CR 32, 458-461 = <u>Oeuvres</u>, I, 276-280.

1858   Sur la résolution de l'équation du cinquième degré, CR 46, 508-515 = <u>Oeuvres</u> II, 5-12.

1859a   Sur la théorie des équations modulaires, CR 48, pp. 940-947, 1079-1084, 1095-1102 and CR 49, pp. 16-24, 110-118, 141-144 = <u>Oeuvres</u>, II, 38-82.

1859b   Sur l'abaissement de l'équation modulaire du huitième degré, Ann. di Mat. II, 59–61 = Oeuvres, II, 83–86.

1862   Note sur la théorie des fonctions elliptiques, in Lacroix: Calcul différentiel et Calcul intégral, Paris 6th ed. 1862 = Oeuvres, II, 125–238.

1863   Sur les fonctions de sept lettres, CR 57, 750–757 = Oeuvres, II, 280–288.

1865/66   Sur l'équation du cinquième degré CR 66 and CR 67 passim = Oeuvres, II, 347–424.

1877–1882   Sur quelques applications des fonctions elliptiques CR 85 – CR 94 passim = Oeuvres, III, 266–418.

1900   Lettre de Charles Hermite à M. Jules Tannery sur les fonctions modulaires, ms. = Oeuvres, III, 13–21.

1905–1917   Oeuvres, 4 vols, Paris.

L. O. Hesse (1811–1874)

1844   Ueber die Wendepunkte der Curven dritter Ordnung, J. f. M., 28, 97–106 = Werke, (no. 9), 123–136.

1855a   Ueber Determinanten und ihre Anwendung in der Geometrie, insbesondere auf Curven Vierter Ordnung, J. f. M., 49, 243–264 = Werke (no. 24) 319–343.

1855b   Ueber die Doppeltangenten der Curven Vierter Ordnung J. f. M. 49, 279–332 = Werke (no. 25) 345–404.

1897   Ludwig Otto Hesse's Gesammelte Werke, 1st pub. 1897, reprinted Chelsea, 1972.

D. Hilbert (1862-1943)

1900a   Ueber das Dirichlet'sche Prinzip J. D. M. V., VIII, 184-188, trans. in Birkhoff Merzbach, 399-402.

1900b   Mathematische Probleme, Göttinger Nachrichten 253-297 (translated in Bull A. M. S. 1902, 437-479 reprinted in Mathematical developments arising from Hilbert Problems, Proc. Symp. Pure Math. A. M. S. 1972, vol. I, 1-34.)

E. Hille (1894-1980)

1976   Ordinary Differential Equations in the Complex Domain, Wiley.

A. Hurwitz (1859-1919)

1881   Grundlagen einer independenten Theorie der elliptischen Modulfunktionen ... Math. Ann. 18, 528-592 = Math. Werke, Birkhäuser, Basel, 1932, I, 1-66.

1886   Über einige besondere homogene linearen Differential-gleichungen Math. Ann. 26, 117-126 = Math. Werke, I (no. 9) 153-162.

C. G. J. Jacobi (1804-1851)

1829   Fundamenta Nova Theoriae Functionum Ellipticarum = Ges. Werke, I, (2nd ed.) 49-239.

1834   De functionibus duarum variabilium quadrupliciter periodicis, quibis theoria transcendentium Abelianarum innititur, J. f. M. 13, Ges. Werke, II, (2nd ed.) 23-50.

1850   Beweis des Satzes, dass ein Curve nten Grades im

Allgemein $\frac{1}{2}n(n-2)(n^2-9)$ Doppeltangenten hat,

J. F. M. 40, 237-260 = <u>Ges. Werke</u>, III, 517-542.

1969   <u>C. G. J. Jacobi's Gesammelte Werke</u>, 2nd edition, 8 vols.

Chelsea.

W. S. Jevons (1835-1882)

1871   Helmholtz on the Axioms of Geometry, Nature, 4,

481-482.

D. M. Johnson

1979   The Problem of Invariance of Dimension in the Growth of

Modern Topology, Part I, A. H. E. S. 20.2, 97-188.

1981   ibid. Part II, A. H. E. S. 20.2/3, 85-267.

C. Jordan (1838-1922)

1868a   Note sur les équations modulaires, C. R. 66, 308-312 =

<u>Oeuvres</u>, I, 159-163.

1868b   Mémoire sur les groupes de mouvement, Ann. di Mat. II

167-215, 322-345 = <u>Oeuvres</u>, IV, 231-302.

1869   Commentaire sur Galois, Math. Ann. I, 142-160 = <u>Oeuvres</u>,

I, 211-230.

1870   <u>Traité des substitutions et des équations algébriques</u>,

Paris.

1871   Sur la résolution des équations différentielles

linéaires, C. R. 73, 787-791 = <u>Oeuvres</u>, I, 313-317.

1876a   Sur les équations du second ordre dont les intégrales

sont algébriques, C. R. 82, 605-607 = <u>Oeuvres</u>, II, 1-4.

1876b   Sur la détermination des groupes formé d'un nombre fini des substitutions linéaires, C. R. 83, 1035–1037 = Oeuvres, II, 5–6.

1876c   Mémoire sur les covariants des formes binaires, I, J. de Math. (3), II, 177–232 = Oeuvres, III, 153–208.

1876/77   Sur une classe de groupes d'ordre fini contenus dans les groupes linéaires, Bull. Math. Soc. France, 5, 175–177 = Oeuvres, II, 9–12.

1877   Détermination des groupes formé d'un nombre fini de substitutions, C. R. 84, 1446–1448 = Oeuvres, II, 7–8.

1878   Mémoire sur les équations différentielles linéaires à intégrale algébrique, J. f. M., 84, 89–215 = Oeuvres, II, 13–140.

1879   Mémoire sur les covariants des formes binaires, II, J. de Math. (3) V, 345–378 = Oeuvres, III, 213–246.

1880   Sur la détermination des groupes d'ordre fini contenus dans le groupe linéaire, Atti Accad. Nap 8 no. 11 = Oeuvres, II, 177–218.

1915   Cours d'Analyse III, 3rd ed. Paris.

1961   Oeuvres de Camille Jordan, 4 vols, ed J. Dieudonné, Paris.

N. M. Katz

1976   An overview of Deligne's work on Hilbert's Twentyfirst Problem, in Mathematical Developments arising from Hilbert Problems, P. S. P. M. AMS XXVIII, vol. 2, 537–557.

L. Kiepert (1846-1934)

1876   Wirkliche Ausführung der ganzzahligen Multiplication der elliptischen Functionen, J. f. M. 76, 21-33.

B. M. Kiernan

1972   The Development of Galois Theory from Lagrange to Artin, A. H. E. S. 8, 40-154.

C. F. Klein (1849-1925)

1871   Über die sogenannte Nicht - Euklidische Geometrie, I, Math. Ann, 4, = Ges. Math. Abh. I, (no. XVI) 254-305.

1872   Vergleichende Betrachtungen über neuere geometrische Forschungen (Erlanger Programm) 1st pub. Deichert, Erlangen = Ges. Math. Abh. I (no. XXVII) 460-497.

1873   Über die sogenannte Nicht-Euklidische Geometrie, II, Math. Ann. 6 = Ges. Math. Abh., I (no. XVIII) 311-343.

1875/76   Über binäre Formen mit linearen Transformationen in sich selbst, Math. Ann. 9, = Ges. Math. Abh. II (no. LI) 275-301.

1876   Über [algebraisch integrirbare] lineare Differential-gleichungen, I, Math. Ann. 11 = Ges. Math. Abh. II (no. LII) 302-306.

1877a   Über [algebraisch integrirbare] lineare Differential-gleichungen, II, Math. Ann. 12 = Ges. Math. Abh. II (no. LIII) 307-320.

1877b   Weitere Untersuchungen über das Ikosaeder Math. Ann. 12 = Ges. Math. Abh. II (no. LIV) 321-379.

1878/79a   Über die Transformationen der elliptischen Funktionen und die Auflösung der Gleichungen Fünften Grades Math. Ann. 14 = Ges. Math. Abh. III (no. LXXXII) 13–75.

1878/79b   Über die Erniedrigung der Modulargleichungen Math. Ann. 14 = Ges. Math. Abh. III (no. LXXXIII) 76–89.

1879a   Über die Transformation siebenter Ordnung der elliptischen Funktionen Math. Ann. 14 = Ges. Math. Abh. III (no. LXXXIV) 90–134.

1879b   Über die Auflösung gewisser Gleichungen vom siebenten und achten Grade Math. Ann. 15 = Ges. Math. Abh. II (no. LVII) 390–425.

1880/81   Zur Theorie der elliptischen Modulfunktionen Math. Ann. 17 = Ges. Math. Abh. III (no. LXXXVII) 169–178.

1881   Über Lamésche Funktionen Math. Ann. 18 = Ges. Math. Abh. II (no. LXII) 512–520.

1882a   Über Riemanns Theorie der algebraischen Funktionen und ihrer Integrale, Teubner, Leipzig, = Ges. Math. Abh. III (no. XCIX) 499–573.

1882b   Über eindeutige Funktionen mit linearen Transformationen in sich, I, Math. Ann. 19 = Ges. Math. Abh. III (no. CI) 622–626.

1882c   Über eindeutige Funktionen mit linearen Transformationen in sich, II, Math. Ann. 20 = Ges. Math. Abh. III (no. CII) 627–629.

1883   Neue Beiträge zur Riemannschen Funktionentheorie, Math. Ann. 21 = Ges. Math. Abh. III (no. CIII) 630–710.

1884   Vorlesungen über das Ikosaeder und die Auflösung der Gleichungen vom fünften Grade. Teubner, Leipzig

(English translation <u>Lectures on the Icosahedron</u> by
G. G. Morrice 1888, Dover reprint 1956).

1894   <u>Vorlesungen über die hypergeometrische Funktion</u>,
Göttingen.

1921   <u>Gesammelte Mathematische Abhandlungen</u>, I, Springer
Berlin.

1922   ibid., vol. II.

1923a   ibid., vol. III.

1923b   Göttinger Professoren (Lebensbilder von eigener Hand):
Felix Klein, Mitteilung des Universitätsbundes
Göttingen, 5, 1, 11-36.

1967   <u>Vorlesungen über die Entwicklung der Mathematik im 19.
Jahrhundert</u>. Chelsea reprint, 2 vols. in 1.

M. Kline

1972   <u>Mathematical Thought from Ancient to Modern Times</u>.
Oxford U. P.

L. Koenigsberger (1837-1921)

1868   Die Multiplication und die Modulargleichungen der
elliptischen functionen, Teubner, Leipzig.

1871   Die linearen transformationen der Hermite'schen
ϕ-function Math. Ann. 3, 1-10.

A. Krazer (1858-1926)

1909   Zur Geschichte des Umkehrproblems der Integrale,
J. D. M. V., 18, 44-75.

A. Krazer, W. Wirtinger

1913   Abelsche Funktionen und allegemeine Thetafunktionen.
Enc. Math. Wiss, II. B. 7.

L. Kronecker (1823-1891)

1857   Über die elliptischen Functionen, für welche complexe
Multiplication stattfindet, Monatsber. Königl. Preuss.
Akad. der Wiss. zu Berlin, 455-460 = Werke, II,
(no. XVII) 177-184.

1858a   Über Gleichungen des siebenten Grades, Monatsber, etc.
Berlin, 287-289 = Werke, II, (no. V), 39-42.

1858b   Sur la résolution de l'équation du cinqième degré,
(Extrait d'une Lettre adressée à M. Hermite), CR 46,
1150-1152, = Werke, II, (no. VI), 43-48.

1859   Sur la théorie des substitutions (Extrait d'une Lettre
adressée à M. Brioschi) Ann. di mat, II, 131 = Werke II
(no. VII) 51-52.

1863   Über die Auflösung der Pell'schen Gleichung mittels
elliptischer Functionen, Monatsber. etc. Berlin 44-50 =
Werke II (no. XXI), 219-226.

1881/82   Grundzüge einer arithmetischen Theorie der algebraischen
Grössen.Festschrift Kummer, J. F. M., 92, 1-122 = Werke II
(no. XI), 237-388.

1883   Über bilineare Formen mit vier Variabeln, Abh. Königl.
Preuss. Akad. des Wiss. zu Berlin II, 1-60 = Werke, II,
(no. XVII), 425-496.

1897   <u>Leopold Kronecker's Werke</u>, II.

1929   ibid., IV. The complete <u>Werke</u>, in 5 vols, ed. K. Hensel,
Teubner, Leipzig, reprinted Chelsea 1968.

E. E. Kummer (1810-1893)

1834   De generali quadam aequatione differentiali tertii
ordinis, Programme of the Liegnitz Gymnasium = J. f. M.
100, 1-9 (1887) = <u>Coll. Papers</u>, II, 33-39.

1836   Über die hypergeometrische Reihe... J. f. M., 15, 39-83
and 127-172 = <u>Coll. Papers</u>, II, 75-166.

1975   <u>Collected Papers</u>, 2 vols, edited A. Weil, Springer.

J. P. S. Kung and G.-C. Rota

1984   The invariant theory of binary forms, Bull. A. M. S.
10.1, 27-85.

G. Lamé (1795-1870)

1837   Mémoire sur les surfaces isothermes, etc. J. de Math.
2, 147-183.

E. Lampe (1840-1918)

1892-3   Nachruf für Ernst Eduard Kummer, J. D. M. V. vol. 3,
13-28, = Kummer, <u>Coll. Papers</u>, I, 1975.

S. Lang

1976   <u>Modular Functions</u>, Springer.

P. A. Laurent (1813-1854)

1843   Extension du théorème de M. Cauchy, relatif à la
convergence du développement d'une fonction suivant les
puissances ascendantes de la variable, reprinted in
J. Peiffer, Les premiers exposés globaux de la théorie
des fonctions de Cauchy, 1840-1860, Thesis, Paris.

A. M. Legendre (1752-1833)

1786   Mémoire sur les intégrations par les arcs d'ellipse
Hist. Acad. Paris, 616-644.

1814   Exercises de calcul intégral, vol. II.

1825   Traité des fonctions elliptiques et des Intégrales
Euleriennes. 3 vols. Paris.

1823   Elémens de Géométrie (12th ed.) Paris.

J. Lehner

1966   A short course in automorphic functions Holt, Rinehart,
and Winston, New York.

J. Liouville (1809-1882)

1845   Sur diverses questions d'analyse et de physique
mathématique, J. de Math. 10, 222-228.

N. I. Lobachevskii (1792-1856)

1829   Ueber die Anfangsgründe der Geometrie. Kasan Bulletin.

1837   Géométrie Imaginaire, J. f. M. 17, 295-320.

1840   Geometrische Untersuchungen etc., Berlin, trans.
G. B. Halsted as Geometrical researches on the theory

of parallels, appendix in Bonola [1912].

J. Lützen

1984   Joseph Liouville, Mathématicien pur et appliqué,

            A Scientific Biography, vol. I. Matematisk Institut,

            Odense Universitet, Preprint no. 4.

Ju. Manin

1974   Cubic Forms, trans. M. Hazewinkel, North Holland,

            Amsterdam.

E. Mathieu (1835-1890)

1860   Le nombre de valeurs que peut acquérir une fonction

            quand on y permute ses variables de toutes les manières

            possibles, J. de Math. 5, 9-42.

1861   Mémoire sur l'étude des fonctions des plusieurs

            quantitées... J. de Math. 6, 241-323.

1875   Sur le fonction cinq fois transitive de 24 quantitées.

            J. de Math. 18, 25-46.

M. Matsuda

1980   First Order Algebraic Differential Equations - A

            Differential Algebraic Approach, Springer L. N. I. M.

            804.

K. O. May (1915-1977)

1972   Gauss, Carl Friedrich, D. S. B. V, Scribners.

434

H. Mehrtens

1979   Das Skelett der modernen Algebra, <u>Disciplinae Novae</u>
            ed. C. J. Scriba, Joachim Jungius Gesellschaft, Hamburg.

F. Meyer (1848-1898)

1890/91   Bericht über den gegenwärtigen Stand der Invarianten
            Theorie J. D. M. V. 1, 79-288.

G. A. Miller, H. F. Blichfeldt, L. E. Dickson

1916   <u>The Theory and Application of Finite Groups</u>, 1st ed.
            New York, Dover reprint 1962.

F. Minding (1806-1885)

1839   Wie sich entscheiden lässt, ob zwei gegebener Krummen $\cdots$
            J. f. M. 19, 370-387.

G. Mittag-Leffler (1846-1927)

1876   <u>En Metod alt kommer i analytiske berittning af de</u>
            <u>Elliptiska functionerna</u>, Helsingfors.
1880   Sur les équations différentielles linéaires à coefficients
            doublement périodiques CR 90, 299-300.

A. F. Monna

1875   <u>Dirichlet's Principle. A mathematical comedy of errors</u>
            <u>and its influence on the development of analysis.</u>
            Oosthoek, Scheltema and Holkema, Utrecht.

435

F. Müller

1872   Ueber die Transformation vierten Grades der elliptischen

            functionen, Berlin.

1873   Beziehungen zwischen dem Moduln der elliptischen

            functionen, etc., Schlömilch's Zeitschrift, 18, 280-288.

E. Netto (1864-1919)

1882   Die Substitutionen theorie und ihre Anwendung auf die

            Algebra, Teubner, Leipzig.

E. Neuenschwander

1978a   The Casorati-Weierstrass Theorem, H. M. 5, 139-166.

1978b   Der Nachlass von Casorati (1835-1890) in Pavia, A. H. E. S.

            19.1, 1-89.

1979   Riemann und das "Weierstrassche" Prinzip der

            Analytischen Fortsetzung durch Potenzreihen, J.D.M.V.,

            82 (1980), 1-11.

1981a   Lettres de Bernhard Riemann à sa famille, Cahiers du

            Séminaire d'Historie des Mathématiques, 85-131.

1981b   Studies in the History of Complex Function Theory, II,

            Bull. A. M. S. 5.2, 87-105.

C. A. Neumann (1832-1895)

1865   Vorlesungen über Riemann's Theorie der Abel'schen

            Integrale, Teubner, Leipzig.

1865   Das Dirichlet'sche Princip in seiner Andwendung auf die

            Riemann'schen Flachen, Teubner, Leipzig.

436

I. Newton (1642-1727)

1968   The Mathematical Papers of Isaac Newton, ed.

D. T. Whiteside, II, 10-89, Cambridge.

E. Papperitz (1857-1938)

1889   Ueber die Darstellung der hypergeometrischen

Transcendenten durch eindeutige Functionen, Math. Ann.

34, 247-296.

J. F. Pfaff (1765-1825)

1797   Disquisitiones Analyticae, I, Helmstadt.

H. T. H. Piaggio (1884-   )

1962   An Elementary Treatise on Differential Equations and

their applications, London.

E. Picard (1856-1941)

1879a   Sur une propriété des fonctions entières. C. R. 88,

1024-1027.

1879b   Sur une généralisation des fonctions périodiques et sur

certaines équations différentielles linéaires C. R. 89,

140-144.

1880a   Sur une classe d'équations différentielles linéaires

C. R. 90, 128-131.

1880b   Sur les équations différentielles linéaires à

coefficients doublement périodiques C. R. 90, 293-295.

1881   Sur les équations différentielles linéaires à coefficients

doublement périodiques J. f. M. 90, 281-302.

J. Plemelj (1873-   )

1964   Problems in the sense of Riemann and Klein, Interscience
            Tract no. 16, Wiley.

J. Plücker (1801-1868)

1834   System der analytischen Geometrie, Berlin.

1839   Theorie der algebraischen Curven, Berlin.

L. Pochhammer

1870   Ueber hypergeometrische Functionen n^{ter} Ordnung,
            J. f. M. 71, 316-352.

1889   Ueber ein Integral mit doppeltem Umlauf, Math. Ann.
            35, 470-494.

H. Poincaré (1854-1912)

1880   Extrait d'un Mémoire inedit de Henri Poincaré sur les
            fonctions Fuchsiennes. Acta Math. 39 (1923),
            58-93 = Oeuvres, I, 578-613.

1881a   Sur les fonctions Fuchsiennes, C. R., 92, 333-335 =
            Oeuvres II, 1-4.

1881b   ibid., C. R. 92, 395-398 = Oeuvres, II, 5-7.

1881c   Sur une nouvelle application et quelques propriétés
            importantes des fonctions Fuchsiennes, C. R., 92,
            859-861 = Oeuvres, V, 8-10.

1881d   Sur les applications de la géométrie non-Euclidienne à
            la théorie des formes quadratiques, Association
            Française pour l'avancement des sciences, 10th session,

Algiers, 16 April, 1881 = Oeuvres, V, 267-274.

1881e   Sur les fonctions Fuchsiennes, C. R., 92, 957 = Oeuvres, II, 11.

1881f   Sur les fonctions abéliennes, C. R. 92, 958-959 = Oeuvres, IV, 299-301.

1881g   ibid., C. R. 92, 1198-1200 = Oeuvres, II, 12-15.

1881h   ibid., C. R., 92, 1214-1216 = Oeuvres, II, 16-18.

1881i   ibid., C. R., 92, 1484-1487 = Oeuvres, II, 19-22.

1881j   Sur les groupes Kleinéens, C. R., 93, 44-46 = Oeuvres, II, 23-25.

1881k   Sur une fonction analogue aux fonctions modulaires, C. R. 93, 138-140 = Oeuvres, II, 26-28.

18811   Sur les fonctions Fuchsiennes, C. R., 93, 301-3 = Oeuvres, II, 29-31.

1881m   ibid., C. R., 93, 581-582 = Oeuvres, II, 32-34.

1881n   Sur la théorie des fonctions Fuchsiennes, Mém. Caen for 1882, 3-29 = Oeuvres, II, 75-91.

1882a   ibid., C. R., 94, 163-166 = Oeuvres, II, 35-37.

1882b   Sur les groupes discontinus, C. R. 94, 840-843 = Oeuvres, II, 38-40.

1882c   Sur les fonctions Fuchsiennes, C. R. 94, 1038-1040, = Oeuvres, II, 41-43.

1882d   ibid., C. R., 94, 1166-1167 = Oeuvres, II, 44-46.

1882e   Sur une classe d'invariants relatifs aux équations linéaires, C. R. 94, 1042-1045 = Oeuvres, II, 47-49.

1882f   Sur les fonctions Fuchsiennes, C. R. 95, 626-628 = Oeuvres, II, 50-52.

1882g   Sur les fonctions uniformes qui se reproduisent par des substitutions linéaires, Math. Ann. 19, 553-564 = Oeuvres II, 92-105.

1882h   ibid., (Extrait d'une lettre adressée à M. F. Klein.)

Math. Ann. 20, 52-53 = Oeuvres, II, 106-107.

1882i   Théorie des Groupes Fuchsiens, Acta Math. 1, 1-62 =

Oeuvres, II, 108-168.

1882j   Sur les Fonctions Fuchsiennes, Acta Math. 1, 193-294 =

Oeuvres, II, 169-257.

1883a   Sur les groupes des équations linéaires, C. R., 96,

691-694, = Oeuvres, II, 53-55.

1883h   ibid., C. R. 96, 1302-1304 = Oeuvres, II, 56-58.

1883c   Sur les fonctions Fuchsiennes, C. R. 96, 1485-1487 =

Oeuvres II, 59-61.

1883d   Mémoire sur les Groupes Kleinéens Acta Math. 3, 49-92 =

Oeuvres, II, 258-299.

1884a   Sur les groupes des équations linéaires, Acta Math., 4,

201-311 = Oeuvres, II, 300-401.

1884b   Mémoire sur les fonctions zétafuchsiennes, Acta Math. 5,

209-278 = Oeuvres, II, 402-462.

1886   Sur les intégrales irregulières des équations linéaires,

Acta Math. 8, 295-344 = Oeuvres, I, 290-332.

1907   Sur l'uniformisation des fonctions analytiques, Acta

Math. 31, 1-63 = Oeuvres, IV, 70-139.

1908   Science et méthode, Paris.

1916-1954   Oeuvres, 11 vols., Paris.

J. V. Poncelet (1788-1867)

1832   Théorèmes et problèmes sur les lignes du troisième ordre,

Quetelet, Corr. Math., 7, 79-84.

J. C. Pont

1974   La Topologie Algébrique des origines à Poincaré,

            P. U. F. Paris.

E. G. C. Poole

1960   Introduction to the Theory of Linear Differential

            Equations, Dover reprint of 1936 edition by Oxford U. P.

V. Puiseux (1820-1883)

1850   Recherches sur les fonctions algébriques, J. de Math. 15
1851   Suite, ibid., 16.

W. Purkert

1971   Zur Genesis des abstrakten Körperbegriffs, I, N. T. M.

            8.1, 23-37.

1973   ibid., II, N. T. M. 10.2, 8-26.

1976   Ein Manuskript Dedekinds über Galois - theorie, N. T. M.

            13.2, 1-16.

J. M. Radner

1979   Bernhard Riemann's probationary lecture, Cambridge M. Phil.

            Thesis (unpublished).

H. Reichardt

1976   Gauss und die nicht-euklidische Geometrie, Teubner,

            Leipzig.

J. L. Richards

1979   The Reception of a Mathematical Theory: Non-Euclidean

             Geometry in England 1865-1883 in B. Barnes, S. Shapin (eds.)

             Natural Order: Historical Studies of Scientific Culture,

             Sage Publications, Beverley Hills.

Bernhard Riemann (1826-1866)

1851   Grundlagen für eine allgemeine Theorie der Functionen

             einer veränderlichen complexen Grösse (Inaugural

             dissertation), Göttingen, = Werke, 3-45.

1854a   Über die Darstellbarkeit einer Function durch einer

             trigonometrische Reihe, K. Ges. Wiss. Göttingen, 13, 87-132

             = Werke, 227-271.

1857a   Beiträge zur Theorie der durch Gauss'sche Reihe

             F(α, β, γ, x) darstellbaren Functionen, K. Ges. Wiss.

             Göttingen = Werke, 67-83.

1857b   Selbstanzeige der vorstehenden Abhandlung, Göttingen

             Nachr. no. 1 = Werke, 84-87.

1857c   Theorie der Abelschen Functionen, J. f. M. 54, 115-155 =

             Werke, 88-144.

1859   Ueber die Anzahl der Primzahlen unter einer gegebene

             Grösse, Monatsberichte Berlin Akademie, 671-680 = Werke,

             145-153.

1865   Ueber das Verschwinden der Theta-functionen, J. f. M.,

             65 = Werke, 212-224.

1867   Ueber die Fläche vom kleinsten Inhalt bei gegebener

             Begrenzung K. Ges. Wiss. Göttingen, 13, 21-57 = Werke,

             301-337.

1854b   Ueber die Hypothesen welche der Geometrie zu Grunde
             liegen, K. Ges. Wiss. Göttingen, 13, 1-20 = <u>Werke</u>,
             272-287.

1953a   Zwei allgemeine Lehrsatze über lineare Differential-
             gleichungen mit algebraischen Coefficienten, MS. =
             <u>Werke</u>, 379-390.

1953b   Fragmente über die Grenzfälle der elliptischen
             Modulfunctionen, MS. = <u>Werke</u>, 455-465.

1953c   Convergenz der p-fach unendlichen Theta-Reihe, MS. =
             <u>Werke</u>, 483-486.

1953d   Zur Theorie der Abel'schen Functionen, MS. = <u>Werke</u>,
             487-504.

1953e   Nachträge = <u>Werke</u> after 559.

1953   <u>Bernhard Riemann's Gesammelte Mathematische Werke und
             Wissenschaftliche Nachlass</u>, ed. R. Dedekind and
             H. Weber, with Nachträge, ed. M. Noether and
             W. Wirtinger. Dover reprint of 2nd ed. 1902.

Gustav Roch (1837-1866)

1864   Ueber die Anzahl der willkürlichen Constanten in
             algebraischen Functionen, J. f. M. 64, 372-376.

1866   Ueber die Doppeltangenten an Curven vierter Ordnung,
             J. f. M., 66, 97-120.

G. Rosenhain (1816-1887)

1851   Mémoire sur les fonctions de deux variables et à quatre
             périodes, qui sont les inverses des integrales ultra-
             elliptiques de la première classe, Paris, Mem. Savans.
             Etrang. 11, 361-468.

D. E. Rowe

1983 A forgotten chapter in the History of Klein's Erlanger

Programm, H. M. 10.4, 448-454.

G. Salmon (1819-1904)

1879 A Treatise on Higher Plane Curves, 3rd ed., Chelsea

reprint.

L. Schläfli (1874-1895)

1870 Über die Gauss'sche hypergeometrische Reihe, Math. Ann.,

3, 286-295 = Ges. Math. Abh., III, 153-162.

1956 Gesammelte Mathematische Abh. III, Basel.

L. Schlesinger (1864-1931)

1895 Handbuch der Theorie der Linearen Differential-

gleichungen, vol. I. Teubner, Leipzig.

1897 ibid., vol. II, 1.

1898 ibid., vol. II, 2.

1904 Über das Riemannsche Fragment zur Theorie der Linearen

Differentialgleichungen und daran anschliessende neuere

Arbeiten, in Verhandlungen des Dritten Internationalen

Mathematiker Kongresses in Heidelberg ed. A. Krazer,

Teubner, Leipzig.

1909 Bericht über die Entwicklung der Theorie der linearen

Differentialgleichungen seit 1865, J. D. M. V. XVIII,

133-267.

444

A. Schlissel

1976/77   The Development of Asymptotic Solutions of Linear

Ordinary Differential Equations, 1817-1920, A. H. E. S.,

16, 307-378.

E. Scholz

1980   Geschichte des Mannigfaltigkeits-begriffs von Riemann

bis Poincare, Birkhäuser, Boston.

1982   Herbart's Influence on Bernhard Riemann, H. M. 9.4,

413-440.

B. Schoeneberg

1974   The Theory of Elliptic Modular Functions, Springer.

F. Schottky (1851-1935)

1877   Ueber die conforme Abbildung mehrfach zusammenhängender

ebener Flächen, J. f. M., 83, 300-351.

H. A. Schwarz (1843-1921)

1869a   Ueber einige Abbildungsaufgaben, J. f. M. 70, 105-120 =

Abh. II, 65-83.

1869b   Conforme Abbildung der Oberfläche eines Tetraeders auf

die Oberfläche einer Kugel, J. f. M. 70, 121-136 = Abh.

II, 84-101.

1869c   Zur Theorie der Abbildung, Programme of the ETH Zurich =

Abh., II, 108-132.

1870a   Ueber einen Grenzübergang durch altenirendes Verfahren,

Vierteljahrschrift Natur. Gesellschaft Zurich, 15, 272-286
= <u>Abh</u>. II, 133-143.

1870b   Ueber die Integration der partiellen Differential-

gleichung $\dfrac{\partial^2 u}{\partial x^2} + \dfrac{\partial^2 u}{\partial y^2} = 0$ unter vorgeschriebenen Grenz -

und Unstetigkeits-bedingungen, Monatsber. K. A. der Wiss.

Berlin, 767-795 = <u>Abh</u>., II, 144-171.

1872   Ueber diejenigen Fälle, in welchen die Gaussische

hypergeometrische Reihe eine algebraische Function ihres

vierten Elementes darstellt, J. f. M. 75, 292-335 = <u>Abh</u>.

II, 211-259.

1880   Auszug aus einem Briefe an Herrn F. Klein, Math. Ann. 21,

157-160 = <u>Abh</u>. II, 303-306.

1890   <u>Gesammelte Mathematische Abhandlungen</u>, 2 vols., 1st ed.

Berlin 1890, 2nd ed. reprinted in 1 vol., Chelsea, 1972.

J. P. Serre

1973   <u>A Course in Arithmetic</u>, Springer

1980   Extensions Icosaédriques, Seminaire de Théories des Nombres

de Bordeaux, Année 1979-80, no. 19, 19-01, 19-07.

J. A. Serret (1827-1898)

1879   <u>Cours d'Algèbre Supérieure</u>, 4th ed. Paris.

A. Silvestri

1979   Simple Groups of Finite Order in the Nineteenth Century,

A. H. E. S. 20.3, 313-357.

446

H. J. S. Smith (1826–1883)

1859–1865   Report on the Theory of Numbers, Chelsea reprint 1965.

1877   Mémoire sur les équations modulaires Atti della

             R. Accad. Lincei = Coll. Math. Papers, II, (no. XXXV)

             224–241.

1965   Collected Mathematical Papers, (1st ed. Oxford 1894)

             2 vols., Chelsea.

L. A. Sohnke (1842–1897)

1834   Aequationes modulares pro transformatione functionum...

             J. f. M. 12, 178.

D. M. Y. Sommerville (1879–1934)

1911   Bibliography of non-Euclidean Geometry, reprint Chelsea.

T. A. Springer

1977   Invariant Theory, Springer Verlag, L. N. I. M. 585.

J. B. Stallo

1888   The Concepts and Theories of Modern Physics, Belknap

             Press, Harvard University reprint 1960.

K. G. C. von Staudt (1798–1867)

1856–1860   Beiträge zur Geometrie der Lage, 3 vols, Nurnberg.

J. Steiner (1796-1863)

1848   Ueber allgemeine Eigenschaften der algebraischen Curven, Berlin Bericht, 310-316.

1852   Ueber solche algebraische Curven, welche einen Mittelpunkt haben,... J. f. M. 47, 7-105.

L. Sylow (1832-1918)

1872   Théoremes sur les groupes de substitutions, Math. Ann. 5, 584-594.

J. Tannery (1848-1910)

1875   Propriété des Intégrales des Équations différentielles linéaires à coefficients variables, Ann. Sci. de l'École Normale Supérieure ser. 2, 4.1, 113-182.

1876   La géométrie imaginaire et la notion d'espace, Rev. Phil. 2, 433-451, 2, 553-575.

J. Thomae (1840-1921)

1870   Über die höheren hypergeometrischen Reihen, insbesondere über die Reihe ... M. Ann. 2, 427-444.

1874   Integration einer linearen Differentialgleichungen zweiter Ordnung durch Gauss'sche Reihen, Schlömilch's Zeitschrift, 19, 273-285.

L. W. Thomé (1841-1910)

1872   Zur Theorie der linearen Differentialgleichungen, J. f. M., 74, 193-217.

1873   Zur Theorie der linearen Differentialgleichungen, J. f. M.,
           75, 265–291.

1884   Zur Theorie der linearen Differentialgleichungen, J. f. M.,
           96, 185–281.

J. Tissot

1852   Sur un déterminant d'intégrales définies J. de. Math. 17,
           177–185.

R. Tobies

1981   Felix Klein, Biographien...50, Teubner, Leipzig.

I. Toth

1977   La revolution non-euclidienne, La Recherche, 75.8,
           143–151.

1984   Three errors in the Grundlagen of 1884: Frege and
           non-Euclidean Geometry, Frege Conference 1984, ed.
           G. Werschung, Akademie-Verlag, Berlin.

H. Valentiner (1850–1930)

1889   De endelige Transformations – Gruppers Theori, Danish
           Academy publications, series V, vol. 6.

J. Wallis (1616–1703)

1656   Arithmetica Infinitorum, Oxford.

S. von Waltershausen

1856   *Gauss zum Gedächtniss*, Leipzig.

W. C. Waterhouse

1980   The early proofs of Sylow's Theorem, A. H. E. S. 21,
        279-290.

H. Weber (1842-1913)

1875   Neuer Beweis des Abel'schen Theorems, Math. Ann. 8,
        49-53.

1876   *Theorie der Abel'schen Functionen vom Geschlecht 3*,
        Berlin.

L. Wedekind

1876   Beiträge zur geometrischen Interpretation binärer Formen,
        Math. Ann. 9, 209-217.

1880   Das Doppelverhaltnis und die absolute Invariante binärer
        biquadratischer Formen, Math. Ann, 17, 1-20.

K. T. W. Weierstrass (1815-1897)

1841   Zur Theorie der Potenzreihen, MS. = *Werke*, I, 67-74.

1842   Definition analytischer Functionen einer Veränderlichen
        vermittelst algebraischer Differentialgleichungen, MS. =
        Werke I, 75-84.

1854   Zur Theorie der Abel'schen functionen, J. f. M., 47,
        289-306 = *Werke* I, 133-152.

450

1856a   Über die Theorie der analytischen Facultäten,

        J. f. M., 51, 1-60 = Werke, I, 153-221.

1856b   Theorie der Abel'schen functionen, J. f. M., 52,

        285-339 = Werke, I, 297-355.

1894   Mathematische Werke, 7 vols., Olms, Hildesheim.

1902   Vorlesungen über die Theorie der Abelschen Transcendenten =

        Werke, IV.

A. Weil

1974   Two lectures on number theory, past and present,

        L'Enseignement Mathématique (2) 20, 87-110 = Oeuvres, III,

        279-302.

1976   Elliptic functions according to Eisenstein and Kronecker,

        Springer.

1979   Oeuvres Scientifiques, Collected Papers, Springer.

E. T. Whittaker (1873-1956) and G. N. Watson (1886-1965)

1973   A Course of Modern Analysis (1st edition 1902, 4th 1927,

        reprinted). C. U. P.

A. Wiman (1865-　　)

1913   Endliche Gruppen Linearen Substitutionen Enc. Math. Wiss.

        IB 3f.

E. Wirtinger (1865-1945)

1904   Riemanns Vorlesungen über die hypergeometrische Reihe und ihre Bedeutung, in Verhandlungen des dritten internationalen Kongresses in Heidelberg ed. A. Krazer, Teubner, Leipzig.

1913   Algebraische Functionen und ihre Integrale, Enc. Math. Wiss., II, B2.

H. Wussing

1969   Die Genesis des abstrakten Gruppenbegriffes, Berlin.
1979   C. F. Gauss, Teubner, Leipzig.

A. P. Youschkevitch

1976/77   The Concept of Function up to the Middle of the 19th Century, A. H. E. S. 16, 37-85.

452

Additions to the Bibliography

Since the manuscript was submitted to Birkhäuser I have become
aware of the following items which are of interest.

A. L. Cauchy 1981 <u>Equations différentielles ordinaires. Cours inedit.
Fragment.</u> Introduction by Ch. Gilain. Johnson Reprint.

B. A. Dubrovin 1981 Theta functions and non-linear equations,
English trans. in Russian Math. Surveys <u>36:2</u> 11-92.

J. Dutka 1984 The Early History of the Hypergeometric Function
A.H.E.S. <u>31.1</u>, 15-34.

W. Kaufmann-Bühler. The Life of the hypergeometric function - a
biographical sketch, to appear in the Mathematical Intelligencer.

D. Mumford, 1983. <u>Tata Lectures on Theta, vols. I and II.</u> Birkhäuser
Boston.

M. F. Singer, 1981. Liouvillian Solutions of n-th order homogeneous
linear differential equations, Amer. J. Maths <u>103.4</u> 661-682;
other papers by the same on the algebraic solution of differential
equations are to appear, he has kindly sent me preprints.

An English translation of Wussing's book has appeared:
H. Wussing, 1984. <u>The Genesis of the Abstract Group Concept.</u> MIT Press.

458